T0260205

Big Data Analytics for Internet of Things

Big Data Analytics for Internet of Things

Edited by

Tausifa Jan Saleem
National Institute of Technology
Srinagar, India

Mohammad Ahsan Chishti
Central University of Kashmir
Ganderbal, Kashmir, India

The right of Tausifa Jan Saleem and Mohammad Ahsan Chishti to be identified as the author(s) of the editorial material in this work has been asserted in accordance with law.

Registered Office
John Wiley & Sons, Inc., 111 River Street, Hoboken, NJ 07030, USA

Editorial Officesw
111 River Street, Hoboken, NJ 07030, USA

For details of our global editorial offices, customer services, and more information about Wiley products visit us at www.wiley.com.

Wiley also publishes its books in a variety of electronic formats and by print-on-demand. Some content that appears in standard print versions of this book may not be available in other formats.

Library of Congress Cataloging-in-Publication Data

Names: Saleem, Tausifa Jan, editor. | Chishti, Mohammad Ahsan, editor.
Title: Big data analytics for Internet of things / edited by Tausifa Jan
  Saleem, Mohammad Ahsan Chishti.
Description: First edition. | Hoboken, NJ : Wiley, 2021. | Includes
  bibliographical references and index.
Identifiers: LCCN 2020049761 (print) | LCCN 2020049762 (ebook) | ISBN
  9781119740759 (hardback) | ISBN 9781119740766 (adobe pdf) | ISBN
  9781119740773 (epub)
Subjects: LCSH: Big data. | Internet of things.
Classification: LCC QA76.9.B45 B4995 2021 (print) | LCC QA76.9.B45
  (ebook) | DDC 005.7–dc23
LC record available at https://lccn.loc.gov/2020049761
LC ebook record available at https://lccn.loc.gov/2020049762

Cover Design: Wiley
Cover Image: © Blue Planet Studio/iStock/Getty Images Plus/Getty Images

Set in 9.5/12.5pt STIXTwoText by SPi Global, Pondicherry, India

10   9   8   7   6   5   4   3   2   1

Contents

List of Contributors

Nailah Afshan
Department of Computer Science
and Engineering
Islamic University of Science
and Technology
Pulwama, India

Nufazil Altaf
School of Business Studies
Central University of Kashmir
Kashmir, India

Asha Ambhaikar
Department of Computer Science
and Engineering
Kalinga University
Naya Raipur
Chhattisgarh, India

Shashwati Banerjea
Department of Computer Science
and Engineering
Motilal Nehru National Institute
of Technology Allahabad
Prayagraj
Uttar Pradesh, India

Molood Barati
School of Engineering
Computer and Mathematical Sciences

Auckland University of Technology
Auckland, New Zealand

Dhruba Kumar Bhattacharyya
Department of Computer Science
and Engineering
School of Engineering
Tezpur University
Tezpur
Assam, India

Mohammad Ahsan Chishti
Department of Information
Technology
Central University of Kashmir
Kashmir, India

Shoumen Palit Austin Datta
MIT Auto-ID Labs
Department of Mechanical
Engineering
Massachusetts Institute of Technology
Cambridge, MA, USA

Rup Kumar Deka
Department of Computer Science
and Engineering
Assam Don Bosco University
Guwahati
Assam, India

Heeba Din
Department of Mass Communication
Islamic University of Science
and Technology
Pulwama, India

Mohammad Eshghi
Computer Engineering Department
Shahid Beheshti University
Tehran, Iran

Amin Fadaeddini
Department of Computer Engineering
Faculty of Engineering
Khatam University
Tehran, Iran

Marie-Laure Furgala
Institut Supérieur de Logistique
Industrielle
KEDGE Business School
Talence, France

Basant Garg
P.S to MOS, Ministry of Commerce
and Industry
Government of India
Udyog Bhawan,
New Delhi, India

Hemant Garg
The PSCADB LTD
Chandigarh, India

Sushil Gupta
Department of Bio-Sciences
Lovely Professional University
Punjab, India

Jugal K. Kalita
Department of Computer Science
College of Engineering and
Applied Science

University of Colorado
Boulder, CO, USA

Ankur Kashyap
Bennett University
Greater
Noida, India

Asif Khan
School of Media Studies
Central University of Kashmir
Kashmir, India

Sarabjeet Kaur Kochhar
Department of Computer
Science
Indraprastha College for
Women
University of Delhi
New Delhi, India

Sachin Kumar
Department of Computer Science
and Engineering
Motilal Nehru National Institute
of Technology Allahabad
Prayagraj
Uttar Pradesh, India

Sunil Kumar
Department of Electrical and
Electronics Engineering
Kalinga University
Naya Raipur
Chhattisgarh, India

María Victoria López López
Deparmento Arquitectura
de Computadores y Automática
Universidad Complutense de Madrid
Madrid, Spain

Babak Majidi
Emergency and Rapid Response
Simulation (ADERSIM)
Artificial Intelligence Group
Faculty of Liberal Arts and
Professional Studies
York University
Toronto, ON, Canada

Eric S. McLamore
Department of Agricultural Sciences
Clemson University
Clemson
SC, USA

Snowber Mushtaq
Department of Computer Science
and Engineering
Islamic University of Science
and Technology
Pulwama, India

Ripal Patel
Department of Electronics and
Communication
Dr. Ambedkar Institute of
Technology
Bengaluru, India

Tanuja Patgar
Department of Electronics and
Communication
Dr. Ambedkar Institute of
Technology
Bengaluru, India

R. Rajathy
Department of Electrical and
Electronics Engineering
Pondicherry Engineering College
Puducherry, India

Ranjeet Kumar Rout
Department of Computer Science
and Engineering
National Institute of Technology
Srinagar, India

Tausifa Jan Saleem
Department of Computer Science
and Engineering
National Institute of Technology
Srinagar, India

Gérald Santucci
INTEROP-VLab
Bureau Nouvelle Région Aquitaine
Europe
Brussels, Belgium

Anjum Sheikh
Department of Electronics and
Communication
Kalinga University
Naya Raipur
Chhattisgarh, India

Omkar Singh
Department of Electronics and
Communication Engineering
National Institute of Technology
Srinagar, India

Shashank Srivastava
Department of Computer Science
and Engineering
Motilal Nehru National Institute
of Technology Allahabad
Prayagraj
Uttar Pradesh, India

C.M. Thasnimol
Department of Electrical and
Electronics Engineering
Pondicherry Engineering College
Puducherry, India

Diana C. Vanegas
Interdisciplinary Group for
Biotechnological Innovation and
Ecosocial Change BioNovo

Universidad del Valle
Cali, Colombia

Syed Rameem Zahra
Department of Computer Science
and Engineering
National Institute of Technology
Srinagar, India

List of Abbreviations

AI	artificial intelligence
ALPR	Automatic License Plate Recognition
ANN	artificial neural network
AWS	Amazon web services
BDA	big data analytics
CBSP	cloud-based security provider
CCP	Cluster Communication Protocol
CHARGEN	Character Generator Protocol
CNN	convolutional neural network
CRF	conditional random field
CSP	cloud service provider
CVM	core vector machine
DBSCAN	density-based spatial clustering of applications
DDoS	distributed denial of service
DNS	Domain Name System
DoS	denial of service
E-DoS	economic denial of sustainability
FCNN	fully convolutional neural network
HTTP	Hyper-Text Transfer Protocol
IaaS	infrastructure as a service
ICMP	Internet Control Message Protocol
IDS	Intrusion Detection System
IoRT	internet of robotic things
IoT	Internet of Things
IP	Internet Protocol
IPS	Intrusion Prevention Systems
IRAS	Intrusion Responsive Autonomic System
IWD	intelligent water drop
KDD	Knowledge Discovery in Databases

KNN	K-nearest neighbor
LAN	local area networks
LAND	local area network denial
M-IoU	mean intersection over union
ML	machine learning
NaaS	networking as a service
NTP	Network Time Protocol
OCSVM	one class support vector machine
PaaS	platform as a service
PCA	principal component analysis
PDC	Phasor Data Concentrator
PMU	phasor measurement unit
POS	part of speech
QoS	quality of service
RDBMS	Relational DataBase Management System
RNN	recurrent neural network
SaaS	software as a service
SCADA	supervisory control and data acquisition
SCIT	self-cleansing intrusion tolerance
SDN	software-defined networking
SDNFV	software-defined network function virtualization
SLA	service-level agreement
SNMP	Simple Network Management Protocol
SOAP	Simple Object Access Protocol
SQL	structured query language
SSDP	Simple Service Discovery Protocol
SVM	support vector machine
SVR	support vector regression
SYN	synchronize
TCP	Transmission Control Protocol
TTL	time-to-live
UPnP	universal plug and play
VAE	variational auto-encoder
WAMS	wide area monitoring system
WAN	wide area networks

1

Big Data Analytics for the Internet of Things

An Overview

Tausifa Jan Saleem[1] and Mohammad Ahsan Chishti[2]

[1] *Department of Computer Science and Engineering, National Institute of Technology Srinagar, India*
[2] *Department of Information Technology, Central University of Kashmir, Kashmir, India*

Internet of Things (IoT) is an emerging idea that has the prospective to completely reform the outlook of businesses. The goal of the IoT is to transmute day-to-day objects to being smart by utilizing a broad range of sophisticated technologies, from embedded devices and communication technologies to data analytics. IoT is bound to transform the ways of our everyday working and living. The number of IoT devices is anticipated to amount to several billion in the next few years. This unpredictable growth in the number of devices connected to IoT and the exponential rise in data consumption manifest how the expansion of big data seamlessly coincides with that of IoT. The growth of big data and the IoT is swiftly accelerating and affecting all areas of technologies and businesses. The main objective of data analytics in IoT is to identify trends in the data, extract concealed information, and to dig out valuable information from the raw data generated by IoT systems. This is extremely crucial for dispensing elite services to IoT users. In this regard, investigating the technological advancements in the said area becomes indispensable. To this purpose, this book uncovers the recent trends in big data analytics for IoT applications so that novel, optimized, and efficient designs of IoT use-cases are formulated.

This book contains high-quality research articles discussing various aspects of IoT data analytics like enabling technologies of IoT data analytics, types of IoT data analytics, challenges in IoT data analytics, etc. This is critically important for keeping researchers up-to-date with the eco-system they have to deal with. IoT is being used as a field for garnering huge business profits. It is extremely important to squeeze out the best decisions or wisdom from the data that is being fed into the systems of business organizations. The book involves discussions of ways for extracting valuable insights from Big Data. The techniques that are suitable for digging out best decisions from the humungous IoT data to gain control of IoT devices are unleashed in the book. The book discusses almost every aspect of IoT data analytics.

The following topics are explored in this book:

- Enabling technologies for IoT Big Data Analytics
- Machine Learning Techniques for IoT Data Analytics
- Types of IoT Data Analytics
- IoT Data Analytical Platforms
- Challenges in IoT Data Analytics
- Deep Learning Architectures for IoT Data Analytics
- Personalization in IoT
- How IoT makes cities smarter
- Role of IoT and Big Data in Environmental Sustainability
- Synchro-phasor Data Management in Power Grids
- Autonomous Vehicle Identification in Smart Transportation
- Cloud-based Water Management System
- Security and Privacy Requirements in IoT
- Mitigation of DDOS attacks
- Opportunities provided by Data Fusion
- Role of IoT and Big Data in Journalism
- Role of IoT and Big Data in Finance

The book comprises of sixteen chapters. Following provides a glimpse of their contribution:

The second chapter entitled "Data, Analytics and Interoperability between Systems (IoT) is Incongruous with the Economics of Technology: Evolution of Porous Pareto Partition (P3)" aspires to inform that tools and data related to the affluent world are not a template to be "copied" or applied to systems in the remaining (80%) parts of the world which suffer from economic constraints. The chapter suggests that we need different thinking that resists the inclination of the affluent 20% of the world to treat the rest of the world (80% of the population) as a market. The 80/20 concept evokes the Pareto theme in P3, and the implication is that ideas may float between (porous) the 80/20 domains (partition).

The third chapter entitled "Machine Learning Techniques for IoT Data Analytics" discusses the various supervised and unsupervised machine learning approaches and their highly significant role in the smart analysis of IoT data. A detailed taxonomy of various machine learning algorithms together with their strengths, challenges and shortcomings is discussed. Following this, a review of application areas and use cases for each algorithm is presented in the chapter. It is quite helpful in having a better understanding of the usage of each algorithm and helps in choosing a suitable data analytic algorithm for a particular problem. The chapter concludes that machine learning has a lot of scope in the world of IoT and is proving highly beneficial for efficient analysis of smart data.

The fourth chapter entitled "IoT Data Analytics using Cloud Computing" discusses the cloud computing framework for IoT data analytics. Moreover, the importance of machine learning in IoT data analytics is also presented in the chapter. The chapter also lists the challenges faced by IoT data analytics when cloud is used as a computing platform.

The fifth chapter entitled "Deep Learning Architectures for IoT Data Analytics" unleashes the opportunities created by Deep Learning in IoT data analytics. Deep Learning has shown phenomenal performance in diverse domains, including image recognition, speech recognition, robotics, natural language processing, human-computer interface, etc. The chapter provides a description of the various Deep Learning architectures. The role of these Deep Learning architectures in IoT data analytics is also presented in the chapter.

The sixth chapter entitled "Adding Personal Touches to IoT: A User-Centric IoT Architecture" focuses on the use of the concept of personalization to achieve the goal of taking the human-computer interaction to the next level. Personalization is a powerful instrument that has the potential of shaping the quality of IoT products and services to keep pace with the constantly evolving customer needs. Use cases and real-life examples are used to demonstrate how using users personal insights spell magic for boosting IoT systems across a variety of domains such as businesses, marketing, recommendation systems and commercial and industrial IoT systems and services. The chapter investigates how personalization is assuming an important, irreplaceable role in the development of IoT systems being deployed across multiple domains and the lives of associated varied strata of users such as the business owners, marketing professionals, business analysts, data analysts, designers and the end-user. The work takes stock of the current scenario and establishes through use cases, and examples that personalization is already being exploited for huge benefits but the concept itself is being given a rather ad-hoc treatment. This is evident as personalization finds no mention in the IoT architecture itself. It is left to dangle on as a last-minute job in most of the IoT systems developed so far. Concerns regarding the usage of personalization viz. privacy and the filter bubble have also been taken into consideration to point out the future directions of work in Big Data Analytics of IoT systems.

The seventh chapter entitled "Smart Cities and the Internet of Things" investigates the development of smart cities from a perspective of the IoT. The chapter uses existing examples of smart cities to forecast what the future holds for cities seeking to utilize the IoT in optimizing their operations and resource usage.

The eighth chapter entitled "A Roadmap for Application of IoT Generated Big Data in Environmental Sustainability" describes the role of IoT generated big data in environmental sustainability. The chapter proposes a roadmap for achieving better environmental sustainability. Moreover, the obstacles that create hindrance in environmental sustainability are also discussed in the chapter.

The ninth chapter entitled "Application of High-Performance Computing in Synchrophasor Data Management and Analysis for Power Grids" discusses the various problems associated with the big data analysis with particular reference to Phasor Measurement Unit's (PMU) data handling and introduces the modern techniques and tools to resolve those pitfalls.

The tenth chapter entitled "Intelligent enterprise-level big data analytics for modelling and management in smart internet of roads" proposes a method based on Fully Convolutional Neural Network for semantic segmentation of vehicle license plates in a complex and multi-language environment. First, the license plates are detected, and then digits in the license plates are segmented. The performance of the proposed algorithm is evaluated using a dataset of real and manually generated data. The impact of various parameters in improving the accuracy of the proposed algorithm is investigated. The experimental results show that the proposed framework can detect and segment the license plates in complex scenarios, and the results can be used in smart highways and smart road applications.

The eleventh chapter entitled "Predictive analysis of intelligent sensing and cloud-based integrated water management system" proposes a water management system with following characteristics; real-time measurement of consumption, monitoring of leakages, ability to control the water supply if there is leakage, a completely automated platform for societies, and apartment complexes to set up their billing system. The proposed system consists of a flow sensor meter installed in the main water inlet pipe that captures information about water usage and communicates through a WiFi network to iOS and Android compatible applications.

The twelfth chapter entitled "Data Security in the Internet-of-Things: Challenges and Opportunities" highlights the IoT security threats and vulnerabilities. The chapter categorizes the IoT security based on context of application, architecture and communication. Furthermore, the chapter discusses the research directions in confidentiality, privacy and IoT data security.

The thirteenth entitled "DDoS Attacks: Tools, Mitigation Approaches, and Probable Impact on Private Cloud Environment" discusses the seriousness of the threats posed by DDoS attacks in the context of the cloud, particularly in the personal private cloud. The chapter discusses several prominent approaches introduced to counter DDoS attacks in private clouds. The chapter presents a generic framework to defend against DDoS attacks in an individual private cloud environment taking into account different challenges and issues.

The fourteenth chapter entitled "Securing the Defense Data for Making Better Decisions using Data Fusion" gives an idea of the problems that arise in the defense related IoT-big data analytics with special attention to its security. Data fusion has been introduced as a probable solution to tackle these problems. The chapter guides the researchers regarding the issues of data fusion, the stages where it could be used and the mathematical techniques that could be adopted to implement it on IoT big data.

The fifteenth chapter entitled "New age Journalism and Big data (Understanding big data & its influence on Journalism)" tries to identify how big data is altering the way journalism is practiced in the twentyfirst century. For the purpose, the chapter takes the case study of award-winning data journalism projects, which have not only used big data for their stories but also using converging big data with new media practices of interactive visualization, revolutionized the practice of journalism. The chapter not only provides a glimpse into how big data is changing journalism but also critically examines the impact, practices and methods involved to lay forward a guide for future research into this genre. The chapter concludes that both IoT and Big Data have tremendous potential to influence the economies of global markets, and at the same time change, the way content (information) is collected and produced for the audiences.

The last chapter entitled "Two decades of big data in finance: Systematic literature review and future research agenda" presents a review on IoT and big data in finance. The chapter identifies the gaps in the current body of knowledge to deliberate upon the areas of future research. The study uses a systematic literature review method on a sample of 105 articles published from 2000 to 2019. The majority of work on big data in finance is dominated by the empirical setup in financial markets, internet finance, and financial services. The chapter contains all-inclusive publications on the big data in finance classified according to various attributes. The chapter would be useful to all the patrons concerned with big data.

2

Data, Analytics and Interoperability Between Systems (IoT) is Incongruous with the Economics of Technology: Evolution of Porous Pareto Partition (P3)

Shoumen Palit Austin Datta[1,2,3,]* Tausifa Jan Saleem[4],
Molood Barati[5], María Victoria López López[6], Marie-Laure Furgala[7],
Diana C. Vanegas[8], Gérald Santucci[9], Pramod P. Khargonekar[10], and
Eric S. McLamore[11]

[1]*MIT Auto-ID Labs, Department of Mechanical Engineering, Massachusetts Institute of Technology, 77 Massachusetts Avenue, Cambridge, MA 02139, USA*
[2]*MDPnP Interoperability and Cybersecurity Labs, Biomedical Engineering Program, Department of Anesthesiology, Massachusetts General Hospital, Harvard Medical School, 65 Landsdowne Street, Cambridge, MA 02139, USA*
[3]*NSF Center for Robots and Sensors for Human Well-Being, Collaborative Robotics Lab, School of Engineering Technology, Purdue University, 193 Knoy Hall, West Lafayette, IN 47907, USA*
[4]*Department of Computer Science and Engineering, National Institute of Technology Srinagar, Jammu & Kashmir 190006, India*
[5]*School of Engineering, Computer and Mathematical Sciences Auckland University of Technology, Auckland 1010, New Zealand*
[6]*Facultad de Informática, Deparmento Arquitectura de Computadores y Automática, Universidad Complutense de Madrid, Calle Profesore Santesmases 9, 28040 Madrid, Spain*
[7]*Director, Institut Supérieur de Logistique Industrielle, KEDGE Business School, 680 Cours de la Libération, 33405 Talence, France*
[8]*Biosystems Engineering, Department of Environmental Engineering and Earth Sciences, Clemson University, Clemson, SC 29631, USA*
[9]*Former Head of the Unit, Knowledge Sharing, European Commission (EU) Directorate General for Communications Networks, Content and Technology (DG CONNECT); Former Head of the Unit Networked Enterprise & Radio Frequency Identification (RFID), European Commission; Former Chair of the Internet of Things (IoT) Expert Group, European Commission (EU); INTEROP-VLab, Bureau Nouvelle Région Aquitaine Europe, 21 rue Montoyer, 1000 Brussels, Belgium*
[10]*Vice Chancellor for Research, University of California, Irvine and Distinguished Professor of Electrical Engineering and Computer Science, University of California, Irvine, California 92697*
[11]*Department of Agricultural Sciences, Clemson University, Clemson, SC 29634, USA*

Big Data Analytics for Internet of Things, First Edition. Edited by Tausifa Jan Saleem and Mohammad Ahsan Chishti.
© 2021 John Wiley & Sons, Inc. Published 2021 by John Wiley & Sons, Inc.

2.1 Context

Since 1999, the concept of the Internet of Things (IoT) was nurtured as a marketing term [2] which may have succinctly captured the idea of data about objects stored on the Internet [3] in the networked physical world. The idea evolved while transforming the use of radio frequency identification (RFID) where an alphanumeric unique identifier (64-bit EPC [4] or electronic product code) was stored on the chip (tag [5]) but the voluminous raw data were stored on the Internet, yet inextricably and uniquely linked via the EPC, in a manner resembling the structure of internet protocols [6] (64-bit IPv4 and 128-bit IPv6 [7]). IoT and, later, *cloud of data* [8] were metaphors for ubiquitous connectivity and concepts originating from ubiquitous computing, a term introduced by Mark Weiser [9] in 1998. The underlying importance of data from connected objects and processes usurped the term big data [10] and then twisted the sound bites to create the artificial myth of "Big Data" sponsored and accelerated by consulting companies. The global drive to get ahead of the "Big Data" tsunami, flooded both businesses and governments, big and small. The chatter about big data garnished with dollops of fake AI became parlor talk among fish mongers [11] and gold miners, inviting the sardonicism of doublespeak, which is peppered throughout this essay.

Much to the chagrin of the thinkers, the laissez-faire approach to IoT percolated by the tinkerers overshadowed hard facts. The "quick & dirty" anti-intellectual chaos adumbrated the artifact-fueled exploding frenzy for new revenue from "IoT Practice" which spawned greed in the consulting [12] world. The cacophony of IoT in the market [13] is a result of that unstoppable transmutation of disingenuous tabloid fodder to veritable truth, catalyzed by pseudo-science hacks, social gurus, and glib publicity campaigns to drum up draconian "dollar-sign-dangling" predictions [14] about "trillions of things connected to the internet" to feed mass hysteria, to bolster consumption. Few ventured to correct the facts and point out that *connectivity without discovery* is a diabolical tragedy of egregious errors. Even fewer recognized that the idea of IoT is *not a point* but an **ecosystem**, where collaboration adds value.

The corporate orchestration of the *digital by design* metaphor of IoT was warped solely to create demand for sales by falsely amplifying the lure of increasing performance, productivity, and profit, far beyond the potential digital transformation could deliver by embracing the rational principles of IoT (Figures 2.1–2.4).

Ubiquitous connectivity is associated with high cost of products (capex or capital expense) but extraction of "value" to generate return on investment (ROI) rests on the ability to implement SARA, a derivative of the PEAS paradigm (see Figures 2.7 and 2.8). SARA – S̲ense, A̲nalyze, R̲espond, A̲ctuate – is not a linear concept. Data and decisions necessary for SARA make the conceptual illustration more akin to The Sara Cycle, perhaps best illustrated by the analogy to the Krebs

By 2022, M2M will be a USD 1.2 trillion opportunity

Total revenue from machine-to-machine, 2011–2022
Source: Machina Research 2012

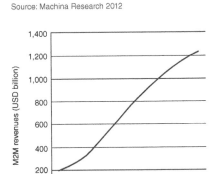

Machina Research

- **Total M2M revenue will grow from USD200 billion in 2011 to USD1.2 trillion in 2022, a CAGR of 18%**
- **Total revenue includes:**
  - device costs where connectivity is integral to the device
  - module costs where devices can optionally have connectivity enabled
  - monthly subscription, connectivity and traffic fees

Figure 2.1 From the annals [15] of the march of unreason: *Internet of things: $8.9 trillion market in 2020, 212 billion connected things.* It is blasphemous and heretical to suggest that this is a *research* [16] outcome.

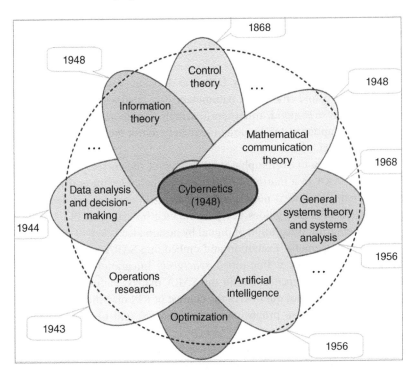

Figure 2.2 A Century of convergence the composition and structure of cybernetics [17]. *Source:* Novikov, D.A. Systems theory and systems analysis. Systems engineering. *Cybernetics.* vol. 47. Springer International Publishing. 2016, pp. 39–44. © 2016, Springer Nature.

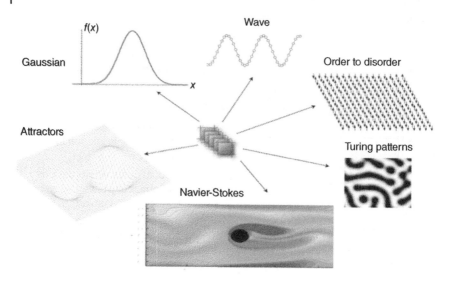

Figure 2.3 Only a few models may capture the behavior of a wide range of systems, underlies the idea of universality [18] (models illustrated in this figure: Gaussian distribution, wave motion, order to disorder transitions, Turing patterns, fluid flow described by Navier–Stokes equations, and attractor dynamics). *Source:* Based on Williams, L.P. (1989). "André-Marie Ampère." *Scientific American*, vol. 260, no. 1, pp. 90–97. © 1989, Scientific American.

[28] Cycle, an instance of bio-mimicry. Data and decisions constantly influence, optimize, reconfigure, and change the parameters associated with, *when* to sense, *what* to analyze, *how* to respond, and *where* to actuate or auto-actuate. Combining SARA with the metaphor of IoT by design may help to ask these questions, with precision and accuracy.

It is hardly necessary to overemphasize the value of the correct questions for each element of SARA in a matrix of connected objects, relevant entities which can be discovered, distributed nodes, related processes, and desired outcomes. Strategic inclusion of SARA guides key performance indicators (KPI). Lucidity and clarity of thoughtful integration of digital by design idea is key to reconfiguring operations management. Execution and embedding SARA is not a systems integration task but rather a fine-tuned *synergistic* integration based on the *weighted combination of dependencies* in the SARA matrix. Failure to grasp the role of data and semantics of queries, in the context of KPI may increase transaction costs, reduce the value proposition for customers, and obliterate ROI or profitability.

This essay meanders, not always aimlessly, around discussions involving data and decision. It also oscillates, albeit asynchronously, between a broad spectrum

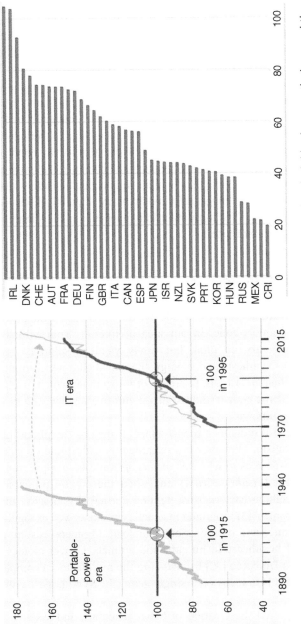

Figure 2.4 (Left) Labor-Productivity Index [19]: Has data failed to deliver? IT was billed as the bridge between *the haves* and *the have-nots*. General process technologies take ~25 years to reach market adoption [20]. *Source:* Syverson, C. (2018). Why hasn't technology sped up productivity?. *Chicago Booth Review.* © 2018, Chicago Booth Review. (Right) Labor Productivity [21] (OECD 2018) is yet another example how the arithmetic of productivity (ratio between volume of output vs input) is misguided, misdiagnosed, mismeasured, and misused as a metric of economic realities. Making Mexico (22.4) appear to be one-fifth as "productive" as Ireland (104.1) suggests formulaic manipulations [22] (GDP per hour worked, current prices, PPP).

of haphazard realities or "dots" which may be more about esoteric analysis rather than focusing on delivering real-world value. In part, this discussion questions the barriers to the rate of diffusion of technologies in underserved communities. Can implementing *simple* tools act as affordable catalysts? Can it lift the quality of life, in less affluent societies, by enabling meaningful use of data, perhaps small data, at the right time, at the lowest cost?

The extremely nonlinear business of delivering tools and technologies makes it imperative to consider the trinity of systems' integration, standards, and interoperability. We advocate that businesses may wish to gradually disengage with the product mindset (sensors, hardware, and software) and engage in the *ecosystem* necessary to deliver **_services_** to communities. The delivery of service to the end-user must be synergized. Hence, system integration may be a subset of synergistic integration. But, before we can view this "whole," it is better to understand the coalition of cyber (data) with the physical (parts). In many ways, this discussion is about cyberphysical systems (CPS) but not for lofty purposes, such as landing on Mars, but for simple living, on Earth.

2.2 Models in the Background

Because it may be difficult to grasp the whole, we tend to focus on the part, and parts, closest to our comfort zone, in our area of interest. This reductionist approach may be necessary *ab initio* but rarely yields a solution, *per se*. Reconstruction requires synthesis and synergy, the global glue which underlies mass adoption and diffusion, of tools, in an age of integration, which, itself, is a khichuri [29] of parts, some known (industrial age, information age, and systems age) and others, parts unknown.

Divide and conquer still remains a robust adage. It may be the philosophical foundation of reductionism. The latter has rewarded us with immense gains in knowledge and the wisdom as to why this *modus operandi* is *sine qua non*. For example, the pea plant (*Pisum sativum*) unleashed the cryptic principles of genetics [30] and unicellular bacteria shed light on normal physiological underpinnings of feedback control [31] common in genetic circuits as well as regulatory networks for maintenance and optimization of biological homeostasis, quintessential for health and healthcare in humans and animals. Cancer biology was transformed by Renato Dulbecco [32] by **_reducing_** the multifactorial complexity of human cancer research to focus on a **_single gene_** (the SV40 large T-antigen) from Papova viruses.

Biomimicry also inspired the creation of better machines and systems [17], using the principles and practice of control theory borrowed from science, strengthened by mathematics and successfully integrated with design and manufacturing, by engineers. An early convergence [33] of control theory with communication may be found in the 1948 treatise "Cybernetics" by Norbert Wiener [34] (who may have borrowed [35] the word "cybernétique" proposed by the French physicist and mathematician André-Marie Ampère [18] to design the then nonexistent science of process control).

In other examples of "divide and conquer," the theoretical duo "Alice and Bob" is at the core [36] of cryptography [37] as well as the game theoretic [38] approach [39] to "prisoner's dilemma" which has influenced business strategies [40] and now it is spilling over to knowledge graph (KG) [41] databases. The simple concept of a lone travelling salesman proposed by Euler in 1759 appears to have evolved [42] as the bread and butter of most optimization engines, which, when considered together with data and information, continues to improve decision support systems (DSS) in manufacturing, retail, transportation, logistics [43], and omnipresent supply chain [44] networks, almost in every vertical which uses DSS.

The purpose of these disparate examples are to emphasize the notion that there are fundamental units of activity or models or set(s) of patterns or certain basic behavioral criteria (for lack of a better descriptive term) that underlie most actions and reactions. When taken apart or sufficiently reduced, we may observe these as isolated units or patterns or models of rudimentary entities. When combined, these simple models/units/patterns/elements can generate an almost unlimited variety of system behaviors observed on grand scales. When viewing the massive scale of systems from the "top," it may be quite counterintuitive to imagine that the observed manifestations are due to a few or a relatively small group of universal "truths" which we refer to as models, units, rules, logic, patterns, elements, or behaviors. To further illustrate this perspective, consider petals (flowers), pineapple (fruit), and pyramids. The variation between and within these three very different examples may boil down to Fibonacci [45] numbers, fractal [46] dimensions, and the Golden [47] Ratio [48] in some form, or the other. In another vein, the number, eight, seems to be central to atoms (octet) and an integral part of the Standard Model in physics (octonions [49]). Number 8 is revered by the Chinese due to its link with words synonymous with wealth and fortune (fa).

If one is still unconvinced and remain skeptical that small sets of underlying elements, generally, may be responsible, albeit in part, for the "big things" we consider diverse, then the "killer" example is that of nucleic acids, deoxyribonucleic acid (DNA) and ribonucleic acid (RNA), made up of only five subunits or molecules (adenine, guanine, cytosine, thymine, and uracil). DNA and RNA serve as the blueprint for all humans, animals, plants, bacteria, and viruses that may ever exist. The infinite diversity of multicellular [50] and unicellular organisms, whose creation is instructed by a *combination of these five molecules* in DNA and

RNA, may vastly exceed 5×10^{30} (5,000,000,000,000,000,000,000,000,000,000 [51]). The known exception to the DNA–RNA dogma may be the case for prions [52] which uses proteins [53] as the transmissible macromolecule.

Parallel examples can be drawn from physical sciences. Large-scale system behaviors can be reduced and mapped to simple models. Combination of these simple models, with widely different microscopic details, applies to, and generates, a large set of possible systems [54] and system of systems. Another example of "hidden complementarities" emerged from cryptic mathematical bridge embedded in natural sciences. It is now established that eigenvectors may be computed [55] using information about eigenvalues. Students are still taught that eigenvectors and eigenvalues are independent and must be calculated separately starting from rows and columns of the matrix. Mathematicians authored papers in related fields [56] yet none "connected the dots" between eigenvectors and eigenvalues. The insight that eigenvalues of the minor matrix encode hidden information may not be entirely new [57] but was neither understood nor articulated. The relationship of centuries-old mathematical objects [58] ultimately came from physicists. Nature inspires mathematical thinking because mathematics thrives when connected to nature. Grasping these connections enables humans to create tools to mimic nature (bio-mimicry).

2.3 Problem Space: Are We Asking the Correct Questions?

The lengthy and winding preface is presented to substantiate the opinion that there may be a disconnect between the volume of data we have generated as a result of the "information age" versus the lackluster gains in performance, as estimated by the productivity [59] index. We may have 2.7 zettabytes [20] (2.7 billion terabytes) of data, but some estimates claim as much as 33 zettabytes [60] of data, at hand (2018). It is projected to reach 175 zettabytes circa 2025.

The deluge of data as a result of "information technology" is far greater in magnitude than the diffusion of electricity [61] a century ago. Productivity increases due to the introduction of electricity and IT offers economic parallels [62] but based on the magnitude of change, the shortfall (in productivity) cannot be brushed aside by attributing the blame to mismeasurement explanations [63] for the sluggish [64] pace. Extrapolating measurements using the tools of classical productivity [65] to determine the impact of IT and influence of data is certainly fraught with problems [66], yet the incongruencies alone cannot explain the shrinkage. In socioeconomic terms, there is a growing chasm between IT and data/information versus productivity, improvement in quality of life, labor, compensation [67], and standard of living.

Despite trillions of dollars invested in data, digital transformation and other IT tools [68] (big data, AI, blockchain), the perforated ROI [69] increasingly points to massive [70] waste. One reason for this "waste" may be due to use of models of data where errors are aggregated under a generalized [71] form or variations [72] of the normal (homoskedastic) distribution. Heteroskedasticity was addressed [73] using ARCH [74] (autoregressive conditional heteroskedasticity [75]) and GARCH [76] models [77] (generalized ARCH). The use [78] of these proven techniques [79] for time series data (for example, sensor data showing water temperature in marine aquaponics [80] or cold chain [81] temperature log of vaccine package during transportation) in financial [82] econometrics [83] may be extended. Applications in predictive [84] modeling and forecasting [85] techniques may wish to adopt these econometric tools (GARCH) as a standard, whenever time series data are used (for example, supply chain [86] management, sensor data in health), but only *if* there is sufficient data (volume) to meet the statistical rigor necessary for successful error correction.

Perhaps, it is best to limit the postmortem analysis of IT failures, snake-oil sales of AI [87], and other debacles. Let us observe from this discussion that in the domain of data, and extraction of value from data to inform decisions and the tools necessary for **meaningful** transformation of data to inform decisions may benefit from re-viewing the processes and technologies with "new" eyes. We must ask, often, if we are pursuing the correct questions, if the tools are appropriate and rigorous. The productivity gap and reports of corporate waste are "sign-posts" on the road ahead, except that the signage is in the incorrect direction, with respect to the intended destination, that is, profit and performance.

2.4 Solutions Approach: The Elusive Quest to Build Bridges Between Data and Decisions

There are no novel proposed solutions in this essay, only new commentary about *approaches* to solutions. The violent discord between volume of data versus veracity of decisions appears to be one prominent reason why the productivity gap may widen to form a chasm. The "background" section discussed how the reductionist approach points to simple models or underlying units or key elements, which, when combined, in some form, by some rules or logic, may generate large-scale systems.

Data models [88] for DBMS are very different from **models in data**. Pattern mining [89] from data [90] is a time-tested tool. What new features can we uncover or learn about data, from patterns? What simpler models or elements are cryptic in data? Are these the correct questions? *If* there are simpler models or patterns in some types of data, can we justify extrapolating these models and *patterns* as a *general feature* of the data? The failure to accept and curate data which may be void of information is of critical importance. The contextual understanding of this issue appears to be uncommon and tools for semantic data curation are nonexistent. Although we have been mining for patterns and models (clustering, classification, categorization, and principal component analysis) for decades, why have not we found simpler models or patterns, yet? Are we using the wrong tools or wrong approaches or looking at wrong places? How rational are we in our search for these general/simple models in view of the fact that models of data from retail or manufacturing or health clinics *should* be quite different? Is model building by humans an irrational approach since humans are innate, irrational organisms endowed with sweeping bias?

Thus, the lowest common denominator of general models/patterns may not be an ingredient for building that experimental "thought" bridge. Increasing volume of data could help GARCH tools but it is a slippery slope in terms of data quality with respect to *informing* DSS and/or the veracity of decisions (output). Data models/patterns as denominators from grocery shopping or dry wall manufacturing or mental health clinics *are* different. In lieu of "universal" common denominators, we may create repertoires of domain-specific common denominators. A comparative analysis between common denominators of retail grocery shopping model from Boston vs. Beijing may reveal the spectrum of nutritional behaviors. If linked to eating habits, perhaps we can extrapolate its *influence* on health/mental health. As this suggestion reveals, we may be able to explore very tiny subsets of models.

Domain-specific denominator models (DSDM) are not new. It requires an infrastructure approach to data analytics which needs multitalented teams to explore almost every cross section and combination of very large volumes of data, from specific domains, to identify obvious correlations as well as unknown/nonobvious relationships. If there is any doubt about the quality of the raw data, then quality control may mandate data curation. The latter alone, makes the task exponentially complex. Curation may introduce reasonable doubt in evaluating any outcome because the possibility exists that curation algorithms and associated processes were error-prone or untrustworthy (post-curation jitters).

Another demerit for DSDM and the idea of denominator models, in general, may be rooted in the "apples vs oranges" dilemma. Denominator models that underlie science and engineering systems are guided by natural laws, deemed *rational*. The quest for denominator models in data (retail, finance, supply chain, health, and agriculture) are influenced, infected, and corrupted by irrational [91] human behavior. Rational models of irrational behavior [92] may coexist elsewhere but remains elusive for data science due to volatility and the vast *spectrum of irrationality* that may be introduced in data by human interference.

Perhaps, the concept of DSDM, ignoring its obvious caveats, may be applied to select domains for specific purposes, for example, healthcare, where deliberate human interference to introduce errors in data is a criminal offense. Case-specific model building, and pattern recognition, may benefit from machine learning (ML) approaches. The latter fueled a plethora of false [93] claims but real success is still a *work in progress* because the bridge between data and decisions will be perpetually *under construction*. Productivity gap and corporate waste are indicators that existing approaches (see Figure 2.5) are flawed, failing, or have [94] failed. We need new roads. The boundary of our thought horizon "map" is in Figure 2.5. The tools are incremental variations [95] garnished with gobbledygook alphabet soup. Unable [96] to create any breakthrough, the return of seasonal "winters of AI" indicates the struggle to shed new light in this field since the grand edification [97] during the 1950s. Unable to cope with data challenges, hard facts [85], and difficult progress, the field offered a perfect segue for con artists and hustlers to inculcate falsehoods and deceive [98] the market. ML was substituted [99] by mindless drivel from ephemeral captains of industry and generated hype [100] from corporate [101] marketing machines as well as greedy academics.

2.5 Avoid This Space: The Deception Space

Data consumers have been led astray by vacuous buzz words manufactured mostly by consulting groups. Part of the productivity gap may be due to fake news, propaganda [94], and glib strategy from smug consultants to coerce large contracts with cryptic "billable hours" to help "monetize" false promises due to "big" data, fabricated [102] claims [103] of "intelligence" in artificial intelligence (AI) [104], and deliberately conniving misrepresentations [105] of "blockchain" as a panacea [106] for all problems [107] including basic food safety and security. Callous and myopic funding agencies invested billions in academic [108] industry partnerships to fuel banal R&D efforts orchestrated by corporate collusion [109] and perhaps [110] criminal [111] practices. Abominable predatory practices on display in Africa are disguised under the "smart cities" marketing campaign to mayors of **African cities, which cannot even provide clean drinking water to its residents.** Vultures from the industry [112] are selling mayors of African cities surveillance technology and AI in the name of cameras for smart city safety and security. These behemoths are cognizant as to how autocrats use data as an ammunition to plan and justify abuse of its citizens, through algorithms of repression.

2.6 Explore the Solution Space: Necessary to Ask Questions That May Not Have Answers, Yet

Uploading data from nodes along a variety of supply chains is an enormous undertaking given trillions of interconnected processes and billions of nodes with extraordinarily diverse categories of potential data streams, with different security

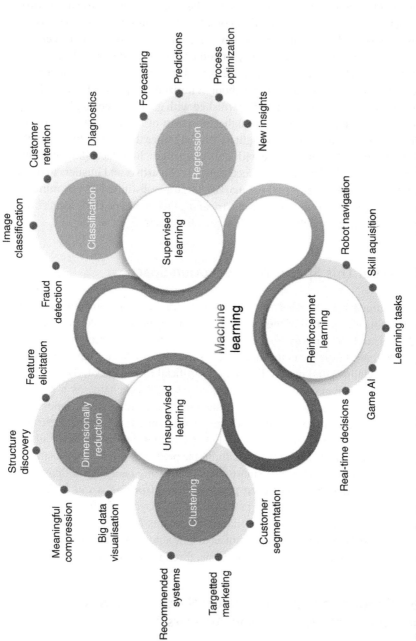

Figure 2.5 It appears that we have been mining for patterns and other simpler models (such as clustering, classification, categorization, regression, and principal component analysis). But, have we found a set(s) of simpler models or patterns, yet, to test the concept of domain-specific denominator models (DSDM)?

mandates, for example, (i) sensor data about heavy metal (mercury) contamination in water used for irrigation, (ii) near real-time respiratory rate (RR) of patient with chronic obstructive pulmonary disease (COPD) under remote monitoring telemedicine in rural nursing home, and (iii) automated check-out scan data from retail grocery store sales, of fast-moving consumer goods, contracted for replenishment (penalty for out of stock) under vendor managed inventory (VMI). The *e-tail* revolution is creative supply chain optimization and reducing retail information asymmetry.

Transforming data and data analytics to inform decision support for *small* cross-sections of examples cited, here, may be theoretically easy in *"power point"* diagrams which "connects" nodes and integrate decision feedback to optimize processes, *using pixels*. The reality may be different. Aggregating data from various nodes, sub-nodes, devices, and processes, on a platform, to enable collective evaluation of dependencies, which could influence outcomes/decisions, may be beneficial, or germane for certain domains, for example, healthcare [113] and clinical [114] environments where patient safety [115] must be of paramount importance.

Agreement on any *one* standard platform is unlikely to succeed. But an anastomosis of platforms is probably rational, if interoperable. An open platform of platforms with secure, selective, interoperable data exchange, between platforms, may be valuable. Synthesis and convergence of data acquisition and analytics begins to catalyze information flow, decision support, and **meaningful** [116] use of data [117]. This suggestion is a few decades old, but still far from practice. The drive to connect data was accelerated by the introduction of the concept of the IoT [118]. Platform [119] efforts [120] are addressing data [121] upload from devices and sensors [122] but nowhere near a turn-key implementation (Figure 2.6).

2.7 Solution Economy: Will We Ever Get There?

There are not any silver bullets and one shoe does not fit [124] all. If we focus on the data to decision process, alone, in any vertical or domain, the variations of analysis and analytics may be astronomical. Initial investments necessary for these endeavors almost guarantee that the extracted value from data (and relevant information) may not be democratized or made functionally available to those who cannot pay the high cost. In principle, the outcome from data to decisions, when appropriate, may be sufficiently distributed and democratized to provide value for communities under economic constraints. Any **meaningful** solution, therefore, is not a scientific or engineering outcome, alone, but must be combined with the economics of technology [125] which must be a catalyst for implementation and adoption by the masses, if transaction costs [126] can be sustained by the community of users, in less affluent geographies.

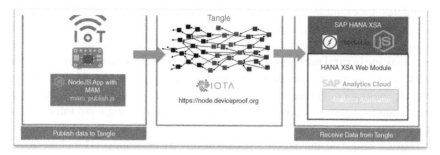

Figure 2.6 National and international consortiums, in partnership with large and small software companies, are addressing data acquisition and aggregation. In this example, Tangle appears to be a data aggregation platform (example shows data from a temperature sensor) which can serve data analytics engines to extract information (if data contain information). Replication of "Tangle" for various verticals (retail, health, and logistics) and the ability to use open-data distribution services [123] may facilitate interoperability between data "holding" services like Tangle. When coupled with supply chain track and trace systems, a retail store (Target, Tesco, Metro, and Ahold) can use Tangle data to inform a customer that the One-Touch blood glucose testing strip (healthcare product manufactured by J&J) will arrive at the store (third-party logistics provider and distribution transportation service) on Monday by 2 a.m. and placed on store shelf by 530 a.m. (retail store replenishment planning) or delivered to the customer on Tuesday before 9 a.m. (online fulfillment services). Any data can be uploaded/downloaded from Tangle.

The economic principle for impoverished environments may be rooted in micro-finance [127] and micro-payments [128] with *low* transaction costs (*the downside*: misinformation [129] can be propagated and disseminated at *low* cost, too). By eliminating classical "product sales," the focus shifts to delivery of "service" which is a *package* of the product plus other resources (retail mobile banking, infrastructure, telecommunications, cybersecurity [130], security [131], and customer service). Users pay (pennies) *only when they use the service*. Pay-a-penny-per-use or pay-a-price-per-unit [132] (PAPPU) is a *metaphor* for economic instruments which may lower the barrier to entry into markets with billions of users.

The economic incentive for democratization of data is the potential to unleash and create new markets for data, information, and decision support, for billions of new consumers (users). The **reward** in the lucrative service economy model depends on harvesting the economies of scale where each user (market of billions) may pay one or more "pennies" (micro-payment for pay-per-use services). The **risk** in the service economy is the collection of that "penny" (per use) at the last step of the *seamless* service delivery process, if the user is satisfied with the quality of service (QoS) metrics. The plethora of partners necessary to create and sustain the ecosystem to deliver the *seamless* service is a herculean task. Sharing a fraction of that "penny" with the partners in the ecosystem is not a trivial challenge. If the QoS delivery metrics suffer due to poor performance of any one partner (component), the end-user "penny" may be unpaid if the QoS metric fails to reach a predetermined value (time, duration, speed, rate, and volume). The inability of one provider (weakest link) in the service supply chain can be financially detrimental to all other supply chain partners due to loss of that penny, *albeit, only for that transaction* (unless the partner has a chronic problem, then, it must be excluded from the ecosystem and the entire value network [133]). Delivery of service is a real-time convergence of operations management which includes (but is not limited to) multiple value chains which must integrate [134] the physical supply chain and the financial supply chain with the service supply chain and customer relationship management (brand expectation).

Determining the cost of execution, to deploy the example in Figure 2.6, may be one way to study feasibility. Simulating models to explore financial engineering of "what if" scenarios, may project the potential for adoption of services in the context of various economies of scale and PAPPU models. The reward for unchaining the economics of technology is in adoption, by the next billion users.

2.8 Is This Faux Naïveté in Its Purest Distillate?

Decision scientists must build a compass to help extract value from data. One compass will not suffice to guide domain-specificity. Existing tools may limp along with *snail-ish* advances (Figure 2.5) yet it may remain inaccessible to the masses because the tools may not be feasible for mass adoption. The struggle to transform data into information is still in quest of a Renaissance.

The path from *data-informed* to *information-informed* to *knowledge-informed* decision remains amorphous. Transforming information to knowledge is in the realm of unknown unknowns. Making sense of data is handicapped due to (i) an apparently insurmountable semantic barrier, (ii) scarcity of tools to facilitate location-aware and context-aware discovery of data at the edge or point of use, and (iii) lack of standards and interoperability between objects, platforms, and devices for data and analytics sharing.

Users in less affluent nations may not want to idle away while the architects of Renaissance are still in short supply. In the near term, it is necessary that we continue to work on dissemination of data which can deliver at least some value, sooner, rather than later. Decision support based on sensor data analytics may provide economic benefits [135] and incentives, if we can share the digital dividends with the masses, for example, in health [26] and agriculture, including every facet of food, required daily, globally.

Tangle, a tool [136] to share sensor data using masked authenticated messaging [137] (MAM) may offer hope. Can nano-payments for sensor data address some of the feasibility challenges [138] and pave the way for human-centric economy of things [139] using IoT as a design metaphor? SNAPS [140] is a tiny step in that general direction: distributing low-cost tools to enable data-informed decision support (DIDAS) for less complex problems. Assuming the Pareto principle holds true, perhaps 80% of the problems may be addressed, and even resolved, with simple tools to deliver solutions as a *service*, at the right-time, at the point of use.

2.9 Reality Check: Data Fusion

The inflated view of the sensor-based economy [141] is carefully [142] crafted [143] to create [144] new markets [145] and momentum [146] for sales [147] of sensors and data services [148] aimed to amplify the IoT [149] hype to fortify the deception game. It is promoting the desired effect by spawning mass hysteria and skillfully obfuscating the hard facts which then paves the ground for hordes of consultants to act as "trusted advisors" to make sense of this "revolution" which is supposedly going to change the future of work, life, and living. One glaring outcome of delusional [150] propaganda [151] is the near-trillion dollar [48] waste related to investment in technology with a failed ROI. Trillions of sensors and devices that *could* connect to the Internet (basis for the cosmic scale of IoT) is due to the scale of unique identification [152] made possible by adopting a 128-bit structure in the internet protocol. The unique address spaces in IPv6 [153] is *29 orders of magnitude higher* than IPv4 if one compares [154] 4.3×10^9 address spaces for the 64-bit IPv4 versus 3.4×10^{38} unique address spaces for the 128-bit IPv6. New possibilities [155] and applications [156] may arise due to the flexibility of IPv6 to directly connect to the Internet (rather than sub-nesting under/via gateway nodes).

The difference between promise and perils in deploying the concept of IoT as a design metaphor is rooted in grasping the difference between connectivity, discovery, and actionable insight. Just because something is *connected* does not mean value emerges, automatically, without a **connected ecosystem**. If a visitor's tablet can discover the printer in an office and use it to print a meeting agenda, then we have extracted some type of value between the connectivity of the tablet and the printer, which were able to "discover" each other, and that discovery enabled the gain in efficiency (printing the agenda). In its basic form, this is an example of very simple data fusion which leads to an actionable output and provides "information" for the meeting attendees in terms of the printed agenda. Connecting trillions of entities to the Internet is futile unless discovery and data fusion enable semantically meaningful extraction of data to move up the DIKW [157] value chain [158] where data precede information, knowledge, and wisdom.

The PEAS paradigm resembles OODA (Figure 2.8) because "observations" refer to scanning (sensing) the environment and "orientation" informs the image of the environment by encapsulating both descriptive and predictive analytics ("decide" includes prescriptive analytics). The integration of data fusion and analytics with ABS (Agent-Based System) is critical in the era of IoT. The networked society faces a deluge [159] of data, yet the human ability to deal with data, analytics, and synthesis of information may be inefficient. How can devices discover data and facilitate processes without intervention by humans? Automated on/off action taken by a domestic thermostat and HVAC based on temperature sensors may be quite primitive when considering autonomous objects in air (UAV), land, and water.

Raw sensor data unless discovered and combined with "perceptions" from the environment, may be context-deprived and over/under utilized, which lowers the value of the data with respect to the desired goals. The perception from the environment is not unique but a "learning" task for the system. It may reuse the experience (learning), when relevant and appropriate, at a different instance (Figure 2.7). Can this "learning" become mobility-enabled and "teach" other devices, for example, by transmitting a *tutor* virion to another computer or drive or system? Can this device communicate in natural language (NL) and/or respond/ understand the semantics in human queries?

Taken together, unleashing the value of data may require coordination of ABS in every facet of our interaction with machines, objects, and processes which may benefit from feedback. ABS is an old [160] concept [161] but resistant to succinct definition [162] because agent activity must remain agile and adapt to the operating objective (PEAS) and problem context (OODA). Equation-based models

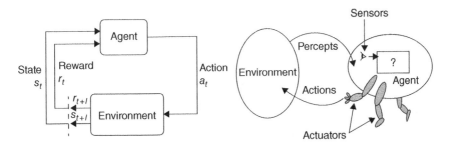

Figure 2.7 SNAPS is one layer in a proposed analytics platform (layer cake) which consists of a portfolio of tools aligned with the concept of PEAS [23], a mnemonic borrowed from agent-based systems (ABS) **(R)**, to address systems performance through convergence of percepts, environment, actuators, and sensors. *Source:* Modified from Russell, Stuart and Norvig, Peter (2010) *Artificial Intelligence: A Modern Approach.* 3rd ed, Prentice Hall. **(L)** Reinforcement Learning [24] (Figure 2.5), a ML technique, compared with PEAS. *Source:* Modified from Sutton, Richard S., and Andrew G. Barto, 2018. *Reinforcement Learning: An Introduction.* 2nd edition, MIT Press.

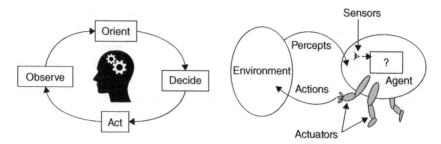

Figure 2.8 PEAS, a mnemonic borrowed from agent-based systems **(R)**, addresses systems' performance through convergence of percepts, environment, actuators, and sensors. The OODA [25] loop **(L)** and PEAS [26] contribute to advance DIKW (data, information, knowledge, and wisdom), which begins with data fusion [27]. *Sources:* Angerman, W.S. (2004). *Coming full circle with boyd's ooda loop ideas: an analysis of innovation diffusion and evolution* (Thesis); Kamenetz, A. (2011). *Esther duflo bribes India's poor to health. Fast Company* (8 August); Castanedo, F. (2013). *A Review of Data Fusion Techniques.* The Scientific World Journal.

(EBM) create rigid, hard-coded software. ABS design induces agility, may enable "drag & drop" variant configuration to adjust (on-demand) to volatility, uncertainty, and ambiguity, inherent in most environments. In the context of democratization of data and benefits for the masses, agents can be highly personalized and "belong" to people, for example, personal agents, as discussed [163] elsewhere, with respect to cybersecurity. A similar modus operandi can be adopted for other use-cases where data fusion [164] can be dynamic and composable (composed when necessary, depends on context) not only for use-cases but also for individual user-specific case/application (healthcare treatment plan) (Table 2.1).

The role of software agents to "discover" and then determine which data and/or data fusion may be meaningful or relevant (user-specific), is an old idea, still waiting to be effectively applied. Connecting data must be contextual. The *established* contextual relationship must be discovered and "understood" by agents or group of agents. Another old idea is to *pre-establish* the context based on KGs (Knowledge Graphs). The thinking that W3C standard resource description framework [166] (RDF) triples are the solution for KGs is incomplete. This myopia, is, in part, one reason why the semantic web [167] failed to flourish. The brilliant idea of representing subject-predicate-object (SPO) as a relational RDF graph is certainly useful and applicable in many instances but the approach *bites the dust* when the reductionist 1:1 granular relationship fails to represent reality. The latter is painfully obvious, especially in medicine and healthcare, where the rigidity of the RDF standard structure and RDF schema may be an anathema. The "force-fitting" of RDF to healthcare applications [168] oversimplifies scenarios to the point where it may, inadvertently, introduce errors, simply due to exclusion, which may prove to be fatal.

Table 2.1 Generally, agents are computational entities (software) designed to perform specific tasks, autonomously.

Agent characteristics	Definition
Autonomy	Operates without the direct intervention of humans or others
Sociability	Interacts with other agents, that is, communicates with external environment such as sensors, fusion systems, and human operators
Reactivity	Perceives its environment and responds in a timely fashion
Pro-activity	Exhibits goal-directed behavior by taking the initiative
Learnability	Learns from the environment over time to adjust knowledge and beliefs
Mobility	Moves with code to a node where data reside
Anthromorphicity	Externally behaves like human

Agents embedded in devices (sensors) may have logic capabilities to perform artificial reasoning tasks (ART) and/or optimization [165] in multi-agent systems (MAS).
Source: Based on Rafferty, E.R.S., et al. (2019). Seeking the optimal schedule for chickenpox vaccination in Canada: using an agent-based model to explore the impact of dose timing, coverage and waning of immunity on disease outcomes. *Vaccine* (November 2019).

One proposal suggests adapting [169] RDF by creating relationships between sets/subsets (rather than points and vertices as in classical SPO) using the set theoretic [170] approach. It is easy to grasp why "set" of symptoms and potential set of causes may make more sense in medicine and healthcare. The overlapping (Venn diagram) subset of relationships may be indicative of likely causes for symptoms. Generic symptoms, for example, fever, can be due to a plethora of causes and why a rigid 1:1 relationship in RDF could turn lethal in healthcare applications. The finer granularity of RDF is a disadvantage, yet it is key to merging attribute lists about an entity sourced from different data sources. The latter enables better search and discovery across diverse domains, the hallmark of globalization of enterprise systems.

An even older idea [171] which is recently [172] enjoying scientific [173] as well as public attention [174] is the labeled property graph (LPG). It is suitable for use-cases which may be focused on providing stores for single applications and single organizations, such as, DSDM. LPG proponents are less committed to standardization, interoperability, and sharing. It is in contrast to the W3C ethos and RDF which favors standardization, interoperability, and sharing, which makes it useful in discovery using graph pattern searches. Optimization of local (domain specific) searches [175] using graph-traversal algorithms [176] are better suited for property graph (PG) databases. KG networks embedded between sets/subsets may give rise to amorphous "linked" clouds, which could be industry [177] specific and may be domain specific [178] as well as user specific. Imagine if data from each patient could be used by an ***automated knowledge graph engine*** to create precision, patient-specific, personalized KGs. Extracting relationships and contextualizing the relevance of symptoms may improve the accuracy of diagnosis. When viewing KGs in a population study (epidemiology), it may be easier to detect outlier events or cases that did not fit the expected patterns.

Therefore, DSDM may be represented as domain-/user-specific KG networks. Agents may be invaluable in working within this environment to discover relationships and contexts (specificity reduces search space), as well as discover data sources, and perhaps, based on embedded logic, decide whether the features or attributes calls for data fusion.

For any agent-based approach to succeed, it is critical that the agent framework and standards are interoperable with the KG network and the data domains where the agent is searching. The opposing tendencies of RDF vs LPG in terms of standardization, interoperability, and sharing may limit agent-mediated "cross-investigation" of domains, discovery, and data. Therefore, it begs to question the expectation that one agent must perform in all domains. Perhaps, the success of agent search and discovery depends on semantically annotated structured data. The latter depends on ontological structure. W3C proposed [179] OWL standard web ontology language [180] and recent variations (VOWL [181]) may contribute to interoperability. The old idea of Internationalized Resource Identifier (IRI) as a complement to the Uniform Resource Identifier (URI) [182] to identify resources (to facilitate discovery) is a valid principle but yet to be adopted in practice. The plethora of old ideas (referred here) suggests that the value of these ideas may have to be revisited. We need new "blood" and new "eyes" to reimagine new ways to address interoperability. However, in reality, today, on top of this wobbly incompatible infrastructure, we are layering the "snake oil AI" and unleashing an incorrigible torrent of half-truths.

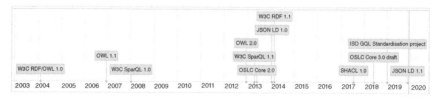

Timeline 2.1 Long march of graph-related [183] specifications. Is standardization a sufficient solution?

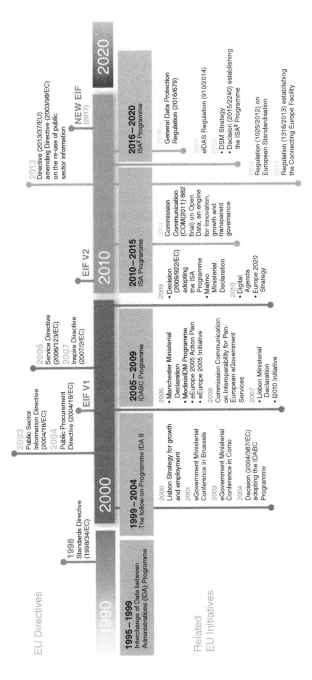

Timeline 2.2 EU's elusive [184] quest for interoperability: is 30 years not enough? *Source:* Carbone, L., et al. (2017) State of play of interoperability: report 2016.

2.10 "Double A" Perspective of Data and Tools vs. The Hypothetical Porous Pareto (80/20) Partition

Africa (>1 billion) and Asia (>4 billion), if combined, may soon represent 80% of the world's population. Global corporations view this "80%" world as a "market" which promises new markets, new customers, and new wave of consumerism. It has little to do with lifting the lives of people. Discussion about the physics and mathematics of data, therefore, is *a tempest in a tea cup*. For ~7 billion people, the trials and tribulations of data and data analytics, we have discussed here, can be dismissed with an eye roll. It is useless for pressing daily applications for ~7 billion of the ~7.8 billion people in the world.

Thus far, what we have discussed, on one hand, may be an exploration of the tessellated facets in our search for meaning, and on the other hand, it is a discussion which may find parallels with the "six blind men and the elephant" syndrome [185] apparently divorced from complementarity [186] or synergy. It is as if the "commerce" from 20% of the global population, relevant or not, is thrust upon the remainder of the world market. In 80% of the cases for 80% of the global population, the daily decisions about FEWSH (bare necessities of life: food, energy, water, sanitation, healthcare) do not require AI, ML algorithms, or optimization of "state space" for hundreds of variables (Figures 2.9 and 2.10).

In 80% of the cases for 80% of the global population, the daily decisions about FEWSH require data, *small data*, data in near real time, and data that impact and enhance the user experience. In that context, the clamor about *democratization of data* is tantamount to chest-thumping. The data in these use-cases may be related to a subset of FEWSH (food, farm, agriculture, water, healthcare). If the tools are there to acquire these data, then the data are available. Therefore, is democratization really an issue? Is it a politically correct word that the 20% world prefers to use as a hand-waving advocacy of problems that are divorced from reality on the ground? Is "democratization" a "theme" song for advocacy groups in OECD nations who are displaying the symptoms of the *six blind men and elephant* syndrome?

Table 2.2 offers a glimpse of one problem in healthcare. The tools to acquire the data are in short supply. Measuring the risk of osteoporosis is a prerequisite for prevention and treatment, if affordable. Arm-chair "scenarios" of medical IoT will want to connect DEXA (DXA) scan data with sales of milk and exposure to sunlight as a "wellness" indicator. From the "double A" perspective, it may be a futile "power point" exercise because milk *may not be available* for the age group [200] generally at high risk of osteoporosis in the AA nations. In most parts of Africa and Asia, there is an opulence of sunlight.

Just because there is an "IoT" scenario, does not mean it is worthwhile or valid for users in "double A" nations. Just because there are data, does not mean there is information. Can we reduce incidence rates of osteoporosis simply by adding

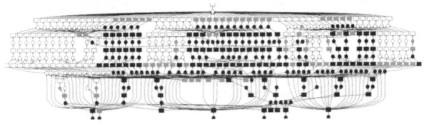

Figure 2.9 Blind men and elephant. Each man guesses his own part of the elephant but blinded by hype [187] they cannot perceive the "whole" elephant. A metaphor for focus on parts, which occludes the system. Cartoon (bottom): penchant for decision trees by *power-point* rather than search for low hanging fruits. *Source:* Modified from Morrison, A. (2019). Is data science/machine learning/AI overhyped right now?

more DXA machines per capita? A recent (2013) study using seven national electronic healthcare records (EHR) databases revealed that Denmark (14.2 DXA units per million) showed age- and sex-standardized incidence rates (IRs) of hip/femur fractures 2× higher than those observed in the United Kingdom (8.2 DXA units/million), Netherlands (10.7 DXA units/million), and Spain (8.4 DXA units/million), while Germany (21.1 DXA units/million) yielded IRs in the middle range (Table 2.3).

2.11 Conundrums

On one extreme we have presented sophisticated ideas for making sense of data. On the other hand, we doubt whether the toothless call for *democratization of data* from the affluent 20% of the world can help to lift the lives of people on the other side of the porous Pareto partition (80% of the world). It is not a true "Pareto"

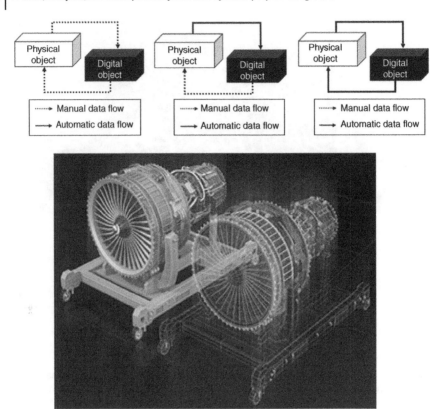

Figure 2.10 Digital Duplicate (left), Digital Shadow [188] (center), Digital Proxy (right), and Digital Twin (bottom) are variations of digital models of physical objects, integrated with data flow. But, do we know if it is *meaningful* for data-related needs for 80% of the world? It is unlikely to be solved by Digital Twins [189] or flamboyant gimmicks peddled by fake pundits on the pages [190] of *Forbes*. However, the R&D related to these tools may trickle through the "pores" from the 20% side of the partition to the other side (80%) of the *porous* Pareto [191] partition and occasionally [192] may be helpful. *Source:* Modified from Fuller, A., et al. (2019). Digital twin: enabling technology, challenges and open research. October 2019.

scenario, but the 80/20 nature of this problem evokes the Pareto principle as an analogy, hence, Pareto partition. The R&D outcome of the 20% may contribute certain elements to the 80% side. Thus, the "partition" is a metaphorical porous membrane, with bidirectional porosity.

But operations must be contextual. For example, is it necessary to deal with data and data models in this scholastic [203] manner (Figure 2.11) for all problems? The 80/20 global partition may be prominent in agriculture, healthcare, and energy. In case of the latter, what is the value of smart metering or load balancing

Table 2.2 Availability [193] of DEXA (dual energy X-ray absorptiometry) scan machines to measure bone mineral density (BMD), a fair prognosticator for osteoporosis.

The number and provision of central DXA units available in the EU27 (Data on reimbursement and waiting time [10])

	DXA units/million	Waiting time (d)	Cost (€)	Reimbursement
Austria	28.7	14	30 [11]	Yes
Belgium	53	14	34 [12]	Partial
Bulgaria	1.2	0	59 [9]	None
Cyprus	23.9	20	75 [9]	Yes (depending on income)
Czech Republic	5.2	40[*]	32 [9]	Yes
Denmark	14.6	30	187 [13]	Yes
Estonia	8.9	14	14 [14]	Yes
Finland	16.8	1	146 [28]	Yes
France	29.1	14	41 [9]	Yes (conditional)
Germany	21.1	0	36 [9]	Yes
Greece	37.5	11[*]	115 [9]	Yes
Hungary	6.0	15[*]	7 [15]	Yes
Ireland	10.0	140[*a]	99 [16]	Yes (conditional)
Italy	18.6	83[*]	81 [9]	Yes (conditional)
Latvia	4.9	10[*]	18 [29]	Yes
Lithuania	3.4	6[*]	28 [30]	No
Luxembourg	2.0	30	59 [31]	Yes
Malta	9.7	105[*]	184 [32]	Yes
Netherlands	10.7	14[*]	84 [17]	Yes
Poland	4.3	1	10 [9]	Yes (conditional)
Portugal	26.9	8	5 [33]	Yes
Romania	2.4	7	5 [34]	Yes
Slovakia	10.7	18[*]	32 [9]	Yes
Slovenia	27.1	11[*]	29 [35]	Yes (conditional)
Spain	8.4	105[*]	109 [9]	Yes
Sweden	10.0	60	152 [18]	Yes
UK	8.2	11[*]	51 [36]	Yes[a]

The European standard [194] is 11 (DEXA) DXA units/million. In an updated estimate, the poorest country in EU27 offers four machines/million [195] whereas Bulgaria's neighbor Greece boasts 37.5 DXA units/million. In comparison, Indonesia [196] has only 0.13, India [197] 0.18, and Morocco [198] 0.6 DXA units/million. For the health of the people in these nations, how can democratization of data lower their risk of osteoporosis? Are we asking the correct questions? Are we pursuing the wrong reasons? Are we arm-chair analysts helping the (BMD) medical device industry [199] accelerate their sales campaigns to AA nations? Can data provide relevant answers? *, average of range; a, data; d, days.

Source: Hernlund, E., et al. "Osteoporosis in the European Union: medical management, epidemiology and economic Burden: a report prepared in collaboration with the international osteoporosis foundation (IOF) and the European federation of pharmaceutical industry associations (EFPIA)." *Archives of Osteoporosis*, vol. 8, no. 1–2, 2013, p. 136. © 2013, Springer Nature.

Table 2.3 Plague of unethical profitability makes US pharmaceutical [201] business model in healthcare an abomination which is inappropriate for mimicry in any part of the world.

Drug	Prescribed for	UK price	US price	Price rise
NEXIUM per 20 mg tablet	Acid reflux	£0.66	£7.40	1120%
ACTIMMUNE, 12 vials	**Genetic diseases, osteopetrosis**	**£5,400**	**£42,990**	800%
DARAPRIM per tab	HIV, cancer, malaria patients	£2.30	£619	**26,900%**
NASONEX, 50 mg	**Nasal allergies**	**£7.68**	**£224**	2900%
CINRYZE, 2 vials	HAE, genetic disorder	£1,336	£3,645	200%
HARVONI per tab	**Hepatitis C virus**	**£464**	**£928£270%**	
SOVALDI per 400 mg tab	Hep C in children under 12	£416	£855	**200%**
DIAZEPAM, per tab	**Anxiety, relaxation, muscle spasms**	**£0.02**	**£3.05**	**15,200%**
OVEX, 100 mg tablet	Threadworm parasite	£2.54	£300	**11,800%**
LIPITOR, per 10 mg tab	**Statin**	**£0.46**	**£4.50**	980%
VIAGRA, per 25 mg tab	Male impotency	£4	£61	1500%
ZOCOR per 10 mg tab	**Statin**	**£0.64**	**£4.20**	656%
CYMBALTA per 30 mg capsule	Anti-depressant	£0.80	£9.48	1200%
EPIPEN, 300 mg	**Allergies**	**£52.90**	**£523**	1000%
HUMALOG INSULIN	Diabetes	£16.61	£215.30	1300%
HIP REPLACEMENT OPERATION		**£7,313**	**£26k-£37k**	350%
KNEE REPLACEMENT OPERATION		£6,315	£24,801	**390%**
CATARACT OPERATION		**£803**	**£5,780**	720%

Source: Dr James Nolan [202].

algorithms when there is not enough energy, at an affordable cost, to supply the basic tenets of economic growth? How many farmers in "80%" world can afford to use drone-on-demand [208] systems? Why should people from the majority sector (80%) need useless marketing tools [209] when daily healthcare for the less fortunate can solve a myriad of problems with *just-in-time little bits of data,* for example, daily blood glucose level from a diabetic (versus the *always-on* real-time

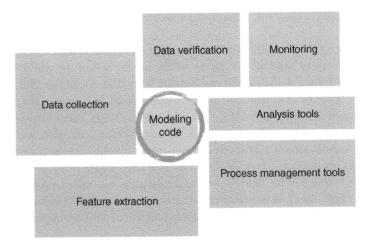

Figure 2.11 From monitoring an event to using the data to inform a decision, there are a plethora of steps [204] in the standard operating procedure (SOP) for the "20%" deploying data to drive decisions. However, irrespective of socioeconomic issues, in future, all aspects of feature selection and feature engineering may emerge as a pivotal or rate-limiting step in dealing with diverse data sources. In this context, automated feature extraction and other feature-related steps may be a very significant step. *In combination, automated feature engineering and automated knowledge graph engines may usher new dimensions in data and data analytics, if automated data curation could improve data quality* (Figure 2.12) [207]. *Source:* Mendez KM, Broadhurst DI, Reinke SN. The application of artificial neural networks in metabolomics: a historical perspective. *Metabolomics* 2019; 15(11):142. doi: 10.1007/s11306-019-1608-0.

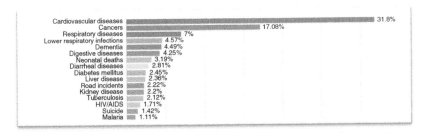

Figure 2.12 Share of deaths, by cause (2017) percent of total deaths [205]. Data refers to specific cause of death, which is distinguished from risk [206] factors for death (water and air pollution, diet, sanitation). *Source:* Ritchie, H. (2018). What do people die from? Licensed under CC BY 4.0

monitoring of blood glucose) or monitoring individuals for silent myocardial ischemia [210], a leading contributor to death. This is the debate where the economics of technology and its ***relevance to the community*** are crucial issues which may ***enable*** adoption or ***disable*** the dissemination of technology, which could have contributed to economic growth, workforce development, and sustainable job creation.

Worldwide, Africa [212] accounts for 9 out of every 10 child deaths due to malaria, 9 out of every 10 child deaths due to AIDS, and for half of the world's child deaths due to diarrheal disease and pneumonia. More than one billion children are severely deprived of at least one of the essential goods and services they require to survive, grow, and develop [213] – these include nutrition, water, sanitation facilities, access to basic healthcare services, adequate shelter, education, and information. As a result, almost 9.2 million children under five die every year and 3.3 million babies are stillborn. Most of the 25,000 children under five that die each day are concentrated in the world's poorest countries in sub-Saharan Africa and South Asia. There, the child mortality rate is 29 times greater than in industrialized countries: 175 deaths per 1000 children compared with 6 per 1000 in industrialized countries.

These facts (paragraph above) and Figure 2.13 offer a vastly contrasting view to that of data tools and democratization of data as essential for lifting the lives of the people living on the majority side of the porous Pareto partition. Simple forms of small amount of data, sufficiently informing ordinary tasks, may be suitable for delivery of global public goods and services to the majority of the 80% world. It is absolutely ludicrous to think that "big" data, AI/ML, blockchain, or smarmy publicity [214] stunts may help, in this context. What we need is the concept [215] of "bit dribbling" perhaps coupled with PAPPU systems to help people improve their quality of life without the constant quest for charity.

Technology may play a central role to reach the billions who need services but not in the form of business [216] which is stable in the West and copied by the thoughtless Eastern schools, especially in India. Technical tools will generate data. The ability to use that data, judiciously, may be key to the value of data, for impoverished nations. Coupling social need with technical catalysts must be optimized in the context of the community and not according to *Wired* or *MIT Tech Review* or *HBR*. Advanced R&D is the bread and butter of progress, but the application of advanced tools must be contextual to the services that the community can sustain. Just because auto maker Koenigsegg claims the *Agera* model was built with a "less is more" philosophy does not mean it is a pragmatic standard of transportation suitable for Calcutta, India. In the realm of systems engineering courses and education, dynamic optimization (DO, Figure 2.14) illustrates a similar perspective. The principle is worth teaching, worldwide, but the practice must be relevant to the case. Do we all need DO in everyday life and living? Is it necessary for all types of edge analytics to process data using convolutional neural networks (CNN) on a mobile device or phone?

The conundrum of **not** applying the tools we think we have mastered is counterintuitive to the problem-solving ethos in the 20% world. We are ever ready to

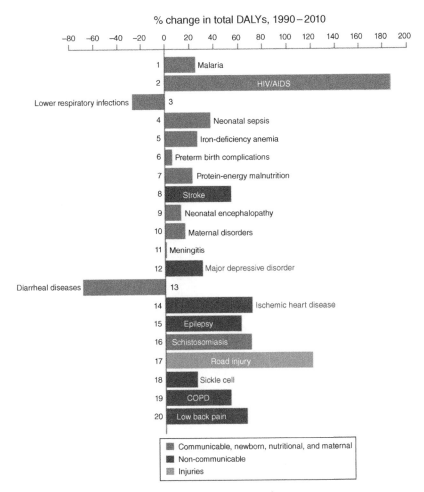

Figure 2.13 Shifts [211] in leading causes of DALYs for females, Ghana (1990–2010). The leading 20 causes of DALYs are ranked from top to bottom in order of the number of DALYs they contributed in 2010. DALYs (Disability-adjusted life years): The sum of years lost due to premature death (YLLs, Years of life lost due to premature mortality) and years lived with disability (YLDs, Years of life lived with any short-term or long-term health loss causing disability). DALYs are also defined as years of healthy life lost. *Source:* Institute for Health Metrics and Evaluation, Human Development Network, The World Bank. *The Global Burden of Disease: Generating Evidence, Guiding Policy – Sub-Saharan Africa Regional Edition.* Seattle, WA: IHME, 2013. © 2013, Institute for Health Metrics and Evaluation.

use the latest and greatest gadgets from the bleeding edge to derive and drive the best possible perfection and performance. The quagmire of lies aside, we do have real tools which offer notable advantages. But, the volume of the 80% of the world and the economic handicap in these communities must be assimilated in order to

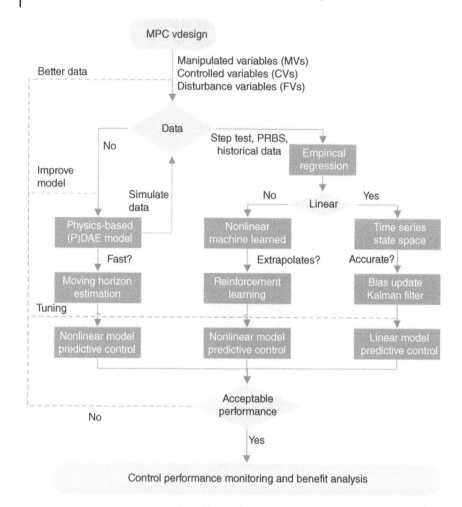

Figure 2.14 Dynamic optimization [217] is a central component of systems engineering where applications of numerical methods for solution of time-varying systems are used to improve performance and precision of engineering design and real-time control applications, which may have a broad spectrum of use, for example, from optimizing the artificial pancreas to fuel cells. Principles of DO may be taught [218] worldwide but DO, systems level data science [219], and Bayesian [220] statistics are excellent tools yet, often, less useful for 80% of the tasks for 80% of the world on the other side of the porous Pareto partition. *Source:* Modified from Hedengren, J.D. Drilling automation and downhole monitoring with physics based models.

change our thinking. The acronym KPI is for "performance" which is euphemistic for profit in the affluent world. It may not be in the best interest of the people. For 80% of the world, perhaps KPI should stand for "key people indicator" and ascertain whether a tool or the service improves the life of people.

Improving lives, however, is relative to the life you aim to improve, a life with disabilities [221] versus life with social [222] void are active domains in robotics. Robotics is useful but the robot propaganda, written mostly by hacks [223] and driven by media [224] sales, is a sign of the times. Essential robotics and robots for tasks that are dangerous, dirty, and dull (repetitive) are a welcome relief, for all involved.

The idea of the automated robotic factory was popularized by Philip K. Dick's fiction "Autofac" published [225] in 1955 (*Galaxy* magazine). The "lights-out" automated manufacturing facility FANUC [226] (factor automated numerical control) has been in operation since 2001, in Japan, but it is an exception. Even though "lights-out" robotics made significant strides in heavy industry, it is far from the Orwellian scenarios promoted through chicanery [227] and buffoonery [228] by discombobulating the masses. The promise of robotics must be balanced with the degree of **trust** [229] **in automated execution** and the system must be **acutely aware** of the perils due to cyberthreats. Cybersecurity is quintessential for automation.

In other instances where human life is at risk (for example, transportation, manufacturing, and mining), the trust in automated action (robot) is as good as the planning for "what ifs" **when** the auto execution goes awry. But, that is a deterministic perspective where what could go wrong is anticipated, albeit with some degree of uncertainty. However, if a mobile robot crashes with a holonic manufacturing podium, it may generate a cascade of events where the outcome may be nondeterministic. The critical question in such a scenario is the extent to which a nondeterministic outcome can be tolerated and the *acceptable* cost of risk despite the "open-ended" uncertainty. Few can even approach to answer this/these questions [230] because it verges on the domain of unknown unknowns.

But, that may not deter simulation aficionados from pursuing stochastic (*what if*) models to capture distribution of randomness in nondeterministic outcomes. Heuristics approaches may surface to suggest contingency measures. This is "video gaming" of automation [231] which could turn deadly in reality. The *executive* robot may be suited for "3D" tasks (dull, dirty, and dangerous) but unsuitable for relinquishing human oversight and control if lives are at risk. However, even worse are evil acts perpetrated by humans to bury [232] and ignore [233] the failure of automation, in the pursuit of profit.

Robotic tools in the 20% world are engaged in sophisticated activity which may be subjected to high oversight. In general, the 80% world is not a customer for such implementations in terms of mass consumption. Automation replacing or reorganizing jobs is not a new event (for example, auto industry) because technology [234] shifts the cycle of jobs and with it, the economy. Rapid changes in skill sets and the volatility of job categories influence other domains, namely, K-16 education, training, skills development, (capital, labor market, and employment) and communication (hopefully, the truthful variety). The rate of change in certain ecosystems is dreadfully slow (for example, education system) whereas the evolution of the job market may resemble the rapid pace of bacterial growth, albeit slower than viral growth rates. The diffusion of robotics will take time and only if the building blocks of automation can be popularized, globally, in a manner that Lego blocks may have inspired young minds to compose, create, and construct.

The 80% world can benefit from robotics, for example, by reducing global disease [235] burden in emerging economies. In India, children are still used to clean sewers (flexible enough to reach cramped spaces, similar to chimney sweepers [236] in seventeenth-eighteenth century UK). Can robotic tools replace the children? Robotics can improve lives and public health [237] rather than fear-mongering and flagrant deception [238] how robots will replace human jobs. Similar to enantiomeric profiling of chiral drugs [239] and opioids excreted in sewer water [240], the post-pandemic world must monitor wastewater for pathogens [241] as a surveillance strategy (prevention tool) or an early warning system prior to the onslaught of detectable clinical [242] symptoms in the general population.

2.12 Stigma of Partition vs. Astigmatism of Vision

The "partition" suggestion does not disguise the reality of the "ours" vs. "theirs" view of a divided world. It is unfortunate but necessary to serve as a constant thorn in our conscience and sow discomfort. The "partition" thinking originates from the corporate pursuit of developing a smörgåsbord of bleeding edge tools and then coerce 80% of the world to buy such products and services ("next billion users"). To add insult to injury, corporations from the 80% world are salivating to acquire rights to these products and bring it to their market (for example, Tata (TCS), TechMahindra, Wipro, Infosys, and other "body-shops" in India [243]). There exists a *nano-cosm* of people in the 80% world who could be a part of the 20% world. Because they are an influential minority and hold the financial power in the 80% world, they are aligning their astigmatic vision, greed, and "profit" objectives with the 20% world.

This mismatch may be at the heart of this global dilemma and creates the necessity to consider the porous Pareto partition in terms of people and service for the end-user. People in AA nations are not buying facial recognition software systems. The abuse [244] is perpetrated by governments. People in the 80% world are not seeking quantum computing to process exabytes of data. People are seeking simple information, for example, for their health (data for blood cholesterol level) or from their farm (data about concentration of heavy metal contaminants in irrigation water used for fresh produce, such as, tomatoes). These services help people, the end-user, the consumer. This discussion is about what science, engineering, and technology businesses can do for people where the KPI is user-centricity and human-centric [245] well-being, a fact which is immensely clear in the post-pandemic world.

This mismatch between the business to consumer (B2C) services versus the business to business (B2B) services is not new. The 80% world is always looking to the 20% world when planning strategic moves for climbing "up" the supply chain. The fact that the tools from the 80% of the world may not fit in the 20% world is obvious from "frugal innovation" calls [246] by others. Yet, imagination, invention [247], and innovation [248] from R&D in the 20% world are often helpful in the lives of the 80% world. What may be often lost is the *translation* of the advances from the 20% world, for people-centric applications, in the 80% world. This discussion is not singing the praises about the investment in research that only the 20% of the world can afford to push forward because we know [249] the facts. The world is indebted for the strides made possible due to entrepreneurial innovation in such havens such as Massachusetts and California. This discussion is about exposing the lies [250] but not slowing the leaps of vision from the 20% world which may help create low-cost [251] tools to serve the 80% world. The world needs R&D from the 20% world, hence, "porous" partition may facilitate the flow of innovation. Not "as is" but with contextual modifications to better serve communities in the 80% world, at a self-sustaining cost (for example, the PAPPU model, pay-a-penny-per-use or pay-a-price-per-unit).

Mental health is one problem where "porosity" is most welcome because most of the world are affected by generic [252] as well as specific issues, which contributes to economic [253] drain. Inherited bipolar and unipolar disorders [254] do not discriminate on the basis of race, color, religion, or national origin. The neurochemical, neuroendocrine, and autonomic abnormalities associated with these disorders need biomedical research to elucidate the neurobiological basis of these diseases. The latter is not feasible for the 80% world. Harvesting data [255] from external symptoms and pattern analysis [256] may offer a low-cost substitute, to inform the nature of treatment required. People in the 80% world may find it useful.

However, this discussion is not a *to-do* list. It is not a roadmap. It may be a compass, oscillating asynchronously from esoteric thoughts to bare necessities. We are immersed in this duality. One cannot exist without the other. The role of the "partition" is to help focus on the issues that are unique to the environment and community that we wish to serve. It is not a partition of R&D or people or products but a partition for ***delivery of service***.

The idea of democratization of data is a bit buzzy but gimmicks are key to marketing. August institutions, including MIT, are complicit in sponsoring potentially puffy pieces to keep the hype [257] alive. But, the fact remains that enabling data to inform decision is a bedrock of measurement, central to all, irrespective of economic status. The porous Pareto partition is a catalyst to focus on services for the 80% world, where less could be more and serves our sense of égalité.

Dribbling bits of data to inform a person that her RR (Respiratory Rate) is fluctuating, too often, may be a preventative measure (think of the proverb *a stitch in time saves nine*). Informing the person that her RR data are not copasetic, may reduce future morbidity due to COPD (Chronic Obstructive Pulmonary Disease). Providing data and information may be **without** impact on the quality of life in the absence of follow-up (clinic). In terms of data and information, alone, by enabling something simple and even mundane, the people-centric application of technology and data, preferably at the edge [258] (point of use), may help to do more with less. Unbeknownst to us, we are attempting to use the pillars of science, engineering, data, information, and knowledge to build bridges which may serve as a platform to provide service to billions of users. Rather than *gilding the lily*, we are offering a "bare bones" bridge which serves a rudimentary purpose and still may exclude a few. The volume and demand for such low-cost services [**PAPPU** services] may be, eventually, profitable for the business ecosystem.

Supporting a sustainable effort, to lift the quality of life, will depend on the extent of the product **ecosystem** and many other "things" in addition to technical and sensor data as well as the *cohesion of the service supply chain*. Socioeconomic data [259] and related factors are equally significant. Core elements are education of women [260] and trust [261] in women, followed by civic honesty [262], social value [263] as well as inculcating the practice of ethical profitability in social business and entrepreneurial innovation to accelerate the pace of creating pragmatic tools and solutions for remediable [264] injustices.

2.13 The Illusion of Data, Delusion of Big Data, and the Absence of Intelligence in AI

Neither data nor AI [265] is a panacea. Acquisition of data and analysis of data are not a guarantee that there is information in the data or that the information is actionable in terms of delivering value for the user, at an affordable cost. The COVID-19 pandemic has made it clear that the global public goods that define "life-blood" are food, energy, water, sanitation, and healthcare (FEWSH). The 80% world needs contextual, advanced, and affordable array of tools and technologies to leapfrog the conventional practices of FEWSH in the 20% world to vastly improve their crisis response systems.

In this context, energy is one rate-limiting entity and in a "tie" with food and water, in terms of human existence and life. The "hand-me-downs" from the 20% world of energy may not be sustainable. Perhaps the Sahara Desert may be a source of energy for creating a global "battery" field, an idea [266] triggered by an 1877 [267] proposal, in a different context (it was, too, subjected to misrepresentation [268] and mockery [269]). Whether this is a "good" idea or not is *not exclusively* a matter of technological feasibility of implementation or transaction cost of service delivery. The question is, if it is *good for the people*. Global public goods are a matter of context for the community as well as the continent. Exploring the cleavage between entrepreneurial engineering innovation and complexities of social egalitarianism requires willingness to recognize, and adapt, among many *different conceptions of a sense of the future*. It may not be the future deemed appropriate by the 20% world experts. The future is asynchronous and nonlinear (Figures 2.15 and 2.16).

It may be a nonbinary future with multiple paths and unequal connectivity between amorphous nexus of networks representing nonlinear choices, aspirations, and outcomes. "Good" decisions are relative to **that mix** which defies definition yet works as a catalyst for economic rejuvenation. Even this **type** of "good" will (must) change with time and culture because no one version of good can fit all the world [271]. A binary *outcome*, with exceptions, must not be confused with binary *decision-making* because a plethora of nonbinary factors can influence the outcome, which may *appear* as a binary output.

An oversimplified and cherubic example of the latter may resonate with residents of the Boston area. The choice between Mike's [272] and Modern [273], famed confectioners located almost opposite each other on Hanover Street in Boston's North End, is far from binary. The filling in the cannoli, taste of java, and the length of the queue are factored in the decision-making process, which generally presents itself cloaked in a binary-esque outcome. The choice masquerades a slew of nonbinary experiences.

Thus, creativity, imagination, and knowledge (from science, engineering, technology, medicine, and mathematics) may need to connect a few or many "dots" to inform solution development and delivery. It is true that "porosity" may contribute to solutions in the 80% world, and perhaps, less is more, but it will be remiss to leave the reader with the impression that invention/innovation may have to take a second place in the 80% world. In some cases, we must seek *out-of-the-world* or counterintuitive ideas [274] and blend it with incisive insight which may be non-traditional. Far-reaching *convergence* of bio, nano, info, and eco [275] is not an alternative but an imperative to stitch practical solutions, to satisfy, survive, and surpass the criteria dictated by the economics of technology and technology policy [276], which may be necessary to transform grand visions [270] into reality (and uncover new [277] tools, in the process).

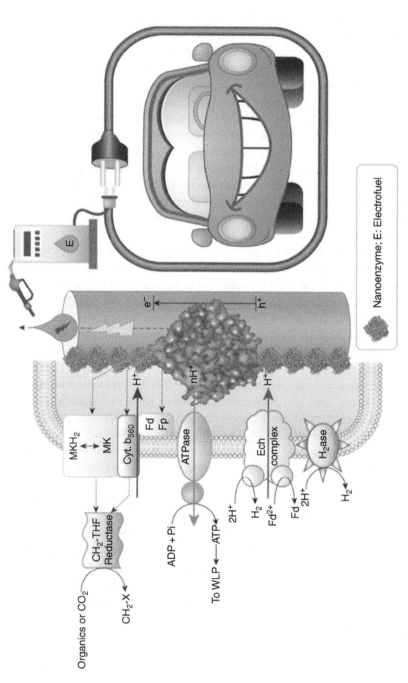

Figure 2.15 The trinity of imagination, invention, and innovation is central for the 80% world to leapfrog the dead weight of old technology and conventional wisdom from the 20% world. *Nanoenzyme–microbe interaction for clean and **affordable** bio-electrofuel production (Figure 1 from Singh et al. Ref.[270]).* Source: Singh, L et al. (2020). Bioelectrofuel synthesis by nanoenzymes: novel alternatives to conventional enzymes. *Trends in Biotechnology,* 38 5, p. S0167779919503129 doi:10.1016/j.tibtech.2019.12.017. © 2020, Elsevier.

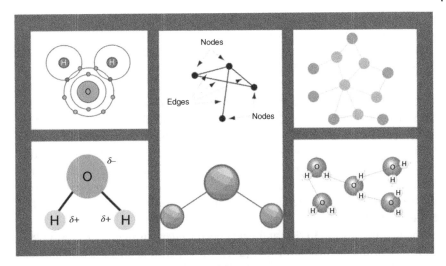

Figure 2.16 Training a neural network to recognize molecules relies on the fact that every molecule may be represented as a *graph* (or a collage of connected graphs, eliciting the idea of a *knowledge graph*). The water molecule may be viewed as a graph with oxygen (O) as the node (vertex). Bonds between oxygen and hydrogen (O−H) serve as the "side" or edge. Most molecules (within reason) may be transformed to a molecular graph and is at the heart of MPNN training to recognize different types of molecules. Then, the *trained* neural network, MPNN, is used to search for similar or *dissimilar* molecules in a repository.

But "grand visions" are often manufactured [278] as incremental mediocrity. Patents for using quantum [279] computing [280] are as absurd as the misuse of the term "cognitive" and the accompanying belligerence [281] in marketing. Those who throw around the term "cognition" may not have consulted a credible expert [282] or explored its meaning/definition {*cognition [n] mental action or process of acquiring knowledge and understanding through thought, experience, and the senses*}. Neither neurology nor modern computational neuroscience comprehends the *combined* electrochemical, cellular, and molecular nature of what it may mean to "acquire knowledge" by animals or humans. Any model, equation, algorithm, or hand-waving "AI" is simply false [283] because it is far beyond our grasp, at this point, to claim anything more than a vague impression of what "*acquire knowledge*" may mean. The other key words in the definition (*thought, experience*) are at depths we do not even dare to know how to measure. Deciphering "processes" based on functional nuclear magnetic resonance imaging (fMRI "activation" maps of real-time blood flow) is mindless drivel due to constraints in spatiotemporal resolution and limited ability of fMRI [284] to reliably detect functional activation. At the current state of instrumentation, resolution is inversely proportional to the ability to detect functional activation. Optimizing both is essential before fMRI data may be even considered precise.

The absurdity cryptic in the claims about cognition, learning, experience, and thought is neither colored by the author's cognitive dissonance nor a figment of our uninformed imagination. Table 2.4 [285] captures the duration of so-called "deep learning" training over "days" on a tensor processing unit [286] (TPU) scale with vast amounts of data (GB) which generated undifferentiated [287] rubbish. By comparing row 1 vs. 10 (bottom), the scores of the relevant match (#1, 0.892) between learning (saved query) vs. challenge (new query) is unimpressively different (#10, 0.765). According to Google BERT, after several days of "deep learning" ***"Blah blah blah blah"*** it was challenged with the query ***"Does this integrate with gmail?"*** which generated value of 0.765, suggesting ***76.5% similarity between the two.***

The fanfare of GPT-3 [288] and ballyhooed *context-awareness* of ELMo [289], BERT [290], and its cousin ALBERT [291], due to permutations and combinations of including masked language model (MLM) and next sentence prediction (NSP), is utterly devoid of intelligence. Except for fanatics feigning ignorance any rational observer may not be incorrect in thinking that throwing data (please see "Data" row in the upper part of Table 2.4) or using generic high volume of data for training (BERT claims to use "Wikipedia" in column 1, row 4) is ineffective. To be effective, these tools (ELMo, BERT, ALBERT, DILBERT, ROBERTA, and GPT-3) must use training data *relevant to the context* of the target (search). Context-awareness in the absence of data curation is as fake as claiming that a marble bowl is made of gold. Nevertheless, these advances in search techniques are immense strides [292] but the tools still are not "intelligent" but dumb as doorknobs. The doorknob does not turn unless one turns it or actuates it, manually or mechanically. Wikipedia as an experimental ***control*** is a plausible idea. Using curated data for training (ANN) may improve accuracy of search and better guide ***informed users*** to extract notions of connections and relationships with BERT-esque tools as ***supplementary aids*** (Tshitoyan et al. [293] provides supporting evidence, Table 2.5). These tools are of limited value for non-mission-critical applications, for example, recommendations (movies, books, and restaurants), weather for entertainment (IBM's Weather Channel), and fault-tolerant uses (open garage door, on/off sprinklers). Nonessential human-centric uses (congestion routing, temperature control, and voice message to email) may qualify if the outcome is almost correct in 80% of the cases. Actual use with humans-in-the-loop (healthcare, emergency response, and security) may be scuppered if based on any credible risk versus reward analytics, except for offering *nonbinding and non-executable* suggestions or alerts for human decision makers.

The paramount significance of curated contextual data in training any model (including artificial neural networks, ANNs) cannot be overemphasized.

Table 2.4 Google BERT thinks *"blah blah blah"* is **76.5% similar** to *"Does this integrate with gmail?"*

	BERT	RoBERTa	DistilBERT	XLNet
Size (millions)	**Base:** 110 **Large:** 340	**Base:** 110 **Large:** 340	**Base:** 66	**Base:** ~110 **Large:** ~340
Training time	**Base:** 8×V100×12days* **Large:** 64 TPU Chips×4 days (or 280×V100×1 days*)	**Large:** 1024×V100×1 day; 4–5 times more than BERT	**Base:** 8×V100×3.5 days; four times less than BERT	**Large:** 512 TPU Chips×2.5 days; five times more than BERT
Performance	Outperforms state-of-the-art in October 2018	2–20% improvement over BERT	3% degradation from BERT	2–15% improvement over BERT
Data	16 GB BERT data (Books Corpus + Wikipedia). 3.3 Billion words	160 GB (16 GB BERT data + 144 GB additional)	16 GB BERT data. 3.3 Billion words.	**Base:** 16 GB BERT data **Large:** 113 GB (16 GB BERT data + 97 GB additional). 33 Billion words.
Method	BERT (Bidirectional transformer with MLM and NSP)	BERT without NSP**	BERT distillation	Bidirectional transformer with permutation-based modeling

(Continued)

Table 2.4 (Continued)

	Saved query	New query	BERT score	USE score	ELMO score	XLNet score
1	How much will this cost?	Is this expensive?	0.892	0.803	0.742	0.720
2	Where is your data stored?	How secure is your product	0.890	0.625	0.765	0.705
3	What temperature is it today?	What time is the game on?	0.880	0.677	0.746	0.730
4	How do I change my password	I cannot find the settings page	0.868	0.671	0.717	0.706
5	Can I sign up for a free trial	Do I need a credit card to get started?	0.868	0.736	0.753	0.759
6	Where can I view my settings	Does this integrate with gmail?	0.866	0.620	0.717	0.691
7	I really do not like this product	I really like this product	0.865	0.747	0.864	0.870
8	What is the Capital of Ireland?	What time is the film in the cinema	0.865	0.578	0.663	0.778
9	Hello, is there anyone there?	What time is the game on?	0.832	0.594	0.680	0.723
10	Blah blah blah blah	Does this integrate with gmail?	0.765	0.519	0.585	0.668

The laughable outcome is not at all surprising despite the hordes of brilliant scientists working to create tools (RNet, XLNet, ELMo, BERT, ALBERT, DILBERT, and ROBERTA) over the past 20 years because it is gnarly to capture *semantics* of language which has evolved over the past 200,000 years. When not to choose the best NLP model. *Sources*: Suleiman Khan (2019). BERT, RoBERTa, DistilBERT, XLNet — which one to use. © 2019, Medium; FloydHub Blog (2019).

Table 2.5 (from Tshitoyan *et al.* Extended Data Table 2.4) – Analogy scores (%) for materials science versus "grammar" from different sources.

Text corpus	Materials	Grammar	All	Corpus size
Wikipedia	2.6	72.8	51.0	2.81B words
Wikipedia elements	2.7	72.1	41.4	1.08B words
Wikipedia materials	2.2	72.8	41.3	781M words
All abstracts	43.3	58.3	51.0	643M words
Relevant abstracts	48.9	54.9	52.0	290M words
Pre-trained model	10.4	47.1	30.8	640k papers

Training using Wikipedia for – metals – is grammatically rich (>72%) but content poor even when using select Wikipedia for materials (2.2% analogy). The smallest corpus (290 M words) used for training using continuous bag of words (CBOW) offers the best performance (48.9%) on materials-related analogies when **curated** for "relevant" abstracts. The best performance for grammar may turn out to be profitable by enabling ELMo, BERT, ALBERT, ROBERTA, and DILBERT (DistilBERT) to be the voices of artificial trainers for the standardized twaddle marketed with impunity and known as *Test of English as a Foreign Language* (TOEFL). The contextual enrichment in this table is similar to the example of enrichment shown in Figure 2.17, suggesting a need for relevance and curation.
Source: Vahe Tshitoyan, John Dagdelen, Leigh Weston, Alexander Dunn, Ziqin Rong, Olga Kononova, Kristin A. Persson, Gerbrand Ceder and Anubhav Jain (2019) Unsupervised word embeddings capture latent knowledge from materials science literature. *Nature* **571**, 95–98. © 2019, Springer Nature.

Individuals and institutions in possession of less than lofty ideals may revert to trickery in an attempt to sow doubt or discombobulate or disqualify the type of outcomes, for example, presented in Table 2.5. It is the age-old deception due to over-fitting [295] which can be also applied to ANN during training and the "fit" may be driven to precision using tools such as recursive feature addition [296] (RFA). For readers seeking a simpler analogy may wish to revisit what we discussed as the "force-fitting" of RDF to healthcare applications (Ref. [168]). The erudition necessary to train ANN with curated data is not easily gleaned from a cursory review. Extensive perusal of scholastic research [297] begins to reveal the minutiae with respect to the nature of the domain-specific data and the context of data curation (see sections 2.3 and 2.4 in Nandy et al., Ref. [297]) that forms the bulk of the *preparatory* work based on rigor and strength of broad-spectrum [298] knowledge. In an earlier section, we referred to *domain-specific* models in a "macro" sense whereas the domain specificity of this example (Nandy et al.) is at the molecular (atomic and/or subatomic) scale.

(Right) Data show large ANN error (−4.91 eV) with respect to DFT for a quintet [Mn(HNNH)6]3+ transition metal complex. The quintet [Mn(HNNH)6]3+

complex highest occupied molecular orbital (HOMO) level is underestimated by 4.9 eV, which is almost **double** the mean absolute error (MAE). This ANN was specifically trained using ΔE_g data models on a set of 64 octahedral homoleptic complexes (OH64). The discrepancy (ANN error) is significant because frontier molecular orbital energetics provide essential insight into chemical reactivity and dictate optical and electronic properties. Small errors could make an immense difference in terms of chemistry of the transition metal complex. In this illustration, the metals are shown as spheres and coordinating atoms as sticks (C atoms, gray; N atoms, blue; H atoms, white). If your healthcare diagnosis and treatment was based on such an ANN outcome, would you trust, accept, and abide by the direction of the treatment suggested by such results? If this outcome is based on data from your electronic health records (EHR) which is known to be erroneous, would you trust poor data quality to inform a poorly performing ANN engine to design your healthcare?

The third piece of evidence that also dispels the marketing myths of AI in favor of viewing through the lens of **artificial reasoning** tools (**ART**, referring to ANN, CNN, RNN, DL, RL), is another variety of neural network [300] with credible capabilities. Message passing neural network [301] (MPNN) for molecules [302] is a tool [303] to unleash data [304] for human-centric applications in health and medicine. This example centers on uncovering and repurposing a previously known molecule as an antibiotic [305] using a plethora of tools including MPNN and collectively referred to as deep learning (DL). Stokes *et al.* and the two other papers (Tshitoyan, Nandy) emphasize data curation and learning, without mentioning the term AI or "artificial intelligence" in the scientific papers. Unfortunately, the marketing and MIT news item [306], as expected, did not shy away from fake sensationalism to bolster the false appeal of AI.

The *learning* that generated the antibiotic (renamed Halicin) is nauseatingly detailed and the *training* (MPNN) was excruciatingly structured, optimized, and re-optimized (using hyperparameter [307] optimization). The old idea of ensembling [308] was applied to improve outcomes *in silico* but predictions were **biologically** tested through rigorous experiments. Even after repeated steps to minimize errors, the authors remain cognizant of the pitfalls: *"It is important to emphasize that machine learning is imperfect. Therefore, the success of deep neural network model-guided antibiotic discovery rests heavily on coupling these approaches to appropriate experimental designs"* (Stokes et al., page 698).

A curated set of 2335 molecules were used as the training set for new antibiotic molecules. The 2335 training dataset included a FDA library of 1760 molecules **pre-selected** (curated) based on their ability to inhibit microbial (*E. coli* BW25113) growth. In other words, molecules with structure and function **known** to possess antimicrobial activity. Training MPNN with this dataset enables the neural network to *learn* the structures in order to select similar (or dissimilar) structures from a larger library of structures. The expectation is that when a "challenge" library is presented to the MPNN, the degree of similarity or dissimilarity, in terms of the output from the MPNN, can be *tuned* by modifying selection parameters. For example, using prediction scores (PS) to categorize molecules from a larger library (in this case, the ZINC database with ~1.5 billion molecules). By selecting higher PS value (>0.7, >0.8, >0.9), the outcome is "enriched" and a subset of molecules (in this case, 107,347,223, reductionism at work) is further subjected to other selection criteria, for example, nearest neighbor analysis (Tanimoto score). Finally, potential molecules (in this case, 23) are biologically screened (microbial assay) to identify the "new" antibiotic candidate(s). One such candidate is Halicin (Stokes et al.), previously identified as the c-Jun N-terminal kinase inhibitor SU3327 and rediscovered as a broad-spectrum antibiotic, renamed Halicin, but still the same molecule as SU3227, albeit repurposed, based on function.

In combination, these three examples offer preliminary evidence that ART such as ANN, MPNN, DL, RL (reinforcement learning), etc., are **excellent** tools. However, ART and related tools are not intelligent, they do not self-operate and the outcome is solely due to the skill and sophistication of the human operators. The steps must be designed with cautious intellectual strategy embracing the **breadth** of diverse knowledge, often dismissed by many institutions. Execution demands **depth** of erudition and incisive foresight to weigh the pros and cons of the criteria used to assess the *quality of curated data* prior to training neural networks with such data.

It is essential to learn the *meaning of context* in order to sufficiently inform the "artificial" part of ART. Models and patterns are like chicken playing tic-tac-toe [309] without context and semantics. Human *knowledge* to equip ART is almost impossible to transfer because we do not have a clue how to abstract **continuous** knowledge and use **discrete** processes to **build** it into an artificial system. Hence, ART may not "possess" an internal model of the external world. The immense variability in terms of features and which features may be *relevant* in which environment makes it difficult to model a state by claiming that feature selection will address all relevant and discrete contexts that the item or object may experience. Even if feature engineering was automated to levels of precision and continuity that could encapsulate all possible permutations and combinations of the behavior of an entity or object, the model may be still inadequate in the hands of different users due to inherent bias. It is not trivial but may not be impossible to model behavior and optimize for some features within a narrow cross section in a retail environment (for example, who may shop at Whole Foods, who may return to Andronico's versus Mollie Stone's).

If reason could inform common sense, then one may prefer ART over AI and champion the value of reasoning in ML techniques using neural networks. Neural networks and ML tools are amplifying, modifying, and regurgitating whatever humans have programmed into the tool. It cannot **learn** beyond the range of data or information provided, until humans decide to change, adjust, or add/subtract parameters/attributes which will influence the **learning** and the output from ART. The obstreperous zeal to move away from the misnomer of modern [310] AI [311] and adopt ART as a generic term may be a *back to the future* moment for rule-based [312] expert systems [313] and principles [314] but coupled with new ML tools [315]. Marketing panders to creators of unstructured data but zettabytes of data anarchy occasionally may offer value irrespective of the clamor for general AI or ambient AI or intuitive AI or cognitive AI. Isn't it possible to deliver value using ART?

Does the acronym matter? AI is a false trigger for technology transitions [316] but it is cheaper [317] and cheaper [318] to promote. It is profitable for conference organizers, narcissistic speakers, greedy social gurus, and other forms of eejits. Irresponsible computation may be draining the energy [319] economy, yet the marketing world is oblivious of the socioeconomic incongruencies in terms of thermodynamics [320] of computation, which is absent from daily discussions. Perpetrating the myth of intelligence in AI may be a moral anathema. ART lacks the cachet and panache, but promotes the rational idea of *learning* tools, which are, and will be, helpful for society.

The state of artificial learning is analogous to receiving a map of the world on a postage stamp and expect the bearer of the map (stamp) to arrive at 77 Massachusetts Avenue, Cambridge, MA, using that postage-stamp-sized map as the only guide. Neurologists shudder [321] at AI, the public are ignorant of the evidence of sham (Tables 2.4–2.6 and Figure 2.17) while marketing accelerates the "show" over substance. Sensationalism amplifies attention and siphon funds away from real-world issues, making it harder for elements of FEWSH [322] to move forward, for the 80% world. The task ahead is to be creative, more than expected, and avoid the oxymoronic implementation of *innovation as usual*. Dynamic combinations [323] and cross-pollination of counterintuitive connections may be worth exploring [324] to find many *different* ways to lift billions of boats, not just a few yachts. Future needs égalitarian resistance to our "default" state of society where we allow, acquiesce, and accept greed [325] as a price and penalty we pay for progress (Figures 2.18).

2.14 In Service of Society

The tapestry of this discussion touches upon models of data, fake propaganda about AI, use of ART, and the lack of managed sanitation services for at least a billion people in the world ravaged by a pandemic. It presents a tortuously

Table 2.6 Even after extensive training using precision data enriched for features using RFA, it is not surprising when gross errors are found in the outcome (analysis).

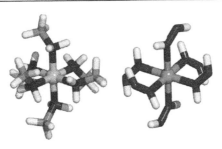

	$[Co(NH_2CH_3)_6]^{3+}$	$[Mn(HNNH)_6]^{3+}$
DFT	−20.00 eV	−18.64 eV
ANN	−19.91 eV	−23.55 eV

Figure 7 (from page 13981 in Nandy et al.) is one example how artificial neural networks (ANN) used in machine learning (ML) exercises and analytics generate erroneous results. ΔE_g data (LEFT) shows ANN error (0.09 eV) with respect to DFT (density functional theory [299]) in a singlet $[Co(NH2CH3)6]3+$ transition metal complex.
Source: Aditya Nandy, Chenru Duan, Jon Paul Janet, Stefan Gugler, and Heather J. Kulik (2018). "Strategies and software for machine learning accelerated discovery in transition metal chemistry." *Industrial and Engineering Chemistry Research*, volume 57, number 42, pages 13973–13986. © 2018, American Chemical Society.

complex series of challenges each with its own bewildering breadth (Figures 2.20 and 2.21).

More than ever, it is thorny to grasp how to balance efforts to continue creativity of thought considering the grim fact that more than 15 million people are infected by SARS-CoV-2 virus and deaths may soon exceed a million [340] people. The pandemic [341] may continue for another few years (2021–2025) and fluctuate in severity (acuity) due to antigenic drift [342] naturally caused by mutations. Lofty, esoteric ideas from any affluent [343] oasis poses moral and ethical dilemma when billions are facing a mirage in their effort to obtain the essentials for survival (FEWSH).

Yet, the pandemic has made molecular epidemiology the third most important job in the world (medical professionals and essential workers are first and second, respectively). Epidemiology is at the heart of public health ***data***. Without metrics to measure performance, we may be forced to rely merely on anecdotes. The plural of anecdotes is not evidence.

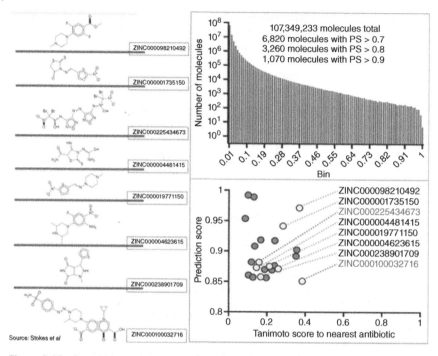

Figure 2.17 Potential candidates (eight molecules) from ZINC database (structures on the left) were scored using nearest neighbor (NN) analysis (yellow circles, bottom right). NN is based on principles derived a thousand years ago [294] (circa 1030). Data are curated at successive steps by enriching for *context* (selecting higher prediction scores, PS, top right) in a manner similar to Table 2.5 (Wikipedia vs relevant abstracts). *Source:* Based on Marcello Pelillo (2014) Alhazen and the nearest neighbor rule. *Pattern Recognition Letters* 38 (2014) 34–37.

2.15 Data Science in Service of Society: Knowledge and Performance from PEAS

We continue exploring the many facets of data but begin to ascend the pyramid how data may optimize the output or performance, an enormously complicated topic with vast gaps of knowledge.

The idea that tools must perform adequately is obvious and simple to grasp. Yet, performance as a scientific and engineering metric is a product of a fabric of dependencies and parameters which must be combined to deliver service, hopefully, useful for the community, for example, public health. In the digital domain, data is the common denominator, for example, data is the central force in epidemiology. The use of data to extract information must be a part of the discussion

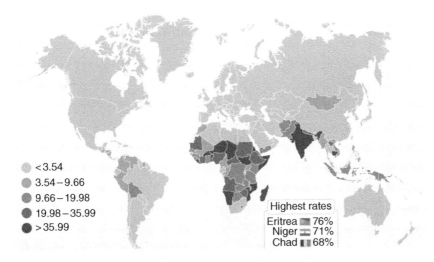

Figure 2.18 Is GINI coefficient [326] a nonstationary end-goal for economic redistribution through ethical social entrepreneurial innovation? A billion [327] people [328] defecate outdoors (figure shows percent of population who are forced to defecate outdoors). The pay-a-penny-per-unit (PAPPU) model could rake in billions if managed sanitation services were developed as a business. If a billion people paid one penny (US) per use per day for their "leased" sanitation service (at home), then the global gross earning for pay-per-use sanitation may be US$3.65 billion annually, an indication of earnings potential and wealth from the business of the poor. The primary assumption is that the individual will choose to pay one penny per day even if their income is only $2 per day (lowest per capita average earnings). This business model depends on availability of many different domains [329] of infrastructure necessary to offer home sanitation as an e-commerce [330] service. The ROI will be realized gradually because earnings will not be US$3.65 billion in the first year. Is the inclination to invest, and wait longer for a ROI, too much to expect from global organizations which could help facilitate delivery of global public goods? [331] *Source:* Andrew, G. (2015). *Open Defecation Around the World.* The World Bank.

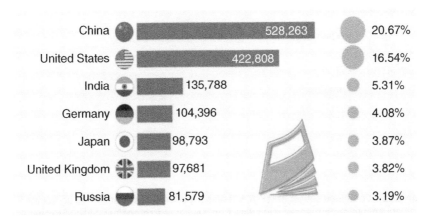

Figure 2.19 Publications in peer-reviewed journals (2018). Face-saving feel-good false positives? Propaganda [332] (does *not* exclude scientists [333]) masks facts [334] and analyses, by ignoring the quality of publications, citations, and investment in R&D (% GDP). Dubious research [335] output tarnishes the image and publications. Are we pointing fingers at a quarter (25.98% = 20.67% + 5.31%) of the global share?

Figure 2.20 GINI coefficient gone awry? Vast slums, adjacent to high-rise residential buildings in a section of Mumbai (Photograph by Prashant Waydande [336]), contradicts the notion that India may be an emerging leader in credible scientific research [337] (Figure 2.19). Are these a few symptoms stemming from grave gender bias, discrimination [338] against females and inequity of women in science and society?

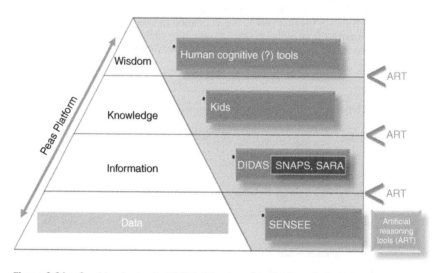

Figure 2.21 Combined subset of P, E, A, S is also a "performance" platform to aggregate tools. Data enable performance (using data from sensor search engines, SENSEE), other databases, logic functions (for example: if this, then that [339] type of operations) and/or ART to deliver near real-time output (mobile app) for end-users (sensor analytics point solution, SNAPS). At the next level, "point" data may auto-actuate to reach SARA (sense, analyze, response, actuate), a part of being data-informed (DIDAS).

with respect to science in the service of society. These tools must be ubiquitous, accessible, composable, and modifiable, on demand.

Information depends on raw data but data may not, always, contain information. Even though data are driving the granularity, the **curation** of the contextual from the granular (data) to extract relevant meaningful information is essential for decision systems including digital transformation. A plethora of real and fake tools and technologies are peddled on this pilgrimage from data to information. On this route, one often finds the snake-oil of AI [283] which is a fake tool and a marketing gimmick. But, under the bonnet of fake AI lies the very valuable techniques from ML and key statistical tools which are the "bread and butter" for many instances, applications, and information-informed outcomes.

We return (please rereview Figures 2.7 and 2.8) to the systems world to revisit PEAS, a mnemonic borrowed from agent systems which consists of percepts (P), environment (E), actuators (A), and sensors (S). Performance is the informed outcome from the PEAS process which is a superset of SARA [344], which we mentioned in the beginning of this essay (see Section 2.1). Performance is an overarching goal of PEAS because solutions evolving from science and engineering in the service of society are not only "point solutions" or "tokens" but a fabric or tapestry, which must integrate, analyze, synthesize, and synergize data and information from various inter-dependent domains to arrive at recommendations or actions that can deliver value to nonexpert users. Point solutions (SNAPS) are useful but "fabrics" may better suit complex cases. PEAS must combine data/information from subsystems (P, E, A, S) for data- (informed) DIDAS and in future, knowledge-informed decision support (KIDS). Data may be inextricably linked to sub-domains and outcomes may be influenced by interrelationships which are predominantly nonlinear. For specific instances we may need to select cross sections of data domains pertinent to specific applications. Once we determine which segments of data are required, the process of "discovery" is key to link the (contextual) data. Logic tools, agents, and ART (in combination) may extract the information. Metrics must evaluate the (semantically viable) outcome, if we wish to certify it as rigorous, reproducible, and credible for data and/or information-informed decision support (DIDAS).

Digital transformation depends on making sense of these relationships and the granularity of nearly noise-free data to support the claim that the relationships in question indeed have dependencies (as opposed to correlation without causation). The ability to discover and connect these interrelated data domains is challenging and yet is the heart of "knowledge" systems. Only those who do not dread the fatiguing climb of the steep path from data to information may reap the harvest of what it is to be information-informed. Ascension to knowledge is difficult by orders of magnitude due to the necessity to discover the relationships/dependencies and then validate the connectivity with credible (curated) data which may be globally distributed in unstructured or structured (databases) forms.

One "connectivity" tool from graph theory is referred to as KG which is a connected graph of data and associated metadata. KG represents real-world entities, facts, concepts, and events as well as relationships between them, yielding a comprehensive representation of relationships between data (healthcare data, financial data, and company data). The metadata about the data may catalyze feature integration, access to models, information assets, and data stores. Interoperability between KG tools is quintessential for data discovery and growth of knowledge networks. Although "knowledge" is the term of choice, the outcome of KGs are still far from knowledge but efficient in establishing connectivity between entities. The term "connectivity graph" is most appropriate. But in the data science parlance it is exaggerated to convey the impression that graph-theoretic tools may approach the "knowledge" level in the DIKW pyramid. It does not. KGs are excellent tools for better connected DIDAS but not KIDS.

The cartoon in Figure 2.22 is a hypothetical example of how KGs and graph networks may serve as a backend for point of care (circle in square) applications, for example, a patient suffering from coronavirus infection who is treated by one or more medical professionals. The physician or nurse may create an *ad hoc* medical profile of the patient to prepare a treatment plan by combining data and/or information, for example, [1] stored data (medical history) from EHR about patient's blood group [2], patient's blood analysis results from pathology lab to determine antibodies (mAbs) in serum [3], available drugs (remdesivir, dexamethasone) for treatment, and [4] whether the ventilator (device) is available, if necessary. Accessing the strands of data in real time at the point of care (on a secure mobile device) is typical of medical decision support. The data/information at the point of care must be relevant to the context of the patient (age, height, weight, symptoms, existing conditions, and allergies). Relationships between data and their dependencies with respect to this *specific* patient must be "understood" by the data discovery process (the importance of semantic metadata) when the medical professional queries the system (technical metadata). Logic, limits, rates, flows, policies, exclusions, etc., which determines "where/when/how/if this then that" must be a part of this system. Data discovery may be a combination of standard data (for example, the range of values for normal blood count, such as, platelets, hemoglobin, etc.) as well as data discovered specifically for the patient's treatment plan.

KGs are useful for relationship mapping and mining. KG maps are useful for discovery of data assets using agents, algorithms, and search engines in knowledge graph networks (KGN) which can be accessed and triggered by external queries. KG is just one element of digital transformation. It is *not a panacea* or a general solution. KGN requires pre-created KG and generating KG requires deep understanding of the nodes that the graph will connect. Individuals creating KGs for domain-specific use must know computational aspects as well domain

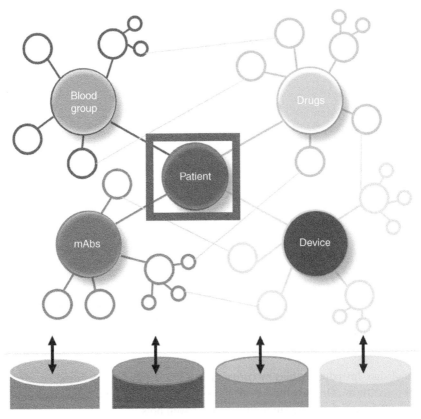

Figure 2.22 Knowledge graphs (KG) may play an increasingly important role if standards may endow KGs with the ability to access, connect, and catalyze data fusion using relevant data (sourced using data discovery agents/tools) followed by reasonable convergence, curation, and analytics of contextual data from distributed databases, based on query or problem specificity. The ability for search and discovery tools to semantically understand the query (language) is one caveat in using KG tools as a backend layer for applications at the point of use. Agents and algorithms for search and discovery of graph networks to access data and information assets may be handicapped by the lack of standards and interoperability between standards in this developing field. Global standards may not be easy to formalize but domain-specific standards developed through agreement between associations may be one mechanism for enabling the dissemination of graph tools for use at the edge (customer, user) in select domains, for example, in a medical subfield (otolaryngology) or water resource optimization (management of reusing wastewater for irrigation) or prevention of food waste (predictive perishability, and shelf-life).

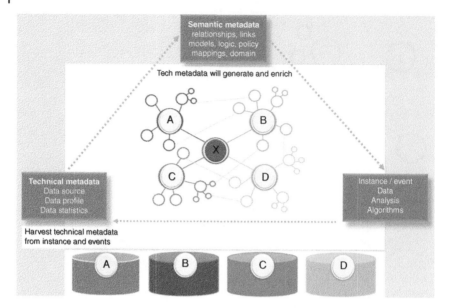

Figure 2.23 Generalized approach to use of knowledge graph networks for application-specific use which assumes interoperability between standards, tools, data, and information (but systems may be incompatible).

knowledge. Introducing bias by connecting select domains may introduce errors and cause harm. On-demand *ad hoc* tools to automate creation of KG and KGNs may suffer from semantic ambiguity of NL-driven bots.

KG abstractions in the PEAS context suggests an overlap between two strategies, both grounded in a multifactorial approach to decision support based on links, relationships, and dependencies. The difference between cartoons in Figures 2.22 and 2.23 versus illustration in Figure 2.24 is one of domain specificity (healthcare, Figure 2.22 or generic, Figure 2.23) compared to the big picture of PEAS (Figure 2.24). In the context of domains, KGs are connecting sub-domains and even granular data/information assets which lie far below the surface (for example, A, B, C, D, Figure 2.23).

Taken together, meaningful use of data and selected types of digital transformation tools are applicable to many domains which can help society. The financial instruments based on the paradigm of PAPPU may be key for ethical profitability if we wish to democratize the benefits and dividends from digital transformation for billions of users. Social business ideas to provide services to help reduce food waste and provide access to primary healthcare may use some version of PAPPU to collect micro-revenue. Suggestions (see items 00, 01, 02, 03, and 04 [345]) about entrepreneurial innovation related to FEWSH may be essential as global public

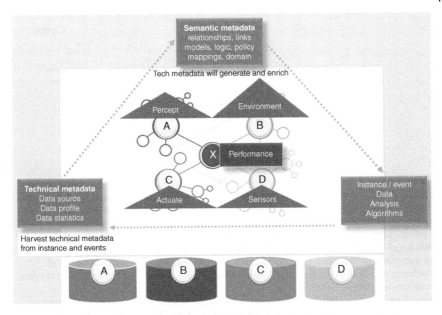

Figure 2.24 Connectivity permeates KG which create relationships between nodes (granular data and information). PEAS is a view from the top of MANY layers of knowledge graphs which may be broadly grouped under percept, environment, actuate, and sensors, representing trillions of use-cases where specific elements within these groups may be involved/weighted/combined in delivering decisions.

goods but they must work "in concert" to deliver the performance – survival in the post-pandemic world and improving the quality of life. Performance as the outcome of the PEAS paradigm (Figure 2.24) works by connecting ***ideas of domains*** (percept, environment, actuate, and sensors) using KGs. Comparing Figures 2.21–2.24 may help to visualize the big picture as an ***abstraction of connectivity*** between PEAS and KG.

These system of systems and multilayer convergence suggested in Figure 2.25 embrace a dynamic broad-spectrum of options and opportunities which may be beyond the grasp of end-users, for decades. It is useful to sketch the "big picture" but implementation of these systems and extracting the synergies are beyond civilian reach. The complexity in the cartoon may be analogous to the problems faced by the "omics" tools in medicine and healthcare (metabolomics, genomics, and proteomics). Physiological changes prior to and during a disease state are almost never a binary outcome. The network view in Figure 2.25 is even more intensified in humans and animals because changes in physiology and metabolism are almost always a network effect and genome-wide associations are only too common. Dissecting data to extract useful information may not be achieved by

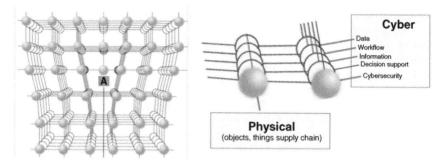

Figure 2.25 PEAS and knowledge graph (KG) networks are elements of the DIKW paradigm which must work in conjunction with cyber-physical systems (CPS), a superset of IoT activities. Actuation (A) could induce commencement of activities limited to a few "point" outcomes ("green" circles in cartoon) or potentially trigger network effects ("yellow" knobs illustrate control elements which may be in a farm, irrigation system, field sensors, or oil/gas pipeline) to modify rate, flow, or simply turn on/off. The actuation may be automatic (if this, then, that) or machine-to-machine (M2M) or some variation of machines and humans in the loop. The *wire frame* layer connecting the lattice represents ecosystems affected directly or indirectly by actuation. The elements may be "cyber" (workflow, cybersecurity, data exchange, information arbitrage, and decision support) as well as "physical" (involving objects – for example – flow of water, detection of molecules and physical in terms of the supply chain of products, and/or services influenced by actuation – for example – recall select batches of lettuce if pathogen is detected by sensor). Actuation is a "performance" outcome, which may engage with and affect multiple nodes, data, and IoT domains.

focusing on one or two aspects. The sluggish pace of progress in harvesting the value of genomics in precision medicine is one example. The herculean task of protein profiling [346] is still a work in progress. Hence, field applications may still need point solutions (SNAPS).

2.16 Temporary Conclusion

Digital transformation is a buzz word unless contextual data are meaningfully used to deliver information of value. Graph structures (KG) are an infrastructure of relationships to connect relevant granular data to improve total performance. The choice of tools from graph theory is based on their mathematical credibility. The combination of KG with PEAS may offer the rigor we seek for measurable progress of data-informed digital transformation to help the 80% and the 20% world. We are still in the early stages and changes will accompany digital transformation (for at least another century) not because it is *due to* digital transformation but because change is a gift of periodicity. Focus, applications, and our demands

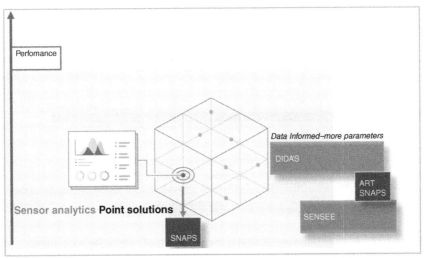

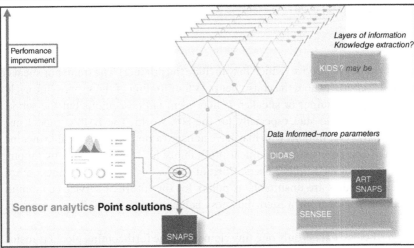

Figure 2.26 Complexity *seems* erudite (red border) but less useful in the real world. Probably 80% of users may benefit from point solutions (SNAPS) generating *usable information* from a cross section of "small" data (time series) which may assist in addressing local point of use problems in near real time.

change. Analyzing, predicting, and understanding these changes may be challenging. Tools in the process of change must evolve, adapt, and serve what is best for our society, industry, and commerce, at that time. With the progress of time, events evolve and unleash their influence on the global ecosystem.

If data cannot deliver performance, then the value of such data is negligible in a world where the "outcome economy" is no longer about the product but the value

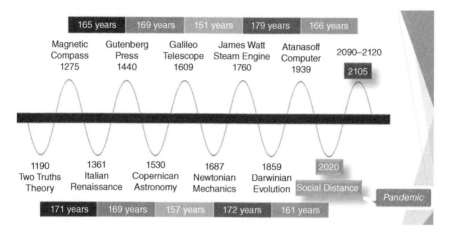

Figure 2.27 A sinusoidal wave illustrates the periodicity of change. We are currently immersed in a social coronary [347] due to the coronavirus pandemic. The Great Transformation [348] of 2021 is expected, soon.

of the meaningful service (which may include the product). The trainer or sneaker is now a "service" where the consumer pays for duration of the service (life cycle) provided by the trainer or sneaker. It may seem tad obsequious but the service economy is here to stay, especially for the 80% world. The PAPPU model may unlock markets of billions who may have nonzero amounts of disposable income. Wealth of the poor may fuel ethical profitability for social business by tapping into markets of next billion users. Current shades of "western" business ethos, roles, and models are unsuitable as pillars for the future of service economies. Expect resistance from *western* business behemoths if they fail to control the 80% world.

Nevertheless, data is only one tiny element in the forthcoming global transformation, which will test our knowledge and our greed. Compassion without knowledge is ineffective. Knowledge without compassion is inhuman [349]. In our pursuit of life, we must think about our duties and responsibilities. In our pursuit of ideas, we must strive to think about the correct questions. Wrong questions will generate wrong answers [350]. Humanity needs compassionate dreamers [351] and an ethical sense of higher purpose guided by humility. Leaders with a higher moral fabric are few and far between. Humanity seeks those who can rise above their personal greed and narcissistic needs. To believe in greater good for the greatest number is that pursuit of "what life expects of me" [352] which often presents itself as a braided lifeline of chance, choice, and character.

Les savants des autres nations, à qui nous avons donné l'exemple, ont cru avec raison qu'ils écriraient encore mieux dans leur langue que dans la nôtre. L'Angleterre nous a donc imités; l'Allemagne, où le latin semblait s'être réfugié, commence insensiblement à en perdre l'usage; je ne doute pas qu'elle ne soit bientôt suivie par les Suédois, les Danois et les Russes. Ainsi, avant la fin du dix-huitième siècle, un philosophe qui voudra s'instruire à fond des découvertes de ses prédécesseurs, sera contraint île charger sa mémoire de sept à huit langues différentes, et. après avoir consume à les apprendre le temps le plus précieux de sa vie, il mourra avant de commencer à s'instruire [353].	The scholars of other nations, to whom we have provided an example, believed with reason that they would write even better in their language than in ours. England has thus imitated us; Germany, where Latin seems to have taken refuge, begins insensibly to lose the use of it: I do not doubt that it will soon be followed by the Swedes, the Danes, and the Russians. Thus, before the end of the eighteenth century, a philosopher who would like to instruct himself about his predecessor's discoveries will be required to load his memory with 7–8 different languages; and after having consumed the most precious time of his life in acquiring them, he will die before having begun to instruct himself [354].

Acknowledgements

Critical review by Robert F Curl (https://www.nobelprize.org/prizes/chemistry/1996/curl/biographical/) and editorial comments by Alain Louchez (http://ipat.gatech.edu/people/alain-louchez) are gratefully acknowledged. Unvarnished opinions and unconventional commentary by the corresponding author may have influenced the timbre of this essay. Some of the opinions of the corresponding author were supported and strengthened by Rodney Brooks (https://www.csail.mit.edu/person/rodney-brooks) based on an earlier and related essay by the corresponding author (https://arxiv.org/abs/1610.07862).

References

1 https://en.wikipedia.org/wiki/Pareto_principle.

2 https://www.postscapes.com/iot-history/.

3 Sarma, S., Brock, D. and Ashton, K. (1999). The networked physical world - proposals for engineering the next generation of computing, commerce, and automatic-identification. MIT Auto-ID Center White Paper. MIT-AUTOID-WH001, 1999. https://autoid.mit.edu/publications-0 https://pdfs.semanticscholar.org/88b4/a255082d91b3c88261976c85a24f2f92c5c3.pdf.

4 Sarma, S., Brock, D., and Engels, D. (2001). Radio frequency identification and the electronic product code. *IEEE Micro* 21 (6): 50–54. https://doi.org/10.1109/40.977758.

5 Sarma, S. (2001). Towards the 5 cents tag. Auto-ID Center. MIT-AUTOID-WH-006, 2001.

6 https://www.rfc-editor.org/rfc/pdfrfc/rfc791.txt.pdf.

7 Deering, S. and Hinden, R. (2017). Internet protocol, version 6 (IPv6) specification. *Internet Engineering Task Force (IETF)* https://www.rfc-editor.org/rfc/pdfrfc/rfc8200.txt.pdf.

8 https://www.bizjournals.com/boston/blog/mass-high-tech/2012/11/mit-aims-to-harness-cloud-data-on-consumer.html.

9 Mühlhäuser, M. and Gurevych, I. (2008). *Introduction to Ubiquitous Computing.* IGI Global https://pdfs.semanticscholar.org/ab0e/b44c7c81a1af3fc2d23fa03f8f04f9e4ca2d.pdf.

10 https://www.bigdataframework.org/short-history-of-big-data/.

11 Kourantidou, M. (2019). Artificial intelligence makes fishing more sustainable by tracking illegal activity. https://theconversation.com/artificial-intelligence-makes-fishing-more-sustainable-by-tracking-illegal-activity-115883.

12 https://www.postscapes.com/iot-consulting-research-companies/.

13 Which market view? The imperfect categories are consumer and industrial IoT. Definitions proposed by domestic and foreign governments (US DHS, US FTC, EU ENISA) are (not surprisingly) neither systematic nor in sync. According to DHS, IoT is defined as "connection of systems and devices (with primarily physical purposes e.g. sensing, heating/cooling, lighting, motor actuation, transportation) to information networks (internet) via interoperable protocols, often built into embedded systems." (p. 2 in "Strategic Principles for Securing the Internet of Things (IoT)", Version 1.0, 15 November 2016. https://www.dhs.gov/sites/default/files/publications/Strategic_Principles_for_Securing_the_Internet_of_Things-2016-1115-FINAL_v2-dg11.pdf). ENISA defines IoT as "a cyber-physical ecosystem of inter-connected sensors and actuators, which enable intelligent decision making." (p. 18 in "Baseline Security Recommendations for IoT" www.enisa.europa.eu/publications/baseline-security-recommendations-for-iot/at_download/fullReport). One version of IoT is in H.R.1668 "Internet of Things Cybersecurity Improvement Act of 2019" (https://www.congress.gov/116/bills/hr1668/BILLS-116hr1668ih.pdf). H.R. 4792 "U.S. Cyber Shield Act of 2019" mentions "internet-connected products" and stipulates that the term 'covered product' means a consumer-facing physical object that can (a) connect to the internet or other network; and (b) (i) collect, send, or receive data; or (ii) control the actions of a physical object or system (https://www.congress.gov/116/bills/hr4792/BILLS-116hr4792ih.pdf).

14 Cisco CEO at CES 2014. Internet of Things is a $19 trillion opportunity. https://www.washingtonpost.com/business/on-it/cisco-ceo-at-ces-2014-internet-of-things-is-a-19-trillion-opportunity/2014/01/08/8d456fba-789b-11e3-8963-b4b654bcc9b2_story.html.

15 www.zdnet.com/article/internet-of-things-8-9-trillion-market-in-2020-212-billion-connected-things/.

16 https://machinaresearch.com/static/media/uploads/machina_research_press_release_-_m2m_global_forecast_&_analysis_2012-22_dec13.pdf.

17 Novikov, D.A. (2016). Systems theory and systems analysis. Systems engineering. *Cybernetics* 47. Springer International Publishing: 39–44. https://doi.org/10.1007/978-3-319-27397-6_4, www.researchgate.net/publication/300131568_Systems_Theory_and_Systems_Analysis_Systems_Engineering.

18 Williams, L.P. (1989). André-Marie Ampère. *Scientific American* 260 (1): 90–97. https://www.jstor.org/stable/24987112.

19 Syverson, C. (2018). Why hasn't technology sped up productivity?. *Chicago Booth Review*. https://review.chicagobooth.edu/economics/2018/article/why-hasn-t-technology-sped-productivity.

20 Wassen, O. (2019). Big data facts - how much data is out there? *NodeGraph* https://www.nodegraph.se/big-data-facts/.

21 OECD (2019). *OECD Compendium of Productivity Indicators 2019*. OECD https://doi.org/10.1787/b2774f97-en.

22 Huff, D. (1954). *How to Lie with Statistics*. Norton http://faculty.neu.edu.cn/cc/zhangyf/papers/How-to-Lie-with-Statistics.pdf.

23 Artificial Intelligence - Intelligent Agents (2017). https://courses.edx.org/asset-v1:ColumbiaX+CSMM.101x+1T2017+type@asset+block@AI_edx_intelligent_agents_new__1_.pdf.

24 Sutton, R.S. and Barto, A.G. (2018). *Reinforcement Learning: An Introduction*, 2nde. MIT Press https://web.stanford.edu/class/psych209/Readings/SuttonBartoIPRLBook2ndEd.pdf.

25 Angerman, W.S. (2004). Coming full circle with boyd's ooda loop ideas: an analysis of innovation diffusion and evolution (Thesis). https://apps.dtic.mil/dtic/tr/fulltext/u2/a425228.pdf.

26 Kamenetz, A. (2011). *Esther duflo bribes India's poor to health. Fast Company* (8 August). https://www.fastcompany.com/1768537/esther-duflo-bribes-indias-poor-health.

27 Castanedo, F. (2013). *A Review of Data Fusion Techniques*. The Scientific World Journal. doi:https://doi.org/10.1155/2013/704504. http://downloads.hindawi.com/journals/tswj/2013/704504.pdf.

28 https://www.the-scientist.com/uncategorized/nature-rejects-krebss-paper-1937-43452.

29 https://www.npr.org/sections/thesalt/2017/07/20/527945413/khichuri-an-ancient-indian-comfort-dish-with-a-global-influence.

30 Mendel, G. 1866. Versuche über Plflanzenhybriden. Verhandlungen des naturforschenden Vereines in Brünn, Bd. IV für das Jahr 1865, Abhandlungen, pp. 3–47. http://www.esp.org/foundations/genetics/classical/gm-65.pdf.

31 Jacob, F. and Monod, J. (1961). Genetic regulatory mechanisms in the synthesis of proteins. *Journal of Molecular Biology* 3 (3): 318–356. https://doi.org/10.1016/S0022-2836(61)80072-7.

32 Dulbecco, R. (1967). The induction of cancer by viruses. *Scientific American* 216 (4): 28–37. https://doi.org/10.1038/scientificamerican0467-28, http://calteches. library.caltech.edu/230/1/cancer.pdf.

33 http://web.mit.edu/esd.83/www/notebook/Cybernetics.PDF.

34 Weiner, N. (1961). *Cybernetics, or Control and Communication in the Animal and the Machine*, 2nde. MIT Press https://doi.org/10.1037/13140-000, https://uberty. org/wp-content/uploads/2015/07/Norbert_Wiener_Cybernetics.pdf.

35 https://www.sissa.it/fa/workshop_old/DCS2003/reading_mat/zuazuaDivS EMA.pdf.

36 https://www.americanscientist.org/article/alice-and-bob-in-cipherspace.

37 Rivest, R.L. et al. (1978). A method for obtaining digital signatures and public-key cryptosystems. *Communications of the ACM* 21 (2): 120–126. https://doi.org/10.1145/359340.359342, https://dl.acm.org/citation. cfm?id=359342.

38 Nash, J.F. (1950). Equilibrium points in N-person games. *Proceedings of the National Academy of Sciences of the United States of America* 36 (1): 48–49. https://doi.org/10.1073/pnas.36.1.48, www.pnas.org/content/pnas/36/1/48. full.pdf.

39 Axelrod, R.M. (2006). *The Evolution of Cooperation*. Basic Books http://www. eleutera.org/wp-content/uploads/2015/07/The-Evolution-of-Cooperation.pdf.

40 Kreps, D.M. et al. (1982). Rational cooperation in the finitely repeated prisoners' dilemma. *Journal of Economic Theory* 27 (2): 245–252. https://doi.org/10.101 6/0022-0531(82)90029-1.

41 https://en.wikipedia.org/wiki/Graph_database#/media/File:GraphDatabase_ PropertyGraph.png.

42 Little, J.D.C. et al. (1963). An algorithm for the traveling salesman problem. *Operations Research* 11 (6): 972–989. https://doi.org/10.1287/opre.11.6.972, https://dspace.mit.edu/bitstream/handle/1721.1/46828/ algorithmfortrav00litt.pdf.

43 Schrader, C.R. (1997). *United States Army Logistics, 1775-1992: An Anthology*, vol. 1. https://history.army.mil/html/books/068/68-1/cmhPub_68-1.pdf.

44 Datta, S. et al. (2004). Adaptive value networks. In: *Evolution of Supply Chain Management: Symbiosis of Adaptive Value Networks and ICT* (eds. Y.S. Chang et al.), 3–67. Springer US https://doi.org/10.1007/0-306-48696-2_1, https://link. springer.com/chapter/10.1007/0-306-48696-2_1.

45 www.cs.cmu.edu/afs/cs/academic/class/15251/Site/current/Materials/Lectures/ Lecture13/lecture13.pdf.

46 https://kluge.in-chemnitz.de/documents/fractal/node2.html.

47 https://trove.nla.gov.au/work/21161891?q&versionId=25229969.

48 Sterling, M.J. (2015). Mathematics and Art. https://www.bradley.edu/dotAsset/ d62c7fce-87ed-4b61-9ce7-23a78b70144d.pdf.

49 Carmody, K. (1988). Circular and hyperbolic quaternions, octonions, and sedenions. *Applied Mathematics and Computation* 28 (1): 47–72. 10.1016/0096-3003(88)90133-6. 10.1016/0096-3003(88)90133-6.

50 Pennisi, E. et al. (2018). The momentous transition to multicellular life may not have been so hard after all. *Science* https://www.sciencemag.org/news/2018/06/momentous-transition-multicellular-life-may-not-have-been-so-hard-after-all.

51 How Many Bacteria Live on Earth? *Sciencing*. https://sciencing.com/how-many-bacteria-live-earth-4674401.html.

52 Prusiner, S.B. (1982). Novel proteinaceous infectious particles cause scrapie. *Science* 216 (4542): 136–144. https://doi.org/10.1126/science.6801762, https://science.sciencemag.org/content/216/4542/136.long.

53 Pattison, I.H. and Jones, K.M. (1967). The possible nature of the transmissible agent of scrapie. *Veterinary Record* 80 (1): 2–9. https://doi.org/10.1136/vr.80.1.2, https://veterinaryrecord.bmj.com/content/80/1/2.

54 *1.4 Universality*. New England Complex Systems Institute. https://necsi.edu/14-universality.

55 Denton, P.B. et al. (2019). Eigenvectors from Eigenvalues. *ArXiv* 1908: 03795. https://arxiv.org/pdf/1908.03795.pdf.

56 Tao, T. and Van, V. (2010). Random matrices: Universality of local eigenvalue statistics. *ArXiv* 0906: 0510. https://arxiv.org/pdf/0906.0510.pdf.

57 Wu, L. et al. (2013). A spectral approach to detecting subtle anomalies in graphs. *Journal of Intelligent Information Systems* 41 (2): 313–337. https://doi.org/10.1007/s10844-013-0246-7, http://citeseerx.ist.psu.edu/viewdoc/download?doi=10.1.1.499.3228&rep=rep1&type=pdf.

58 Denton, P.B. et al. (2019). Eigenvalues: the rosetta stone for neutrino oscillations in matter. *ArXiv* 1907: 02534. https://arxiv.org/pdf/1907.02534.pdf.

59 Committee on Information Technology, Automation, and the U.S. Workforce et al. (2017). *Information Technology and the U.S. Workforce: Where Are We and Where Do We Go from Here?* National Academies Press https://doi.org/10.17226/24649. National Academies of Sciences, Engineering, and Medicine 2017. doi: 10.17226/24649 and https://www.nap.edu/download/24649.

60 www.seagate.com/files/www-content/our-story/trends/files/idc-seagate-dataage-whitepaper.pdf.

61 David, P.A. (1990). The dynamo and the computer: an historical perspective on the modern productivity paradox. *The American Economic Review* 80 (2): 355–361. www.jstor.org/stable/2006600. https://pdfs.semanticscholar.org/dff7/9b2f28cbb79da91becaab803667f30394233.pdf?_ga=2.215911511.1520557695.157068203 5-1238830782.1562127126.

62 Syverson, C. (2013). Will history repeat itself? Comments on is the information technology revolution over? *Intl Productivity Monitor* 25: 37–40. www.csls.ca/ipm/25/IPM-25-Syverson.pdf.

63 Syverson, C. (2016). Challenges to mismeasurement explanations for the U.S. Productivity slowdown. Working Paper, 21974, National Bureau of Economic Research, February 2016. doi:https://doi.org/10.3386/w21974. https://www.nber.org/papers/w21974.pdf.

64 Gordon, R. (2016). The rise and fall of American growth: U.S. Standard of living since the Civil War. https://press.princeton.edu/books/hardcover/9780691147727/the-rise-and-fall-of-american-growth.

65 Solow, R.M. (1956). A contribution to the theory of economic growth. *The Quarterly Journal of Economics* 70 (1): 65. https://doi.org/10.2307/1884513.

66 Information technology and the U.S. workforce: where are we and where do we go from here?. doi:https://doi.org/10.17226/24649 https://www.nap.edu/read/24649/chapter/5#55.

67 Bivens, J. and Mishel, L. (2015). *Understanding the Historic Divergence Between Productivity and a Typical Worker's Pay: Why It Matters and Why It's Real.* Economic Policy Institute https://www.epi.org/files/2015/understanding-productivity-pay-divergence-final.pdf.

68 Blum, A. et al. (2020). *Foundations of Data Science*, 1ste. Cambridge University Press https://www.cs.cornell.edu/jeh/book.pdf.

69 David, J. (2015). Study: nearly 70 percent of tech spending is wasted. *Vox* (31 October). https://www.vox.com/2015/10/31/11620222/study-nearly-70-percent-of-tech-spending-is-wasted.

70 Zobell, S. Why digitaltransformations fail: closing the $900 billion hole in enterprise strategy. *Forbes.* https://www.forbes.com/sites/forbestechcouncil/2018/03/13/why-digital-transformations-fail-closing-the-900-billion-hole-in-enterprise-strategy/.

71 Giller, G.L. (2005). *A Generalized Error Distribution. SSRN Electronic Journal* https://doi.org/10.2139/ssrn.2265027, http://citeseerx.ist.psu.edu/viewdoc/download?doi=10.1.1.542.879&rep=rep1&type=pdf.

72 McDonald, J.B. and Yexiao, J.X. (1995). A generalization of the beta distribution with applications. *Journal of Econometrics* 66 (1–2): 133–152. https://doi.org/10.1016/0304-4076(94)01612-4, https://www.sciencedirect.com/science/article/abs/pii/0304407694016124.

73 Granger, C.W.J. (1983). Co-integrated variables and error-correcting models. UCSD Discussion Paper 83-13.

74 Engle, R.F. (1976). Interpreting spectral analyses in terms of time-domain models. *Annals of Economic and Social Measurement* 5 (1): 89–109. www.nber.org/chapters/c10429.pdf.

75 Engle, R.F. (1982). Autoregressive conditional heteroscedasticity with estimates of the variance of United Kingdom inflation. *Econometrica* 50 (4): 987. https://doi.org/10.2307/1912773, http://www.econ.uiuc.edu/~econ508/Papers/engle82.pdf.

76 Granger, C.W.J. et al. (2001). *Essays in Econometrics: Collected Papers of Clive W.J. Granger.* Cambridge University Press https://dl.acm.org/citation.cfm?id=781849.

77 Granger, C.W.J. (2004). Time series analysis, cointegration, and applications. https://escholarship.org/uc/item/2nb9f668.

78 Engle, R. (2001). GARCH 101: the use of ARCH/GARCH models in applied econometrics. *Journal of Economic Perspectives* 15 (4): 157–168. https://doi.org/10.1257/jep.15.4.157, http://www.cmat.edu.uy/~mordecki/hk/engle.pdf.

79 https://www.nobelprize.org/prizes/economic-sciences/2003/summary/.

80 Fronte, B, Galliano, G. and Bibbiani, C. (2016). From freshwater to marine aquaponic: new opportunities for marine fish species production. Conference VIVUS, 20 and 21 April 2016, Biotechnical Centre Naklo, Strahinj 99, Naklo, Slovenia. https://pdfs.semanticscholar.org/f8b2/fad3fe3f6f92a7b1c8276ad57876d8fd0a70.pdf.

81 Matthias, D.M. et al. (2007). Freezing temperatures in the vaccine cold chain: a systematic literature review. *Vaccine* 25 (20): 3980–3986. https://doi.org/10.1016/j.vaccine.2007.02.052, https://www.nist.gov/sites/default/files/documents/2017/05/09/FreezingReviewArticle-Vaccine.pdf.

82 Ruppert, D. (2011). *Statistics and Data Analysis for Financial Engineering*, 477–504. Springer Texts in Statistics https://doi.org/10.1007/978-1-4419-7787-8_18, https://faculty.washington.edu/ezivot/econ589/ch18-garch.pdf.

83 Duan, J.C. (2001). GARCH model and its application. http://jupiter.math.nctu.edu.tw/~weng/seminar/GarchApplication.pdf.

84 Diebold, F.X. (2019). Econometric data science: a predictive modeling approach. http://www.ssc.upenn.edu/~fdiebold/Textbooks.html. https://www.sas.upenn.edu/~fdiebold/Teaching104/Econometrics.pdf.

85 Datta, S. and Granger, C. (2006). Potential to improve forecasting accuracy: advances in supply chain management. https://dspace.mit.edu/handle/1721.1/41905.

86 Datta, S. et al. (2007). Management of supply chain: an alternative modelling technique for forecasting. *Journal of the Operational Research Society* 58 (11): 1459–1469. https://doi.org/10.1057/palgrave.jors.2602419, https://dspace.mit.edu/handle/1721.1/41906.

87 Narayanan, A. (2019). How to recognize AI snake oil. https://www.cs.princeton.edu/~arvindn/talks/MIT-STS-AI-snakeoil.pdf.

88 Li, L. et al. (2020). A survey of uncertain data management. *Frontiers of Computer Science* 14 (1): 162–190. https://doi.org/10.1007/s11704-017-7063-z, https://link.springer.com/article/10.1007/s11704-017-7063-z.

89 Agrawal, R., Imielinski, T., Swami, A. (1993). Mining association rules between sets of items in large databases. *Proceedings of the 1993 ACM SIGMOD International Conference on Management of Data - SIGMOD '93*, ACM Press, 1993, pp. 207–16. doi:https://doi.org/10.1145/170035.170072.

90 Coenen, F. (2011). Data mining: past, present and future. *The Knowledge Engineering Review* 26 (1): 25–29. https://doi.org/10.1017/S0269888910000378, https://www.researchgate.net/publication/220254364_Data_mining_Past_present_and_future.

91 Kahneman, D. (2012). *Thinking, Fast and Slow*. Penguin Books http://sysengr.engr.arizona.edu/OLLI/lousyDecisionMaking/KahnemanThinkingFast&Slow.pdf.

92 Akerlof, G. and Yellen, J. (1987). Rational models of irrational behavior. *American Economic Review* 77 (2): 137–142. https://notendur.hi.is/ajonsson/kennsla2003/Akerlof_Yellen.pdf.

93 https://spectrum.ieee.org/biomedical/diagnostics/how-ibm-watson-overpromised-and-underdelivered-on-ai-health-care.

94 Minds and machines postponed / canceled - post regarding GE digital layoffs. https://www.thelayoff.com/t/VtbKYAf.

95 http://bit.ly/MRC-BERT-HAN.

96 Rumelhart, D.E. et al. (1986). Learning representations by back-propagating errors. *Nature* 323 (6088): 533–536. https://doi.org/10.1038/323533a0, https://www.iro.umontreal.ca/~vincentp/ift3395/lectures/backprop_old.pdf.

97 McCarthy, J., Minsky, M.L., Rochester, N. and Shannon, C.E. (1955). A proposal for the dartmouth summer research project on artificial intelligence (31 August 1955). http://www-formal.stanford.edu/jmc/history/dartmouth.pdf.

98 Levy, F. (2018). Computers and populism: artificial intelligence, jobs, and politics in the near term. *Oxford Review of Economic Policy* 34 (3): 393–417. https://doi.org/10.1093/oxrep/gry004, https://www.russellsage.org/sites/default/files/gry004.pdf.

99 GE minds + machines (15 November 2016). www.youtube.com/watch?v=OYn9ZtpWCUw.

100 Why robots won't take over the world. https://phys.org/news/2018-04-robots-wont-world.html.

101 Freedman, D.H. What will it take for IBM's Watson technology to stop being a dud in health care?. *MIT Tech Review*. www.technologyreview.com/s/607965/a-reality-check-for-ibms-ai-ambitions/.

102 Salmon, F. (2018). IBM's Watson was supposed to change the way we treat cancer. here's what happened instead. *Slate Magazine* (18 August). https://slate.com/business/2018/08/ibms-watson-how-the-ai-project-to-improve-cancer-treatment-went-wrong.html.

103 Perez, C.E. (2017). Why we should be deeply suspicious of backpropagation. *Medium* (13 October). https://medium.com/intuitionmachine/the-deeply-suspicious-nature-of-backpropagation-9bed5e2b085e.

104 Datta, S. (2016). *Intelligence in Artificial Intelligence*. https://arxiv.org/abs/1610.07862.

105 There's no good reason to trust Blockchain technology. *Wired*. https://www. wired.com/story/theres-no-good-reason-to-trust-blockchain-technology/.

106 Stinchcombe, K. (2018). Blockchain is not only crappy technology but a bad vision for the future. *Medium* (9 April). https://medium.com/@kaistinchcombe/ decentralized-and-trustless-crypto-paradise-is-actually-a-medieval-hellhole-c1ca122efdec.

107 RuuviLab - IOTA masked authentication messaging. *RuuviLab*. https://lab. ruuvi.com/iota.

108 Andreas, K., Fonts, A. and Prenafeta-Boldú, F.X. (2019). The rise of blockchain tech in agriculture and food supply chains. https://arxiv.org/ftp/arxiv/ papers/1908/1908.07391.pdf.

109 Press, G. Big data is dead. Long live big data AI. *Forbes*. www.forbes.com/sites/ gilpress/2019/07/01/big-data-is-dead-long-live-big-data-ai/#5a262cf71b05.

110 Walsh, M.W., and Emily, F. (2019). McKinsey faces criminal inquiry over bankruptcy case conduct. *The New York Times* (8 November). https://www. nytimes.com/2019/11/08/business/mckinsey-criminal-investigation-bankruptcy.html.

111 Black, E. (2001). *IBM and the Holocaust: The Strategic Alliance between Nazi Germany and America's Most Powerful Corporation*, 1ste. Crown Publishers http://posoh.ru/book/htm/ibm.pdf.

112 Allison, S. Huawei's pitch to African mayors: our cameras will make you safe. *The M&G Online*. https://mg.co.za/ article/2019-11-15-00-our-cameras-will-make-you-safe.

113 2007 Joint Workshop on High Confidence Medical Devices, Software, and Systems and Medical Device Plug-and-Play Interoperability: HCMDSS//MD PnP 2007: Improving Patient Safety through Medical Device Interoperability and High Confidence Software. Proceedings: 25–27 June 2007, Cambridge, MA. https://nam.edu/wp-content/uploads/2018/02/3.1-Goldman-Jan-2018-002.pdf.

114 Hatcliff, J. et al. (2011). Medical application platforms – rationale, architectural principles, and certification challenges. www.nitrd.gov/nitrdgroups/ images/8/8b/MedicalDeviceInnovationCPS.pdf.

115 Makary, M.A. and Michael, D. (2016). Medical error—the third leading cause of death in the US. *BMJ* 353 https://doi.org/10.1136/bmj.i2139, https://www.bmj. com/content/353/bmj.i2139.

116 Slight, S.P. et al. (2015). Meaningful use of electronic health records: experiences from the field and future opportunities. *JMIR Medical Informatics* 3 (3): e30. https://doi.org/10.2196/medinform.4457, https://medinform.jmir. org/2015/3/e30/.

117 Meaningful use. *CDC*. (10 September 2019). https://www.cdc.gov/ ehrmeaningfuluse/introduction.html.

118 MIT AUTO-ID LABORATORY. https://autoid.mit.edu/about-lab.

119 Shabandri, B. and Piyush, M. (2019). Enhancing IoT security and privacy using distributed ledgers with IOTA and the tangle. *2019 6th International Conference on Signal Processing and Integrated Networks (SPIN)*, IEEE, 2019, pp. 1069–75. doi:https://doi.org/10.1109/SPIN.2019.8711591 https://assets.ctfassets.net/r1dr6 vzfxhev/2t4uxvsIqk0EUau6g2sw0g/45eae33637ca92f85dd9f4a3a218e1ec/iota1_4_3.pdf.

120 The Coordicide. https://files.iota.org/papers/Coordicide_WP.pdf.

121 https://www.iota.org/.

122 Send IoT data to the IOTA tangle with SAP HANA XSA and analytics cloud. https://blogs.sap.com/2019/10/08/send-iot-data-to-the-iota-tangle-with-sap-hana-xsa-and-analytics-cloud/.

123 https://www.dds-foundation.org/.

124 Shepard, M, Baicker, K., and Skinner, J.S. (2019). *Does one medicare fit all? The economics of uniform health insurance benefits* in tax policy and the economy. *Moffitt* 34 https://doi.org/10.3386/w26472, https://www.nber.org/papers/w26472.pdf.

125 http://bit.ly/Economics-of-Technology.

126 http://bit.ly/COASE5PAPERS.

127 Merelli, A. The inventor of microfinance has an idea for fixing capitalism. *Quartz*. https://qz.com/1089266/the-inventor-of-microfinance-has-an-idea-for-fixing-capitalism/.

128 Georgescu, C. (2012). Simulating micropayments in local area networks. *Procedia - Social and Behavioral Sciences* 62: 30–34. https://doi.org/10.1016/j.sbspro.2012.09.007.

129 Spence, M. (2011). The next convergence: the future of economic growth in a multispeed world. *Farrar, Straus and Giroux*. http://pubdocs.worldbank.org/en/515861447787792966/DEC-Lecture-Series-Michael-Spence-Presentation.pdf.

130 Tabassi, E. et al. (2019). *A Taxonomy and Terminology of Adversarial Machine Learning. NIST IR 8269-draft*. National Institute of Standards and Technology https://doi.org/10.6028/NIST.IR.8269-draft, https://nvlpubs.nist.gov/nistpubs/ir/2019/NIST.IR.8269-draft.pdf.

131 Duddu, V. (2018). A survey of adversarial machine learning in cyber warfare. *Defence Science Journal* 68 (4): 356. https://doi.org/10.14429/dsj.68.12371.

132 Morgan, V., Casso-Hartman, L., Bahamon-Pinzon, D. et al. (2020). Sensor-as-a-service: convergence of sensor analytic point solutions (SNAPS) and Pay-A-Penny-Per-Use (PAPPU) paradigm as a catalyst for democratization of healthcare in underserved communities. *Diagnostics* 10: 22; doi:https://doi.org/10.3390/diagnostics10010022. MIT Library SNAPS -. https://dspace.mit.edu/handle/1721.1/111021.https://dspace.mit.edu/handle/1721.1/123983

133 Datta, S. et al. (2004). Adaptive value networks: convergence of emerging tools, technologies and standards as catalytic drivers. In: *Evolution of Supply Chain Management: Symbiosis of Adaptive Value Networks and ICT* (eds. Y.S.,.C. et al.). Kluwer Academic Publishers https://dspace.mit.edu/handle/1721.1/41908.

134 Silvestro, R. and Paola, L. (2014). Integrating financial and physical supply chains: the role of banks in enabling supply chain integration. *International Journal of Operations & Production Management* 34 (3): 298–324. https://doi. org/10.1108/IJOPM-04-2012-0131, https://www.emerald.com/insight/content/ doi/10.1108/IJOPM-04-2012-0131/full/html.

135 Banerjee, A.V. and Duflo, E. (2011). *Poor Economics: A Radical Rethinking of the Way to Fight Global Poverty*, 1ste. PublicAffairs https://warwick.ac.uk/about/ london/study/warwick-summer-school/courses/macroeconomics/poor_ economics.pdf.

136 Lie, R. (2019). Robertlie/Dht11-Raspi3. 2018. *GitHub*. https://github.com/ robertlie/dht11-raspi3.

137 Handy, P. (2018). Introducing masked authenticated messaging. *Medium* (9 April). https://blog.iota.org/ introducing-masked-authenticated-messaging-e55c1822d50e.

138 Kazimirova, E.D. (2017). Human-centric internet of things. Problems and challenges. https://www.researchgate.net/ publication/319059870_Human-Centric_Internet_of_Things_Problems_and_ Challenges.

139 Calderon, M.A. et al. (2016). A more human-centric internet of things with temporal and spatial context. *Procedia Computer Science* 83: 553–559. https:// doi.org/10.1016/j.procs.2016.04.263, https://www.sciencedirect.com/science/ article/pii/S1877050916302964.

140 McLamore, E.S., Datta, S.P.A., Morgan, V. et al. (2019). SNAPS: sensor analytics point solutions for detection and decision support. *Sensors* 19 (22): 4935. https:// www.mdpi.com/1424-8220/19/22/4935/pdf.

141 The sensor-based economy. *Wired* (January 2017). https://www.wired.com/ brandlab/2017/01/sensor-based-economy/.

142 Pont, S. (ed.) (2013). *Digital State: How the Internet Is Changing Everything*. Kogan Page.

143 Leading the IoT: gartner insights on how to lead in a connected world. https:// www.gartner.com/imagesrv/books/iot/iotEbook_digital.pdf.

144 Evans, D. (2011). The internet of things: how the next evolution of the internet is changing everything. https://www.cisco.com/c/dam/en_us/about/ac79/docs/ innov/IoT_IBSG_0411FINAL.pdf.

145 Puglia, D. (2014). Are enterprises ready for billions of devices to join the Internet?. *Wired* (December 2014). https://www.wired.com/insights/2014/12/ enterprises-billions-of-devices-internet/.

146 Gartner Says 5.8 Billion enterprise and automotive IoT endpoints will be in use in 2020. https://www.gartner.com/en/newsroom/ press-releases/2019-08-29-gartner-says-5-8-billion-enterprise-and-automotive-io.

147 www.cisco.com/c/dam/en/us/products/collateral/se/internet-of-things/ at-a-glance-c45-731471.pdf.

148 The Internet of things: sizing up the opportunity. *McKinsey*. https://www. mckinsey.com/industries/semiconductors/our-insights/ the-internet-of-things-sizing-up-the-opportunity.

149 IoT overview handbook: 2019 background primer on the topics & technologies driving the Internet of things. *Postscapes*. https://www.postscapes.com/iot/.

150 Fleming, M., Clarke, W., Das, S., Phongthiengtham, P., and Reddy, P. (2019). The future of work: how new technologies are transforming tasks (31 October 2019). https://mitibmwatsonailab.mit.edu/research/publications/paper/download/ The-Future-of-Work-How-New-Technologies-Are-Transforming-Tasks.pdf.

151 Andrew, N.G. Why AI is the new electricity. https://www.gsb.stanford.edu/ insights/andrew-ng-why-ai-new-electricity.

152 Cheekiralla, S. and Engels, D.W. (2006). An IPv6-based identification scheme. *2006 IEEE International Conference on Communications* 1: 281–286. https://doi. org/10.1109/ICC.2006.254741, https://ieeexplore.ieee.org/document/4024131.

153 Stallings, W. (1996). IPv6: the new Internet protocol. *IEEE Communications Magazine* 34 (7): 96–108. https://doi.org/10.1109/35.526895, https://ieeexplore. ieee.org/document/526895.

154 Difference between IPv4 and IPv6. *Tech Differences* (4 August 2017). https:// techdifferences.com/difference-between-ipv4-and-ipv6.html

155 Datta, S.P.A. (2012). An unified theory of relativistic identification of information in the systems age: proposed convergence of unique identification with syntax and semantics through internet protocol version 6 (IPv6). *International Journal of Advanced Logistics* 1 (1): 66–82. https://doi.org/10.108 0/2287108X.2012.11006070, https://dspace.mit.edu/handle/1721.1/41902.

156 Datta, S. (2011). Mobile eVote as an IPv6 App. (23 May 2011). https:// shoumendatta.wordpress.com/2011/05/23/mobile-e-vote-ipv6-app/. https:// dspace.mit.edu/bitstream/handle/1721.1/41902/IPv6%20Apps%20and%20SaaS. pdf?sequence=11&isAllowed=y.

157 https://bentley.umich.edu/elecrec/d/duderstadt/Speeches/JJDS6/jjd1341.pdf.

158 Figueroa, A. (2019). Data demystified — DIKW model. *Medium* (24 May 2019). https://towardsdatascience.com/rootstrap-dikw-model-32cef9ae6dfb.

159 https://www.wur.nl/en/Education-Programmes/wageningen-academy-1/ What-we-offer-you/Courses/show-1/Course-Towards-Data-driven-Agri-Food-Business-1.htm.

160 CS 540 lecture notes: intelligent agents. http://pages.cs.wisc.edu/~dyer/cs540/ notes/agents.html.

161 Maes, P. (1997). Intelligent Software. *Proceedings of the 2nd International Conference on Intelligent User Interfaces - IUI'97*, ACM Press, pp. 41–43. doi:https://doi.org/10.1145/238218.238283.

162 Paolucci, M. and Sacile, R. (2005). *Agent-Based Manufacturing and Control Systems: New Agile Manufacturing Solutions for Achieving Peak Performance.* CRC Press http://jmvidal.cse.sc.edu/library/paolucci05a.pdf.

163 Datta, S. (2017). Cybersecurity: agents based approach?. https://dspace.mit.edu/handle/1721.1/107988.

164 White, F.E. (1997). Data fusion group. https://apps.dtic.mil/dtic/tr/fulltext/u2/a394662.pdf.

165 Rafferty, E.R.S., et al. (2019). Seeking the optimal schedule for chickenpox vaccination in Canada: using an agent-based model to explore the impact of dose timing, coverage and waning of immunity on disease outcomes. *Vaccine* (November 2019). doi:https://doi.org/10.1016/j.vaccine.2019.10.065.

166 Lassila, O. and Swick, R.R. (1999). Resource description framework (RDF) model and syntax specification. *W3C working draft* (February 1999). www.w3.org/TR/REC-rdf-syntax/.

167 Berners-Lee, T., Hendler, J., and Lassila, O. (2001). *The Semantic Web.* Scientific American.

168 Shi, L. et al. (2017). Semantic health knowledge graph: semantic integration of heterogeneous medical knowledge and services. *BioMed Research International* 2017: 1–12. https://doi.org/10.1155/2017/2858423, http://downloads.hindawi.com/journals/bmri/2017/2858423.pdf.

169 Chakraborty, A., Munshi, S. and Mukhopadhyay, D. (2013). Searching and establishment of S-P-O relationships for linked RDF graphs: an adaptive approach. 2013 Int Conf on Cloud & Ubiquitous Comp & Emerging Tech. https://arxiv.org/ftp/arxiv/papers/1311/1311.7200.pdf.

170 Chakraborty, A., Munshi, S. and Mukhopadhyay, D. (2013). A proposal for the characterization of multidimensional inter-relationships of RDF graphs based on set theoretic approach. https://arxiv.org/ftp/arxiv/papers/1312/1312.0001.pdf.

171 Buneman, P. (1974). A characterisation of rigid circuit graphs. *Discrete Mathematics* 9 (3): 205–212. https://doi.org/10.1016/0012-365X(74)90002-8, http://homepages.inf.ed.ac.uk/opb/homepagefiles/phylogeny-scans/rigidcircuitgraphs.pdf.

172 Sharma, C. and Roopak, S. (2019). A schema-first formalism for labeled property graph databases: enabling structured data loading and analytics. *Proceedings of the 6th IEEE/ACM International Conference on Big Data Computing, Applications and Technologies - BDCAT'19*, ACM Press, 2019, pp. 71–80 doi:https://doi.org/10.1145/3365109.3368782.

173 Wang, D.-W. et al. (2019). CK-modes clustering algorithm based on node cohesion in labeled property graph. *Journal of Computer Science and Technology* 34 (5): 1152–1166. https://doi.org/10.1007/s11390-019-1966-0.

174 https://aibusiness.com/ending-the-rdf-vs-property-graph-debate-with-rdf/.

175 Tarjan, R. (1972). Depth-first search and linear graph algorithms. *SIAM Journal on Computing* 1 (2): 146–160. https://doi.org/10.1137/0201010.

176 Fleischer, R, and Gerhard, T. (2003). Experimental studies of graph traversal algorithms. In: *Experimental and Efficient Algorithms* (ed K. Jansen, et al), vol. 2647, Berlin Heidelberg: Springer, pp. 120–133 doi:https://doi.org/10.1007/3-540-44867-5_10.

177 Noy, N. et al. (2019). Industry-scale knowledge graphs: lessons and challenges. *Communications of the ACM* 62 (8): 36–43. https://doi.org/10.1145/3331166, https://queue.acm.org/detail.cfm?id=3332266.

178 https://tech.ebayinc.com/engineering/akutan-a-distributed-knowledge-graph-store/.

179 https://www.w3.org/TR/owl-guide/.

180 Sengupta, K. and Hitzler, P. (2014). Web Ontology Language (OWL). In: *Encyclopedia of Social Network Analysis & Mining* (eds. R. Alhajj and J. Rokne). NY: Springer https://doi.org/10.1007/978-1-4614-6170-8.

181 http://vowl.visualdataweb.org/v2/.

182 https://tools.ietf.org/html/rfc3987.

183 www.eads-iw.net/web/nfigay.

184 Carbone, L., et al. (2017) State of play of interoperability: report 2016. *Open WorldCat*. http://dx.publications.europa.eu/10.2799/969314 https://ec.europa.eu/isa2/sites/isa/files/docs/publications/report_2016_rev9_single_pages.pdf.

185 Blind men and the elephant. www.allaboutphilosophy.org/blind-men-and-the-elephant.htm.

186 Bohr, N. (1950). On the notions of causality and complementarity. *Science* 111 (2873): 51–54. https://doi.org/10.1126/science.111.2873.51.

187 Morrison, A. (2019). Is data science/machine learning/AI overhyped right now?. www.quora.com/Is-data-science-machine-learning-AI-overhyped-right-now/answer/Alan-Morrison.

188 Fuller, A., et al. (2019). Digital twin: enabling technology, challenges and open research. October 2019. http://arxiv.org/abs/1911.01276. https://arxiv.org/ftp/arxiv/papers/1911/1911.01276.pdf.

189 https://newsstand.joomag.com/en/iic-journal-of-innovation-12th-edition/0994713001573661267.

190 https://www.forbes.com/sites/bernardmarr/2017/03/06/what-is-digital-twin-technology-and-why-is-it-so-important.

191 Vilfredo, P. (1971). *Manual of Political Economy*. Translated by (ed. A.M. Kelley). MIT Press.

192 McLamore, E.S., Huffaker, R., Shupler, M., Ward, K., Austin Datta, S.P., Katherine Banks, M., Casaburi, G., Babilonia, J., Foster, J.S. (2019). Digital proxy of a bio-reactor (DIYBOT) combines sensor data and data analytics for wastewater treatment and wastewater management systems. (Nature Scientific Reports, in press) Draft copy of "DIYBOT" available from MIT Libraries. https:// dspace.mit.edu/handle/1721.1/123983.

193 Hernlund, E. et al. (2013). Osteoporosis in the European Union: medical management, epidemiology and economic Burden: a report prepared in collaboration with the international osteoporosis foundation (IOF) and the European federation of pharmaceutical industry associations (EFPIA). *Archives of Osteoporosis* 8 (1–2): 136. https://doi.org/10.1007/s11657-013-0136-1, https:// www.ncbi.nlm.nih.gov/pmc/articles/PMC3880487/pdf/11657_2013_ Article_136.pdf.

194 Mithal, A., Dhingra, V. and Lau, E. (2009). The Asian Audit Epidemiology, costs and burden of osteoporosis in Asia 2009. http://www.iofbonehealth.org/.

195 www.iofbonehealth.org/sites/default/files/PDFs/Audit%20Eastern%20Europe_ Central%20Asia/Russian_Audit-Bulgaria.pdf.

196 www.iofbonehealth.org/sites/default/files/PDFs/Audit%20Asia/Asian_ regional_audit_Indonesia.pdf.

197 www.iofbonehealth.org/sites/default/files/PDFs/Audit%20Asia/Asian_ regional_audit_India.pdf.

198 https://www.iofbonehealth.org/facts-statistics.

199 Global bone density test market grows substantially by 2023, asserts MRFR unleashing the forecast for 2017–2023. *Reuters*. www.reuters.com/ brandfeatures/venture-capital/article?id=57229.

200 El Maghraoui, A. et al. (2009). Bone mineral density of the spine and femur in a group of healthy moroccan men. *Bone* 44 (5): 965–969. https://doi.org/10.1016/j. bone.2008.12.025.

201 Angell, M. (2005). The truth about the drug companies: how they deceive us and what to do about it. *Random House Trade Paperbacks* (2005). https://cyber. harvard.edu/cyberlaw2005/sites/cyberlaw2005/images/ NYReviewBooksAngell.pdf.

202 https://www.keele.ac.uk/pharmacy-bioengineering/ourpeople/jamesnolan/.

203 https://people.eecs.berkeley.edu/~pabbeel/.

204 https://github.com/alirezadir/Production-Level-Deep-Learning.

205 Ritchie, H. (2018). What do people die from?. https://ourworldindata.org/ what-does-the-world-die-from.

206 https://www.who.int/nmh/publications/ncd_report_chapter1.pdf.

207 Mendez, K.M., Broadhurst, D.I., and Reinke, S.N. (2019). The application of artificial neural networks in metabolomics: a historical perspective. *Metabolomics* 15 (11): 142. https://doi.org/10.1007/s11306-019-1608-0.

208 http://bit.ly/Farm-IoT-Ranveer.

209 https://www.theaaih.org/.

210 Gutterman, D.J. (2009). Silent myocardial ischemia. *Circulation Journal* 73: 785–797. https://www.jstage.jst.go.jp/article/circj/73/5/73_CJ-08-1209/_pdf.

211 Institute for Health Metrics and Evaluation, Human Development Network, The World Bank (2013). *The Global Burden of Disease: Generating Evidence, Guiding Policy — Sub-Saharan Africa Regional Edition*. Seattle, WA: IHME http://documents.worldbank.org/curated/en/831161468191672519/pdf/808520PUB0E NGL0Box0379820B00PUBLIC0.pdf.

212 World Health Organization (2009). *Global Health Risks: Mortality and Burden of Disease Attributable to Selected Major Risks*. ISBN: 978 92 4 156387 1 https://www.who.int/healthinfo/global_burden_disease/GlobalHealthRisks_report_full.pdf.

213 https://www.who.int/pmnch/media/press_materials/fs/fs_mdg4_childmortality/en/.

214 https://economictimes.indiatimes.com/magazines/panache/jack-dorsey-has-fallen-in-love-with-africa-plans-to-shift-there-for-3-6-months-next-year/articleshow/72275949.cms.

215 The concept of "bit dribbling" may not be new but may be attributed to Neil Gershenfeld (MIT) circa 1999 (*personal communication*). The notion is that very small amounts of data and information (bits) transmitted (dribbled) at the right time, are often enough to serve 80% of cases, in several circumstances.

216 Parker, M. (2018). Shut down the business school. https://www.theguardian.com/news/2018/apr/27/bulldoze-the-business-school.

217 Hedengren, J.D. Drilling automation and downhole monitoring with physics based models. https://apm.byu.edu/prism/index.php/Members/JohnHedengren.

218 https://apmonitor.com/do/index.php/Main/ShortCourse.

219 Osgood, N. (2019). Systems data science. https://www.youtube.com/watch?v=CPUOyqs9G3Q&feature=youtu.be.

220 Bayes, T. (1763). An essay towards solving a problem in the doctrine of chances. *Philosophical Transactions of the Royal Society of London* 53 (0): 370–418. http://www.rssb.be/bsn57/bsn57-6.pdf.

221 Eid, M.A. et al. (2016). A novel eye-gaze-controlled wheelchair system for navigating unknown environments: case study with a person with ALS. *IEEE Access* 4: 558–573. https://doi.org/10.1109/ACCESS.2016.2520093.

222 Robots for lonely hearts – Asian robotics review. https://asianroboticsreview.com/home29-html.

223 Oppenheimer, A. and Fitz, E.E. (2019). *The Robots Are Coming! The Future of Jobs in the Age of Automation*. Vintage Books, a division of Penguin Random House. ISBN: ISBN-13 978-0525565000.

224 Deming, D. (2020). The robots are coming. prepare for trouble. *The New York Times* (30 January 2020). https://www.nytimes.com/2020/01/30/business/artificial-intelligence-robots-retail.html.

225 http://www.philipkdickfans.com/mirror/websites/pkdweb/short_stories/Autofac.htm.

226 https://www.bloomberg.com/news/features/2017-10-18/this-company-s-robots-are-making-everything-and-reshaping-the-world.

227 Ford, M. (2016). *Rise of the Robots: Technology and the Threat of a Jobless Future.* Basic Books.

228 Pugliano, J. (2017). *The Robots Are Coming: A Human's Survival Guide to Profiting in the Age of Automation.* Ulysses Press.

229 The Dutch Safety Board (2009). Crashed during approach, Boeing 737-800, near Amsterdam Schiphol Airport. 25 February 2009. https://catsr.vse.gmu.edu/SYST460/TA1951_AccidentReport.pdf.

230 Pacaux-Lemoine, M.-P. and Flemisch, F. (2016). Layers of shared and cooperative control, assistance and automation. *IFAC-Papers OnLine* 49 (19): 159–164. https://doi.org/10.1016/j.ifacol.2016.10.479.

231 Wickens, C.D. et al. (2010). Stages and levels of automation: an integrated meta-analysis. *Proceedings of the Human Factors and Ergonomics Society Annual Meeting* 54 (4): 389–393. https://doi.org/10.1177/154193121005400425.

232 Hamby, C. (2020). How boeing's responsibility in a deadly crash 'got buried. *The New York Times* (20 January 2020). https://www.nytimes.com/2020/01/20/business/boeing-737-accidents.html.

233 Hamby, C. and Moses, C. (2020). Boeing refuses to cooperate with new inquiry into deadly crash. *The New York Times* (6 February 2020). www.nytimes.com/2020/02/06/business/boeing-737-inquiry.html.

234 Solow, R.M. (1957). Technical change and the aggregate production function. *The Review of Economics and Statistics* 39 (3): 312. https://doi.org/10.2307/1926047, http://www.piketty.pse.ens.fr/files/Solow1957.pdf.

235 http://ghdx.healthdata.org/.

236 Howes, A. (2020). *Arts and Minds: How the Royal Society of Arts Changed a Nation.* Princeton University Press. ISBN: 9780691182643.

237 Endo, N., Ghaeli, N., Duvallet, C. et al. (2020). Rapid assessment of opioid exposure and treatment in cities through robotic collection and chemical analysis of wastewater. *Journal of Medical Toxicology* 16: 195–203. https://doi.org/10.1007/s13181-019-00756-5.

238 https://www.oxfordeconomics.com/recent-releases/how-robots-change-the-world.

239 Kasprzyk-Hordern, B. and Baker, D.R. (2012). Enantiomeric profiling of chiral drugs in wastewater and receiving waters. *Environmental Science & Technology* 46 (3): 1681–1691. https://doi.org/10.1021/es203113y, https://core.ac.uk/reader/161908112.

240 Matus, M. et al. (2019). 24-hour multi-omics analysis of residential sewage reflects human activity and informs public health. *Genomics* https://doi.org/10.1101/728022.

241 Wu, F. et al. (2020). SARS-CoV-2 titers in wastewater are higher than expected from clinically confirmed cases. *Infectious Diseases (except HIV/AIDS)* https://doi.org/10.1101/2020.04.05.20051540.

242 Wu, F. et al. (2020). SARS-CoV-2 titers in wastewater foreshadow dynamics and clinical presentation of new COVID-19 cases. *Infectious Diseases (except HIV/AIDS)* https://doi.org/10.1101/2020.06.15.20117747.

243 Exceptions prove the rule. Not all Indians espouse the "body shop" mantra. Global business and tech leaders (2020) of Indian origin include Mrs Jayashree Ullal (CEO, Arista Networks), Mr Arvind Krishna (CEO, IBM, incoming), Mr Sundar Pichai (CEO, Alphabet/Google), Mr Satya Nadella (CEO, Microsoft), Mr Shantanu Narayen (CEO, Adobe), Mr Rajeev Suri (CEO, Nokia, outgoing), Mr V K Narasimhan (CEO, Novartis) and Mr Ajaypal Singh Banga (CEO, MasterCard). The market cap of these 8 companies (~$3T) may be 20% of the total market cap of the top 50 US companies (~$15T). www.iweblists.com/us.

244 Sabrie, G. (2019). Behind the rise of China's facial-recognition giants (03 September 2019). https://www.wired.com/story/behind-rise-chinas-facial-recognition-giants/.

245 https://www.purdue.edu/rosehub/.

246 Radjou, N., Prabhu, J. (2016). Frugal innovation: how to do more with less. http://naviradjou.com/book/frugal-innovation-how-to-do-more-with-less/.

247 Creighton, J. (2019). 5 Moon-landing innovations that changed life on Earth. https://theconversation.com/5-moon-landing-innovations-that-changed-life-on-earth-102700.

248 https://www.inc.com/bill-murphy-jr/27-innovations-we-use-constantly-but-that-you-probably-didnt-know-were-from-nasa-space-program.html.

249 https://spinoff.nasa.gov/.

250 Marcus, G. (2019). An epidemic of AI misinformation. https://thegradient.pub/an-epidemic-of-ai-misinformation/.

251 Vanegas, D.C., Patiño, L., Mendez, C. et al. (2018). Laser scribed graphene biosensor for detection of biogenic amines in food samples using locally sourced materials. *Biosensors* 8 (2): 42. https://doi.org/10.3390/bios8020042, https://www.mdpi.com/2079-6374/8/2/42.

252 Holmes, E.A. et al. (2016). Applications of time-series analysis to mood fluctuations in bipolar disorder to promote treatment innovation: a case series. *Translational Psychiatry* 6 (1): e720–e720. https://doi.org/10.1038/tp.2015.207, https://www.ncbi.nlm.nih.gov/pmc/articles/PMC5068881/pdf/tp2015207a.pdf.

253 Trautmann, S. et al. (2016). The economic costs of mental disorders: do our societies react appropriately to the burden of mental disorders? *EMBO Reports*

17 (9): 1245–1249. https://doi.org/10.15252/embr.201642951, www.ncbi.nlm.nih. gov/pmc/articles/PMC5007565/pdf/EMBR-17-1245.pdf.

254 Drevets, W.C. et al. (1997). Subgenual prefrontal cortex abnormalities in mood disorders. *Nature* 386 (6627): 824–827. https://doi.org/10.1038/386824a0.

255 Llamocca, P., Junestrand, A., Cukic, M., Urgelés, D., López, V. (2018). Data source analysis in mood disorder research. *XVIII Proceedings of the XVIII Conference of the Spanish Association for Artificial Intelligence*. Granada, Spain: CAEPIA. ISBN: 978-84-09-05643-9 F. Herrera et al. (Eds.), pp. 893–900.

256 Llamocca, P., Urgelés, D., Cukic, M., Lopez, V. (2019). Bip4Cast: Some advances in mood disorders data analysis. *Proceedings of the 1st International Alan Turing Conference on Decision Support and Recommender Systems*, London 2019.

257 Perrault, R., Shoham, Y., Brynjolfsson, E. et al. (2019). *The AI Index 2019 Annual Report*. Stanford, CA: AI Index Steering Committee, Human-Centered AI Institute, Stanford University https://hai.stanford.edu/sites/g/files/ sbiybj10986/f/ai_index_2019_report.pdf.

258 Chen, Y.H. et al. (2019). Eyeriss v2: a flexible accelerator for emerging deep neural networks on mobile devices. (May 2019). http://arxiv.org/abs/1807.07928 ■ https://arxiv.org/pdf/1807.07928.pdf. https://www.rle.mit.edu/eems/wp-content/uploads/2019/04/2019_jetcas_eyerissv2.pdf.

259 Jo, E.S., and Timnit, G. (2019). Lessons from archives: strategies for collecting sociocultural data in machine learning. (December 2019) doi:https://doi. org/10.1145/3351095.3372829. https://arxiv.org/pdf/1912.10389.pdf.

260 Tembon, M.M., and Lucia, F. (2008). Girl's education in the 21st Century: gender equality, empowerment and growth. *The World Bank*. doi:https://doi. org/10.1596/978-0-8213-7474-0.

261 https://www.nytimes.com/2006/10/14/world/asia/14nobel.html.

262 Cohn, A. et al. (2019). Civic honesty around the Globe. *Science* 365 (6448): 70–73. https://doi.org/10.1126/science.aau8712.

263 Grosch, K., Rau, H. (2017). Gender differences in honesty: the role of social value orientation. Discussion Papers, No. 308, University of Göttingen, Center for European, Governance and Economic Development Research (CEGE), Göttingen, Germany. https://www.econstor.eu/bitstr eam/10419/156226/1/882555200.pdf.

264 Sen, A. (2009). *The Idea of Justice*. Harvard University Press https:// dutraeconomicus.files.wordpress.com/2014/02/amartya-sen-the-idea-of-justice-2009.pdf.

265 Artificial intelligence makes bad medicine even worse. *Wired*. www.wired.com, https://www.wired.com/story/artificial-intelligence-makes-bad-medicine-even-worse/.

266 Datta, S. (2008). Arm chair essays in energy. https://dspace.mit.edu/ handle/1721.1/45512.

267 Mackenzie, D. (1877). *The Flooding of the Sahara: An Account of the Proposed Plan for Opening Central Africa to Commerce and Civilization* (ed. S.L. Marston). Searle & Rivington https://ia902205.us.archive.org/23/items/floodingsaharaa01mackgoog/floodingsaharaa01mackgoog.pdf.

268 https://www.jstor.org/stable/1761255?seq=1#metadata_info_tab_contents.

269 https://www.nature.com/articles/019509a0.pdf.

270 Singh, L. et al. (2020). Bioelectrofuel synthesis by nanoenzymes: novel alternatives to conventional enzymes. *Trends in Biotechnology* 38 (5): S0167779919303129. https://doi.org/10.1016/j.tibtech.2019.12.017.

271 Rothman, J. (2020). The equality conundrum. (13 January 2020). www.newyorker.com/magazine/annals-of-inquiry. https://www.newyorker.com/magazine/2020/01/13/the-equality-conundrum.

272 https://www.mikespastry.com/.

273 https://www.modernpastry.com/.

274 Ilic, S. et al. (2018). Deep contextualized word representations for detecting sarcasm and irony. *Proceedings of the 9th Workshop on Computational Approaches to Subjectivity, Sentiment and Social Media Analysis*, Association for Computational Linguistics, pp. 2–7: doi:https://doi.org/10.18653/v1/W18-6202. https://arxiv.org/pdf/1809.09795.pdf.

275 Datta, S. (2008). Convergence of bio, info, nano, eco: global public goods and economic growth. https://dspace.mit.edu/handle/1721.1/41909.

276 Steinmueller, W.E. (2010). Economics of technology policy. In: *Handbook of the Economics of Innovation*, 1ste, vol. 2 (eds. B.H. Hall and N. Rosenberg), 1181–1218. Elsevier https://doi.org/10.1016/S0169-7218(10)02012-5.

277 Wang, L., Chen, Y., Long, F. et al. (2020). Breaking the loop: tackling homoacetogenesis by chloroform to halt hydrogen production-consumption loop in single chamber microbial electrolysis cells. *Chemical Engineering Journal* https://doi.org/10.1016/j.cej.2020.124436.

278 https://patents.justia.com/inventor/andrew-e-fano.

279 https://patents.justia.com/patent/10095981.

280 https://claimparse.com/patent.php?patent_num=10275721.

281 https://www.ibm.com/downloads/cas/KMPVGB4W.

282 https://mcgovern.mit.edu/profile/tomaso-poggio/.

283 Fjelland, R. (2020). https://doi.org/10.1057/s41599-020-0494-4). Why general artificial intelligence will not be realized. *Humanit Soc Sci Commun* 7: 10. www.nature.com/articles/s41599-020-0494-4.pdf.

284 Faro, S.H. and Mohamed, F.B. (eds.) (2006). *Functional MRI: Basic Principles and Clinical Applications*. Springer. ISBN: 978-0-387-23046-7.

285 Suleiman Khan (2019). BERT, RoBERTa, DistilBERT, XLNet — which one to use. https://towardsdatascience.com/bert-roberta-distilbert-xlnet-which-one-to-use-3d5ab82ba5f8.

286 https://cloud.google.com/tpu/.

287 FloydHub Blog (2019). When not to choose the best NLP model. https://blog.floydhub.com/when-the-best-nlp-model-is-not-the-best-choice/.

288 Brown, T.B. et al. (2020). Language models are few-shot learners. *ArXiv* 2005: 14165 [Cs]. http://arxiv.org/abs/2005.14165. https://arxiv.org/pdf/2005.14165.pdf.

289 Peters, M.E., Neumann, M., Iyyer, M., Gardner, M., Clark, C., Lee, K., Zettlemoyer, L. (2018). Deep contextualized word representations. https://arxiv.org/pdf/1802.05365.pdf.

290 Devlin, J., Chang, M.W., Lee, K. and Toutanova, K. (2018). BERT: pre-training of deep bidirectional transformers for language understanding. https://arxiv.org/pdf/1810.04805.pdf.

291 Lan, Z., Chen, M., Goodman, S., Gimpel, K., Sharma, P. and Soricut, R. (2020). Albert: a lite bert for self-supervised learning of language representations. *8th International Conference on Learning Representations (2020)*. https://openreview.net/pdf?id=H1eA7AEtvS.

292 Sun, C. et al. (2019). A deep learning approach with deep contextualized word representations for chemical–protein interaction extraction from biomedical literature. *IEEE Access* 7: 151034–151046. https://doi.org/10.1109/ACCESS.2019.2948155.

293 Tshitoyan, V., Dagdelen, J., Weston, L. et al. (2019). Unsupervised word embeddings capture latent knowledge from materials science literature. *Nature* 571: 95–98. https://doi.org/10.1038/s41586-019-1335-8.

294 Pelillo, M. (2014). Alhazen and the nearest neighbor rule. *Pattern Recognition Letters* 38 (2014): 34–37. https://doi.org/10.1016/j.patrec.2013.10.022.

295 Babyak, M.A. (2004). What you see may not be what you get: a brief, nontechnical introduction to overfitting in regression-type models. *Psychosomatic Medicine* 66 (3): 411–421. https://doi.org/10.1097/01.psy.0000127692.23278.a9, https://people.duke.edu/~mababyak/papers/babyakregression.pdf.

296 Hamed, T. (2017). Recursive feature addition: a novel feature selection technique, including a proof of concept in network security. PhD thesis submitted to University of Guelph, Ontario, Canada. https://atrium.lib.uoguelph.ca/xmlui/bitstream/handle/10214/10315/Hamed_Tarfa_201704_PhD.pdf?sequence=1&isAllowed=y.

297 Nandy, A., Duan, C., Janet, J.P. et al. (2018). Strategies and software for machine learning accelerated discovery in transition metal chemistry. *Industrial and Engineering Chemistry Research* 57 (42): 13973–13986. https://doi.org/10.1021/acs.iecr.8b04015.

298 Virshup, A.M. et al. Stochastic voyages into uncharted chemical space produce a representative library of all possible drug-like compounds. *Journal of the*

American Chemical Society 135 (19): 7296–7303. https://doi.org/10.1021/ja401184g.

299 Lemonick, S. (2019). As DFT matures, will it become a push-button technology? *Chemical & Engineering News* 97 (35) https://cen.acs.org/physical-chemistry/computational-chemistry/DFT-matures-become-push-button/97/i35.

300 Bruna, J., Zaremba, W., Szlam, A. and LeCun, Y. (2014). Spectral networks and locally connected networks on graphs. http://arxiv.org/abs/1312.6203. https://arxiv.org/pdf/1312.6203.pdf.

301 Gilmer, J., Schoenholz, S.S., Riley, P.F., Vinyals, O. and Dahl, G.E. (2017). Neural message passing for quantum chemistry. https://arxiv.org/pdf/1704.01212.pdf.

302 Swanson, K. (2019). Message passing neural networks for molecular property prediction. Master's Thesis. EECS, MIT. https://dspace.mit.edu/bitstream/handle/1721.1/123133/1128814048-MIT.pdf?sequence=1&isAllowed=y.

303 http://chemprop.csail.mit.edu/.

304 Ruddigkeit, L., van Deursen, R., Blum, L.C., and Reymond, J.-L. (2012). Enumeration of 166 billion organic small molecules in the chemical Universe database GDB-17. *Journal of Chemical Information and Modeling* 52 (11): 2864–2875. https://doi.org/10.1021/ci300415d.

305 Stokes, J.M., Yang, K., Swanson, K. et al. (2020). A deep learning approach to antibiotic discovery. *Cell* 180 (4): 688–702.e13. https://doi.org/10.1016/j.cell.2020.01.021, https://www.cell.com/cell/pdf/S0092-8674(20)30102-1.pdf.

306 http://news.mit.edu/2020/artificial-intelligence-identifies-new-antibiotic-0220.

307 Zhang, X. et al. (2019). Deep neural network hyperparameter optimization with orthogonal array tuning. (July 2019). http://arxiv.org/abs/1907.13359. https://arxiv.org/pdf/1907.13359.pdf.

308 Hansen, L.K. and Salamon, P. (1990). Neural network ensembles. *IEEE Transactions on Pattern Analysis and Machine Intelligence* 12 (10) https://pdfs.semanticscholar.org/257d/c8ae2a8353bb2e86c1b7186e7d989fb433d3.pdf.

309 https://thenextweb.com/neural/2020/02/19/study-ai-expert-gary-marcus-explains-how-to-take-ai-to-the-next-level/.

310 Russell, S. and Norvig, P. (2010). *Artificial Intelligence: A Modern Approach*, 3rde. Prentice Hall http://aima.cs.berkeley.edu/.

311 Winston, P.H. and Brown, R.H. (eds.) (1979). *Artificial Intelligence, an MIT Perspective*. MIT Press https://courses.csail.mit.edu/6.034f/ai3/rest.pdf.

312 Buchanan, B.G. and Shortliffe, E.H. (1984). *Rule-Based Expert Systems: The MYCIN Experiments of the Stanford Heuristic Programming Project*. Addison-Wesley http://digilib.stmik-banjarbaru.ac.id/data.bc/2.%20AI/2.%20AI/1984%20Rule-Based%20Expert%20Systems.pdf.

313 Quinlan, J.R. (1987). *Applications of Expert Systems: Based on the Proceedings of the Second Australian Conference*, 1987. Turing Institute Press in association with Addison-Wesley Pub. Co.

314 Giarratano, J.C. and Riley, G. (1989). *Expert Systems: Principles and Programming*, 1ste. Boston: PWS-Kent Publishing Company. ISBN: 0-87835-335-6.

315 Buchanan, B.G. (1989). Can machine learning offer anything to expert systems? *Machine Learning* 4 (3–4): 251–254. https://doi.org/10.1007/BF00130712, https://link.springer.com/content/pdf/10.1007/BF00130712.pdf.

316 Valverde, S. (2016). Major transitions in information technology. *Philosophical Transactions of the Royal Society B* 371: 20150450. http://dx.doi.org/10.1098/rstb.2015.0450, https://www.ncbi.nlm.nih.gov/pmc/articles/PMC4958943/pdf/rstb20150450.pdf.

317 Mearian, L. (2017). CW@50: data storage goes from $1M to 2 cents per gigabyte (+video). *Computerworld* (23 March 2017). https://www.computerworld.com/article/3182207/cw50-data-storage-goes-from-1m-to-2-cents-per-gigabyte.html.

318 Supernor, B. (2018). Why the cost of cloud computing is dropping dramatically. *App Developer Magazine*. https://appdevelopermagazine.com/why-the-cost-of-cloud-computing-is-dropping-dramatically/.

319 Wolpert, D.H. (2018). Why do computers use so much energy?. *Scientific American Blog Network*. https://blogs.scientificamerican.com/observations/why-do-computers-use-so-much-energy/.

320 Wolpert, D.H. (2019). The stochastic thermodynamics of computation. *Journal of Physics A: Mathematical and Theoretical* 2019 https://iopscience.iop.org/article/10.1088/1751-8121/ab0850/pdf.

321 Panesar, S.S., Kliot, M., Parrish, R. et al. (2019). Promises and perils of artificial intelligence in neurosurgery. *Neurosurgery* https://doi.org/10.1093/neuros/nyz471.

322 Hanhauser, E., Bono, M.S. Jr., Vaishnav, C. et al. (2020). Solid-phase extraction, preservation, storage, transport, and analysis of trace contaminants for water quality monitoring of heavy metals. *Environmental Science & Technology* https://doi.org/10.1021/acs.est.9b04695.

323 Sharma, R., Kamble, S.S., Gunasekaran, A. et al. (2020). A systematic literature review on machine learning applications for sustainable agriculture supply chain performance. *Computers and Operations Research* https://doi.org/10.1016/j.cor.2020.104926.

324 Gyrard, A., Gaur, M., Shekarpour, S., Thirunarayan, K. and Sheth, A. (2018). Personalized health knowledge graph. *Contextualized Knowledge Graph Workshop, International Semantic Web Conference*, 2018. https://scholarcommons.sc.edu/cgi/viewcontent.cgi?article=1005&context=aii_fac_pub.

325 Apple agrees to settlement of up to $500 million from lawsuit alleging it throttled older phones. *TechCrunch*. http://social.techcrunch.com/2020/03/02/apple-agrees-to-settlement-of-up-to-500-million-from-lawsuit-alleging-it-throttled-older-phones/.

326 C.,.G. (1909). Concentration and dependency ratios (in Italian). English translation in. *Rivista di Politica Economica (1997)* 87: 769–789.

327 www.reddit.com/r/MapPorn/comments/88mw3q/open_defecation_around_ the_world_2015_960_684/.

328 Nearly a billion people still defecate outdoors. Here's why. *Magazine* (25 July 2017). https://www.nationalgeographic.com/magazine/2017/08/toilet-defecate-outdoors-stunting-sanitation/.

329 Sadhu, B. et al. (2019). The more (antennas), the merrier: a survey of silicon-based mm-wave phased arrays using multi-IC scaling. *IEEE Microwave Magazine* 20 (12): 32–50.

330 Wang, T. and Kang, J.W. (2020). An integrated approach for assessing national e-commerce performance. Trade, Investment and Innovation Working Paper Series, No. 01/20, ESCAP Trade, Investment and Innovation Division, United Nations (UN). January 2020. Bangkok, Thailand. https://www.unescap.org/sites/default/files/publications/Working%20Paper%20No.1_2020.pdf.

331 Andrew, G. (2015). *Open Defecation Around the World*. The World Bank.

332 http://www3.weforum.org/docs/WEF_Global_Risk_Report_2020.pdf.

333 Propaganda by erudite, credible and respectable scientists, are disturbing, devastating, desacralizing. Time, events and publications suggest that "giants" who are *good* are occasionally hypnotized by the slippery slope of metamorphosis from *good* to self-anointed "God" particles. Knowledge, which was once regarded as an oak tree, and supposed to usher in self-deprecation, modesty and humility, now, frequently suffers from bloating, sufficient to spill over the *black hole* of hubris. For example, a trio of brilliant male scientists, in the upper latitudes of North America, are acting as *Nostradamus*, stoked by greed and the quest for immortality, buoyed by corporate largesse, exclusively driven by the desire for wealth creation. A group of complicit organizations and ill-informed media are ever ready to quench the drab voices of reason and restraint, in favour of sensationalizing and amplifying this inane *Nostradamus Effect*. This harms society by generating derelict reports and reduces the credibility of august institutions and organizations which appear as pawns for corporate business development (https://knowledgegraphsocialgood.pubpub.org) often under a camouflage of so-called knowledge for social good. https://knowledgegraphsocialgood.pubpub.org/programcommittee [This statement is the personal opinion of the author].

334 Boroush, M. (2020). *Research and Development: U.S. Trends and International Comparisons. Science and Engineering Indicators 2020*. NSB-2020-3. Alexandria, VA: National Science Board, National Science Foundation https://ncses.nsf.gov/pubs/nsb20203/assets/nsb20203.pdf.

335 Cyranoski, D. (2018). China awaits controversial blacklist of 'poor quality' journals. *Nature* 562: 471–472. www.nature.com. https://doi.org/10.1038/d41586-018-07025-5.

336 https://www.thehindu.com/profile/photographers/Prashant-Waydande/.

337 Indian research quality lags quantity. *Economic Times Blog* (24 December 2019). https://economictimes.indiatimes.com/blogs/et-editorials/indian-research-quality-lags-quantity/.

338 Jaishankar, D. (2013). The huge cost of India's discrimination against women. *The Atlantic* (18 March 2013). https://www.theatlantic.com/international/archive/2013/03/the-huge-cost-of-indias-discrimination-against-women/274115/.

339 Surbatovich, M., et al. (2017). Some recipes can do more than spoil your appetite: analyzing the security and privacy risks of IFTTT recipes. *Proceedings of the 26th International Conference on World Wide Web*, Intl WWW Conferences Steering Com, pp. 1501–10. doi:https://doi.org/10.1145/3038912.3052709. https://www.ftc.gov/system/files/documents/public_comments/2017/11/00026-141804.pdf.

340 https://coronavirus.jhu.edu/map.html.

341 Datta, S. (2020). CITCOM (unpublished) MIT Library. https://dspace.mit.edu/handle/1721.1/111021.

342 Cai, Y. et al. (2020). Distinct conformational states of SARS-CoV-2 spike protein. *Science* https://doi.org/10.1126/science.abd4251.

343 Wiedmann, T. et al. (2020). Scientists' warning on affluence. *Nature Communications* 11 (1): 3107. https://doi.org/10.1038/s41467-020-16941-y, www.nature.com/articles/s41467-020-16941-y.pdf.

344 Giacobassi, C.A., Oliveira, D.A., Pola, C.C., Xiang, D., Tang, Y., Austin Datta, S.P., McLamore, E.S. and Gomes, C. (2020). Sense-Analyze-Respond-Actuate (SARA) Paradigm: Proof of concept systems spanning nanoscale and macroscale actuation for detection of Escherichia coli in water. (in press).

345 PDFs numbered 00, 01, 02, 03 and 04 in https://dspace.mit.edu/handle/1721.1/123984.

346 Yu, L.-R. et al. (2018). Aptamer-based proteomics identifies mortality-associated serum biomarkers in dialysis-dependent AKI patients. *Kidney International Reports* 3 (5): 1202–1213. https://doi.org/10.1016/j.ekir.2018.04.012.

347 Datta, S. (2020) CORONA: a social coronary. https://bit.ly/CORONARY.

348 Poire, N.P. (2011). The great transformation of 2021. ISBN-13 978-0557948901. http://www.lulu.com/shop/norman-poire/the-great-transformation-of-2021/paperback/product-14729057.html.

349 Victor Weisskopf. *MIT*. http://web.mit.edu/dikaiser/www/Kaiser.Weisskopf.pdf.

350 Gary, T. (2018). What if sugar is worse than just empty calories? An essay by gary taubes. *British Medical Journal*: j5808. www.bmj.com/content/bmj/360/bmj.j5808.full.pdf.

351 Marie, C. https://www.iupac.org/publications/ci/2011/3301/jan11.pdf.

352 Frankl, V.E. et al. (2020). *Yes to Life: In Spite of Everything*. Beacon Press. ISBN: ISBN-13 978-0807005552.

353 le Rond d'Alembert, J. (1886). Oeuvres de D'Alembert. ISBN-13 9781103953189.

354 Gordin, M.D. (2015). *Scientific Babel: How Science Was Done before and after Global English*. The University of Chicago Press.

355 https://citations.ouest-france.fr/citation-voltaire/medecins-administrent-medicaments-dont-savent-22351.html.

Market proponents of AI are individuals who blather about neural networks of which they know little, to solve problems using learning tools which they know less, for the society of human beings of whom they know nothing.

(**Adapted from** "Les médecins administrent des médicaments dont ils savent très peu, à des malades dont ils savent moins, pour guérir des maladies dont ils ne savent rien" – **Voltaire** [355])

(Doctors are men who prescribe medicines of which they know little, to cure diseases of which they know less, in human beings of whom they know nothing.)

← **Tweet**

 Geoffrey Hinton
@geoffreyhinton

Suppose you have cancer and you have to choose between a black box AI surgeon that cannot explain how it works but has a 90% cure rate and a human surgeon with an 80% cure rate. Do you want the AI surgeon to be illegal?

3:37 PM · Feb 20, 2020 · Twitter Web App

3

Machine Learning Techniques for IoT Data Analytics

Nailah Afshan[1] and Ranjeet Kumar Rout[2]

[1]*Department of Computer Science and Engineering, Islamic University of Science and Technology, Pulwama, India*
[2]*Department of Computer Science and Engineering, National Institute of Technology, Srinagar, India*

3.1 Introduction

"The most profound technologies are those that disappear. They weave themselves into the fabric of everyday life until they are indistinguishable from it." This is the central statement made by Mark Weiser in his seminal paper in Scientific American in 1991 [1]. The statement fits quite well as far as the technology of IoT is concerned. IoT is among the most prevalent and rising technologies in the current era. The success and emergence of IoT have been possible due to tremendous technological advancements in recent years and rapid convergence of digital electronics, wireless communication, Internet protocols, computing systems, and micro-electro-mechanical systems (MEMS) [2–4]. It actually portrays the next progression phase of the Internet, with great advancement in its data processing abilities. Dynamic Internet revolutionized communication that in turn resulted in the birth of IoT. The future of IoT and its relation with the Internet has been beautifully stated by a senior researcher at HP Labs, Peter Hartwell as, "With a trillion sensors embedded in the environment – all connected by computing systems, software, and services – it will be possible to hear the heartbeat of the Earth, impacting human interaction with the globe as profoundly as the Internet has revolutionized communication." Now it has started to mold and refine our modern lifestyle that includes daily routine of a common man, and all aspects of modern world, a world where devices of different

Big Data Analytics for Internet of Things, First Edition. Edited by Tausifa Jan Saleem and Mohammad Ahsan Chishti.

possible shapes and sizes are fabricated with highly "smart" features allowing them to interact and communicate with other devices as well as with humans, exchange and share their data, make smart decisions on their own, and accomplish important tasks using already defined and set conditions [5]. According to the Cisco Internet Business Solutions Group (IBSG), IoT is simply the point in time when more "things or objects" are connected to the Internet than people. IoT describes a system or an environment where individual things in the physical world together with sensors within or connected to these things, all are connected to the Internet through different wireless and wired Internet connections. It binds together both living and inanimate things, uses sensors for sensing and collection of data, and is set to change what types of devices communicate over an IP Network. It is turning to be a well-known concept covering various horizontal and vertical markets with its vast application areas [6]. The main distinguishing elements of IoT are the smart and self-configuring devices having capability to communicate with each other through global network infrastructure. Such communication between these heterogeneous devices symbolizes IoT as a pioneering technology, enabling universal and extensive computing applications [7]. Hence, tremendous variety of industrial IoT applications [8] in diverse domains have been developed and deployed such as applications in agriculture, medicine, healthcare [9], environmental monitoring, security surveillance, food processing industry, and transportation systems. Thus, IoT is set to transform this world into a smarter, more efficient, and safer place.

To present better services to users, IoT requires data so as to perform intelligently. Thus, different IoT-based systems must be capable of accessing raw data from different constituent resources and then process and analyze it to extract knowledge. Further, IoT itself is generating enormous volumes of data. The size of data present on the Web and the Internet is already quite enormous and still continues to grow at rapid pace. Around 2.5 quintillion bytes of data are generated on daily basis and it is calculated that 90% of the data today was created in last several years [10]. The voluminous data collected from IoT devices (mainly sensors) can have a key role in having a better understanding and control over complicated systems around us, enhancing predictions and decision-making, accuracy, higher efficiencies, productivity, greater automation, and wealth generation. For such value generation, these data need to be analyzed through different algorithms, processed into information which in turn can be converted into knowledge that can be utilized by machines so as to understand the real human world in a better manner [11]. This whole data transformation process can be illustrated in a better manner using "knowledge hierarchy/pyramid." Here, different layers are understood in the context of IoT (see Figure 3.1) [12].

The lowermost layer represents voluminous data generated by different IoT devices which are then filtered and processed into structured and machine-understandable

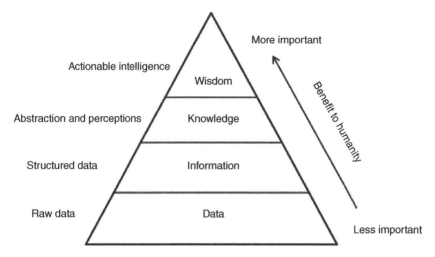

Figure 3.1 Knowledge pyramid in the context of IoT [12]. *Source:* Preethi, N. (2014) Performance evaluation of IoT result for machine learning, *Transactions on Engineering and Sciences*, 2, 11.

information represented by the next higher layer. Then, hidden knowledge from the information is extracted to have a better understanding of the data. Ultimately, using this abstract knowledge, machines can draw some intelligence referred to as wisdom that helps in future predictions as represented by the topmost layer. This wisdom can be then incorporated into some end products and services. Automatic realization of this knowledge pyramid by machines and algorithms is the key point to success of IoT.

To live up to its promises, proper analytics of huge amounts of data generated in IoT is very crucial to derive highly reliable, valuable, and accurate insights. But it is a challenging task as IoT data are quite different from traditional normal one with respect to features which is mainly because of the sensors used in data generation and collection. Some of the prominent features are variety, heterogeneity, rapid growth, and noise [13]. According to statistics, it is predicted that the number of sensors will be increased by 1 trillion in 2030 [14]. This rise in number will further affect the generation and growth of big data. IoT data are generated in a continuous manner, indicating that it would be of enormous volume. Further, the rate of generation of data for different devices varies that poses a serious challenge to its processing. Integration of sensory data generated and gathered from various heterogeneous sources is also posing a great challenge. Another peculiar characteristic of these data is variability and dynamicity that keeps changing with time and location. The characteristics IoT data are illustrated in Figure 3.2 [15].

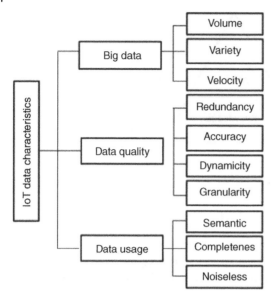

Figure 3.2 IoT data characteristics [15]. *Source:* Mahdavinejad, M.S., Rezvan, M., Barekatain, M., Adibi, P., Barnaghi, P. and Sheth, A.P. (2018) Machine learning for Internet of Things data analysis: a survey, *Digital Communication Network*, 4, 161–175. © 2018, Elsevier.

With regards to extremely challenging nature of IoT data, it is highly impor-tant to incorporate a novel concept known as smart data, obtained from raw IoT data by applying different data processing and analyzing techniques. These smart data can be a better substitute for IoT data capable of delivering action-able information and enhancing decision-making [16, 17]. For this transforma-tion, data science in general and machine learning in particular is the best option. They involve enormous algorithms applicable in diverse fields and have ability to discover hidden trends in data and predicting future data from historical data making and can make IoT systems smarter, more intelligent, and more efficient in decision-making [18]. Integration of IoT and machine learning has a promising future and a great role in making this world a smarter place. Both of these are among the top four emergent technologies as shown below in the latest Gartner's Hype Cycle (see Figure 3.3) [19]. According to this cycle, the emerging technologies for year 2019 fall into five major trends: sens-ing and mobility (IoT), augmented human, postclassical compute and commu-nications, digital ecosystems, and advanced AI and analytics. Thus, IoT, machine learning, and analytics are among the most prevalent technologies and will be having significant impact on business, society, and people over next 5–10 years. The main focus of this chapter is to present taxonomy of different machine learning techniques and demonstrate their implementation in IoT data analytics in order to extract higher level information from it as discussed in the next section.

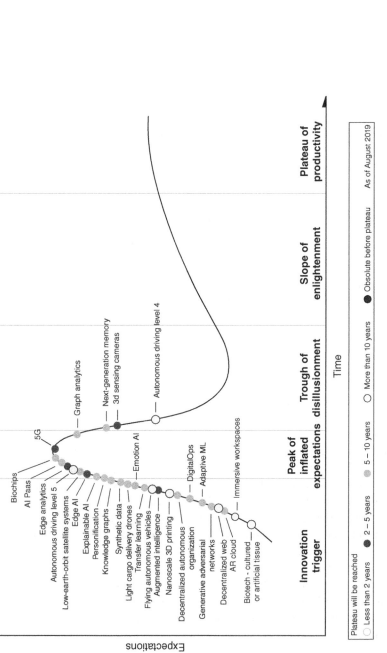

Figure 3.3 Gartner's Hype Cycle for emerging technologies, 2019 [19]. *Source*: Gartner Inc. (2019). Gartner's Hype Cycle Special Report for 2019. http://www.gartner.com/technology/research/hype-cycles/ (accessed 27 April 2020). © 2019, Gartner Inc.

3.2 Taxonomy of Machine Learning Techniques

Machine learning is the state-of-art technology nowadays and was first introduced in late 1950s as a subfield of Artificial Intelligence. In 1959, Arthur Samuel defined machine learning as "Field of study that gives computers the ability to learn without being explicitly programmed" [20]. Later, a more formal and practical definition was given by Tom Mitchell [21] in 1997 as "A Computer program is said to learn from an experience E with respect to some task T and some performance measure P, if its performance on T, as measured by P, improves with experience E." The fields of pattern recognition and computational learning theory actually led to the evolution of machine learning. In the beginning, it was limited to AI applications only but with time, it evolved and its applicability shifted more to computationally robust and viable algorithms. In the last decade, different application domains like computer vision, bioinformatics, spam and fraud detection, biometrics, speech recognition, etc., have been extensively using machine learning approaches for different tasks like clustering, classification, regression, etc. [22]. Different areas related to machines and other domains like mathematics, computer science, statistics, and neuroscience together contribute in the development of numerous techniques and algorithms of machine learning. In this section, we will discuss important basic concepts of machine learning together with some mostly used algorithms for analysis of smart data generated in IoT.

For generating desired results, a learning algorithm needs to work on a set of input samples termed as training set. According to the nature of this training set, the learning algorithms are broadly divided into three groups: supervised, unsupervised, and reinforcement [23–25]. In supervised learning, the samples of input vectors of the training set are associated with their respective target vectors called as labels while as no labels are needed in case of unsupervised learning. Reinforcement learning uses a system of reward and punishment for learning and here suitable action is taken for maximizing payoff in a particular situation. This chapter is mainly concerned with supervised and unsupervised learning as these are extensively used in IoT data analysis. Supervised learning aims to learn prediction of appropriate output vector for each input vector and is further divided into two main classes: classification and regression. In classification, the target labels comprise a finite number of discrete categories whereas continuous variables form the target labels in case of regression [26]. Unsupervised learning is also divided into clustering and feature extraction algorithms. Clustering aims at identifying distinguishable clusters of similar samples from the input dataset and feature extraction is used in the preprocessing phase of input data to transfer it to a more useful representation [24]. Figure 3.4 gives full taxonomy of various machine learning algorithms [27].

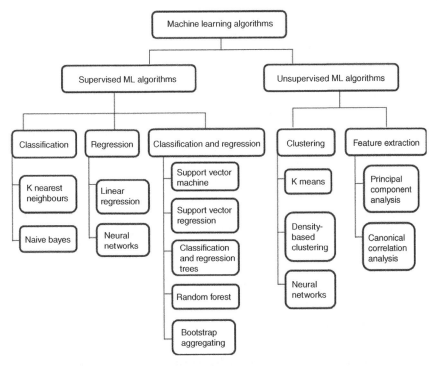

Figure 3.4 Taxonomy of machine learning algorithms used in IoT data analysis [27]. *Source:* Alsharif, M.H., Kelechi, A.H., Yahya, K. and Chaudhry, S.A. (2020) Machine learning algorithms for smart data analysis in Internet of Things environment: taxonomies and research trends, *Symmetry* 12: 88. Licenced under CC BY 4.0.

3.2.1 Supervised ML Algorithm

In supervised learning, the learning is accomplished by training the model on a labeled input dataset called as training dataset. It uses cost function that signifies the measure of error or difference between the actual ground truth and the output calculations made by the algorithm. The aim here is to minimize cost function so that the model is well trained to yield outputs that closely match the ground truth [28]. The algorithm keeps iterating till an acceptable level of accuracy is achieved.

In each of the following subsections, we will be assuming a notation for representing different data parameters. Training set is denoted as $\left\{\left(x_i, y_i\right)\right\}_{i=1}^{N}$, where N is the number of training samples in the training dataset, x_i is the ith sample in the M-dimensional training input vector, and y_i represents the P-dimensional output vector [15]. The M-dimensional input vector is represented as $x = (x_1, x_2, \ldots, x_N)^T$ and $y = (y_1, y_2, \ldots, y_N)^T$ is the corresponding output vector.

3.2.1.1 Classification

Classification is a supervised learning technique in which data are categorized into distinct number of classes where each class is assigned a label [29]. There are numerous types of classification techniques as discussed in the upcoming sections.

3.2.1.1.1 K Nearest Neighbors (KNN) Algorithm KNN is one of the most basic but essential classification as well as regression algorithms. It finds K data points from the training set which are close enough to any data point in the input space so as to classify it. A distance metric is used to find K nearest neighbors of an unknown data point such as Euclidean distance, Mahalanobis distance L_∞ norm, or Hamming distance [30, 31]. To formulate an expression for the algorithm, let $N_k(x)$ denote K nearest neighbors of input data point, "x," "y" be its estimated class label identifier, with the class variable using unassigned variable t. Further, suppose $I(.)$ represents the indicator function attribute: $I(s) = 1$ if s is true and $I(s) = 0$ otherwise. The classification task is depicted as [32]:

$$p\left(t = c \mid x, K\right) = \frac{1}{K} \sum_{i \in N_k(x)} I\left(t_i = c\right),$$

$$y = \arg\max p_c\left((t = c) \mid x, K\right). \tag{3.1}$$

This depicts that the input vector x will be classified and labeled in terms of the labels of its neighbors [33].

Scope in IoT Data Analytics KNN learning algorithm is quite simple, easy to implement, robust against noise, and efficient with respect to time. Despite these benefits, it suffers some shortcomings. It does not work that good with a large dataset, incurs large memory requirement costs, and is quite sensitive to unlabeled training data. It finds good applicability in IoT data analysis. KNN classifiers have been used in monitoring of physiological signals of humans for stress detection [34] and in epileptic patients for seizure activity detection [35] in developing smart healthcare. In [36], the model has been used to optimize the passenger's travel pattern in smart environments. KNN has also been used for developing a model for detection of outliers in IoT streaming data in [37] and also in wireless sensor networks for the purpose of intrusion detection [38].

3.2.1.1.2 Naïve Bayes Classifier Naïve Bayes uses a supervised as well as statistical method for classification and can be employed for binary and multi-class classification problems. It is actually a whole set of algorithms sharing a common principle, i.e. each pair of data points being classified is quite independent of each other. It takes an underlying probabilistic model and then allows capturing uncertainty about it in a principled manner by determining the probabilities of outcomes. The underlying principle of the classifier is Bayes Theorem which uses the concept of conditional probability and assumes that there is no statistical correlation among the data points. The goal here is to determine the probability of features $\{x = (x_1, x_2, \ldots, x_N)^T\}$ present in each class $\{c = (c_1, c_2, \ldots, c_k)\}$ and then return the class that is most probable. Thus, we have to determine the conditional probability $P(c_i | x_1, x_2, \ldots, x_N)$ for each class. In order to do this, we apply Bayes Theorem as follows [15]:

$$P(c_i | x_1, x_2, \ldots, x_N) = \frac{P(x_1, x_2, \ldots, x_N | c_i) P(c_i)}{P(x_1, x_2, \ldots, x_N)}. \tag{3.2}$$

Now, in order to simplify the computation, we use the concept of conditional independence and get the simplified results as:

$$P(c_i | x_1, x_2, \ldots, x_N) \propto P(c_i) \prod_{j=1}^{N} P(x_j | c_i). \tag{3.3}$$

Now, Naïve Bayes further simplifies it by simply picking the c_i which has largest probability provided the features of data point. The final simplified form is:

$$y = \text{argmax}_{c_i} P(c_i) \prod_{j=1}^{N} P(x_j | c_i). \tag{3.4}$$

This is known as Maximum A Posteriori decision rule and y here represents the predicted class identifier for x. Different techniques and distributions are used by different Naïve Bayes classifiers for predicting the prior and conditional probabilities [35].

Scope in IoT Data Analytics Naïve Bayes classifier is interpretable, highly scalable, and computationally efficient. It can work well with high-dimensional data points and can be trained well with a small number of data points. The only negative point here is the assumption that features are mutually independent which is not possible in practical life. This classifier has a lot of scope in smart data analysis and has proved the best model for numerous use-cases like text categorization, spam filtering [39], medical diagnosis [40], etc. It has also been used to evaluate the trust value by combining factors and then compute the resultant quantitative trust of agricultural product [41]. In addition to these applications, it has a great role in IoT security and is being used in network intrusion detection [42]. Prediction of events generated by objects is much needed in many IoT environments, but is very difficult due to dynamicity of IoT. In such scenarios, probabilistic reasoning is of great help as it helps to determine dependent probabilities of events from the ones that are much easier to predict. Thus, the Naïve Bayes model can be employed to determine the probability of future events. It has been used in event prediction in [43] for predicting outbound fight delay events. It has also been used for food monitoring in smart environment by predicting environmental conditions in [44]. In [45], authors have used it for real-time dengue prediction in IoT.

3.2.1.2 Regression Analysis

Regression is a type of supervised learning technique that outputs target values in terms of continuous values. It is a predictive modeling technique that determines relationship between dependent variable (target variable) and independent variable(s) or predictor [46]. It gives conditional expectation of the dependent variable when independent variables are kept fixed. Following subsections present a detailed description of regression technique and the related important concepts.

3.2.1.2.1 Linear Regression
It is the most widely used technique of regression that relates dependent variable and one or more independent variables by determining the best-fit straight line. This is accomplished and understood by learning a specific function $f(x, w)$. This function is actually a mapping represented as $f: \emptyset(x) \to y$. Here, x represents the independent input variable, y is the predicted dependent output, w is the weight vector or matrix $w = (w_1, w_2, \ldots, w_N)^T$, and $\emptyset = (\emptyset_1, \emptyset_2, \ldots, \emptyset_N)^T$ is set of various linear or nonlinear functions of x represented in an amalgamated form. The function $f(x, w)$ is represented as [15]:

$$f(x, w) = \emptyset(x)^T w. \tag{3.5}$$

A wide range of functions such as Gaussian, polynomial, sigmoidal, radial, etc., can be used in $\emptyset(x)$ according to the type of application for which the model is to be used [47].

For training the model and minimizing the cost function, a variety of approaches are used like Bayesian linear regression, least-mean-squares (LMS), regularized least squares, and ordinary least square. LMS approach is the most widely used among these as it is fast, highly scalable (capable of being scaled to accommodate huge datasets), and is capable of learning the parameter requirements online by using stochastic gradient descent [48].

Scope in IoT Data Analytics The development of model using regression is quite straightforward and rapid and very simple when the data to be modeled are less and simple. But linear regression is applicable only for showing linear relationships and its assumption regarding normal distribution of residuals and fixed basis functions is a major shortcoming. Regression analysis is widely used in analysis of IoT for predictive modeling, time series modeling, controlling operations, as well as security. Some of the popular use-cases include remote controlling of IoT-based applications [49], IoT data imputation [50], using IoT to model the prediction of energy usage in smart home appliances [51], attack and anomaly detection in IoT sensors [52], smart farm enhancement [53], etc.

Neural networks is discussed in Section 3.2.2.1.3.

3.2.1.3 Classification and Regression Tasks

3.2.1.3.1 Support Vector Machine Support vector machines (SVMs) are a family of non-probabilistic supervised learning algorithms which can be used for classification as well as regression problems and outliers detection [15]. It is mainly employed for binary classification where the aim of model is to determine a hyperplane or a decision boundary that distinctly divides and classifies the classes of training set. There can be many possible hyperplanes for separating the classes but the aim of SVMs is to find one that is successful enough in maximizing distance (margin) between the data points of individual classes (see Figure 3.5). This helps in classifying future data points with more confidence. The label of an unknown data point is then obtained by the side of the hyperplane on which it lies [54]. In the field of geometry, a hyperplane is a subspace having one dimension less than that of its environmental space. A 3-dimensional space will have 2-dimensional planes as its hyperplanes where as a 2-dimensional space will have 1-dimensional lines as its hyperplanes. A data space consisting of 2D-points that fit in one of two classes is linearly separable and its hyperplane would be an optimal separating straight line.

A separating line is not optimal if it is close enough to data points as it will not be able to generalize correctly and would be very sensitive to noise. So, SVM aims at finding such a line that is as far as possible from all points giving largest minimum distance to the training examples. Twice of this distance is referred to as

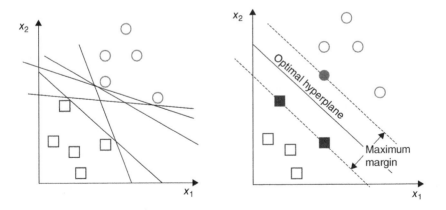

Figure 3.5 Possible hyperplanes.

margin here and accordingly, it is said that a good hyperplane maximizes the margin of training data. The notation for defining hyperplane is:

$$f(x) = \beta_0 + \beta^t x, \tag{3.6}$$

where β is known as the weight vector and β_0 as bias [15]. Conventionally, among all the possible representations of hyperplane, the one chosen is

$$\beta_0 + \beta^t x = 1, \tag{3.7}$$

where x denotes the training examples nearest to hyperplane. Such training examples are called support vectors. This notation is known as **canonical hyperplane**. From geometry, the distance between hyperplane (β, β_0) and point x is determined as:

$$Distance = \frac{|\beta_0 + \beta^t x|}{\|\beta\|}. \tag{3.8}$$

For canonical hyperplane, the numerator is equal to one and we get the distance to the support vectors as

$$Distance_{distance\ support\ vector} = \frac{1}{\|\beta\|}. \tag{3.9}$$

This gives margin M as:

$$M = \frac{2}{\|\beta\|}. \tag{3.10}$$

Finally, the problem of maximizing M is equivalent to minimizing the function $L(\beta)$:

$$\min_{\beta,\beta_0} L(\beta) = \frac{1}{2\|\beta\|^2}, \tag{3.11}$$

provided $y_i(\beta_0 + \beta^t x) \geq 1$ where y_i represents each of the labels of the training examples [15]. SVMs can do a nonlinear classification as well in addition to linear one by finding a hyperplane which is a nonlinear function of the input variable. This is accomplished by using kernel trick where input variable is mapped into higher dimensional feature into high-dimensional feature spaces [55]. Further, SVMs can be used for multi-class classification in addition to binary classification using different methods like Weston and Watkins [56] version, one-vs-all (OVA) SVM, all-vs-all (AVA) SVM, and structured SVM [57].

Scope in IoT Data Analytics   SVM has much better generalization and regularization capabilities and can handle nonlinear data efficiently. It can efficiently deal with high-dimensional data and is too good with respect to memory usage. One major limitation of the model is that probability estimates are not directly provided here and interpretation becomes difficult. As far as IoT data analysis is concerned, SVM has proved quite suitable in a number of applications like classification of traffic data [58], image classification [59], handwriting recognition [60], and protein classification in smart healthcare and medicine [61]. In [62], SVM is used in smart homes for developing a real-time human tracker system by using cameras for human motion prediction and recognition. Authors of [63] used an SVM model and data captured by wireless motion sensors for prediction of the time intervals when visitors are present in a smart home. Authors of [64, 65] also used SVM for recognition of real-time activity in smart homes.

3.2.1.3.2 Support Vector Regression The regression problems can be solved by using the SVM model through a process known as Support vector regression (SVR). This SVR approach is based on only a subset of training points. This combined model is also used in different aspects of IoT like in [66], humidity data prediction and accurate temperature is obtained through this hybrid model.

3.2.1.3.3 Classification and Regression Trees Classification and regression trees (CART) hybrid model (CART) is a scalable and fast learning algorithm which partitions the input domain into axis-aligned cuboid regions R_k. Then the prediction of labels for data points falling into a particular region is done by assigning a classification or regression model to each region [67]. Given an unknown data point (input vector) x, the methodology for estimating the corresponding target label is accomplished through the traversal of binary decision tree. Class for each region may be predicted by classification and then a model based on regression may be used for prediction of constant for the region. Now, we can formulate expressions for both the models. Suppose class variable is denoted by a discrete random variable t, input data point by x and y denotes its predicted class label. The classification task of CART is represented as [15]:

$$p\left(t=c|k\right)=\frac{1}{|R_k|}\sum\nolimits_{i\varepsilon R_k} I\left(t_i=c\right), \tag{3.12}$$

$$y = \arg\max p_c\left(\left(t=c|x\right)\right)=\arg\max p_c\left(\left(t=c|k\right)\right), \tag{3.13}$$

where $I(.)$ represents the indicator function. According to this equation, the input vector x will be assigned the label of mode in its appropriate region [15, 68].

The regression module of the model is expressed as:

$$y = \frac{1}{|R_k|}\sum\nolimits_{i\varepsilon R_k} t_i, \tag{3.14}$$

i.e. the output vector for x is the average of the output vectors t of data points in its corresponding region [15, 68].

The training set provides basis to determine the structure of tree for training CART. This means along with the limiting parameter value, split criterion is determined at each node. It is an NP complete problem to determine an ideal tree topology. The issue of overfitting is addressed by using stopping criteria or pruning technique [24, 25, 69].

Scope in IoT Data Analytics   The main strength of CART is that it is scalable to large datasets, human interpretable, and fast. However, it suffers from irregular labeling of input data points and high sensitivity to the selection of training set [70]. CART has found great applicability in classifying smart citizen behavior. In [36, 69], a coherent and well-structured data mining model is proposed by the author to illustrate the travel patterns of transit riders in China.

3.2.1.3.4 Random Forest   Random forests or random decision forests fall in the category of ensemble learning and use training of an army of trees rather than using a single tree. By constructing a family of decision trees at training time, it is a quite capable approach for operation of classification, regression, and many other problems that function by constructing a family or an army of decision trees during training phase. The output of the model is the most frequent class (classification) or it can be the average prediction (regression) of constituent trees of the forest. A subset of M input features is chosen randomly (with replacement) and used for training each tree [71]. Then there are two possibilities for predicting label of a novel unknown data point. Now onward, there are two possible schemes for estimating the label of a novel data point: (i) mode of the labels estimated by each tree in case of classification tasks (ii) and mean of the labels predicted by each tree in case of regression. The value of M should be reasonable enough as a very small and insignificant value results in random trees having very weak prediction power and very big values result in quite similar arbitrary trees [15].

Scope in IoT Data Analytics   Random forests are considered as highly robust and accurate learning algorithms having great capability to detect anomalies from data. However, it suffers from the issues of overfitting and less human interpretability. In smart cities, they find wide applicability in body part classification and body pose recognition [72]. In [73], random forest technique and IoT are used to model a healthcare monitoring system.

3.2.1.3.5 Bootstrap Aggregating   Bootstrap aggregating, also known as bagging, is an ensemble learning technique mainly focused on making the ML algorithms better by enhancing their accuracy, stability, and robustness with a substantial reduction in the problem of overfitting. In bagging, original training set is used to choose data points randomly with replacements for generating K new M sized training sets and each of them is used for training a machine learning model [15, 74]. Then the approach discussed above in CART is used for predicting label of an unknown data point [15]. Bagging improves the accuracy and results of many ML models like neural networks and CART but can degrade the performance of some stable algorithms like KNN. It may increase the computational complexity and moreover, the individual algorithms do not have any interaction among themselves. It has been widely used in many practical applications including prediction of customer churn [75] and preimage learning [76].

3.2.2 Unsupervised Machine Learning Algorithms

Unsupervised ML algorithms have no labeled data or no prior information about the target vectors for analyzing the data system being investigated. They determine the form and structure of data points from a dataset without referring to any previously identified and known results. The expected outcome is quite opaque here; consequently, they cannot be employed for any classification or regression tasks. Training of the model is a difficult task here. Learning is done by observation rather than by using examples or well-labeled data. Clustering is the most extensively used procedure among all the categories of unsupervised ML algorithms. Following subsections will be discussing some of the unsupervised algorithms in detail.

3.2.2.1 Clustering

3.2.2.1.1 K-Means Clustering The objective of K-means clustering technique is to partition the given unlabeled dataset into K partitions or clusters according to similarity measure among data points of same cluster. Similarity measure is usually determined in terms of distance between the data points [15]. Given a dataset, K-means clustering aims at obtaining a group of K cluster centers, represented as $\{c_1, c_2, \ldots, c_k\}$ such that there is minimization of distances between the data points and their nearest centers [77]. A set of binary indicator variables, $\pi_{nk} \in \{0, 1\}$, is used for allotting data points to cluster centers. Whenever any data point x_n is placed in a cluster with center c_k, then $\pi_{nk} = 1$. The whole problem is modeled as [15]:

$$\min_{c,\pi} \sum_{n=1}^{N} \sum_{k=1}^{K} \pi_{nk} \left\| x_n - c_k \right\|^2, \tag{3.15}$$

$$subject\ to \sum_{k=1}^{K} \pi_{nk} = 1, n = 1, \ldots, N.$$

An iterative approach is used for determining the optimal clusters and then assigning the data points to them.

Scope in IoT Data Analytics   K-means clustering algorithm is scalable, can be used for large datasets, highly flexible, efficient, and converges very fast. However, it needs the value of K in advance, sensitive to outliers, and not suitable to determine clusters with non-convex shapes [15, 27]. IoT data analytics is widely using this algorithm in many applications. In [36], it is used to partition and classify travel pattern regularities. In [78], K-means clustering has been used to develop a model for real-time event processing and clustering to analyze data from sensors through OpenIoT1 middleware which works as an interface for advanced analytical IoT services. It has been used as a base algorithm for optimizing different metrics like fraud detection [79], outlier detection [79, 80], forecasting energy consumption [81], analysis of small dataset [82], stream data analysis [78], and passengers travel pattern analysis [36] in different IoT use cases like that of smart citizen, smart city, smart home, and regulating air traffic.

3.2.2.1.2 Density-Based Spatial Clustering of Applications with Noise (DBSCAN) DBSCAN uses the notion of density as a metric to accomplish clustering. Here, clusters are regarded as highly dense regions of data points which are separated by those of low density or noisy regions in data space. In [15], an algorithm is presented for training DBSCAN.

Scope in IoT Data Analytics   Practically, DBSCAN works well on large datasets, converges fast, and is quite robust against outliers. It is capable of discovering random-shaped clusters [15]. However, DBSCAN loses its grace when there is a large difference in densities in given dataset and is highly affected by the distance metric which is used for finding out the density of a region [27]. DBSCAN is one of the most widely used clustering models, implemented in many real-world IoT applications like that of fraud detection, analyzing passenger travel pattern, labeling data, and so on. It has been used for X-ray crystallography in [83] and authors in [84] used it for detection of anomalies in temperature data. Furthermore, this algorithm was used by authors of [85] for determining the exact number of existing classes and then, categorizing the data to depict the cardinality of current classes, data labeling, and then for data identification. It was also deployed in [79] for deducing the group random shape.

3.2.2.1.3 Neural Networks   Neural networks help in overcoming the shortcoming of other learning algorithms like regression regarding choosing the types of basic functions as selecting optimal basic functions is often difficult. Neural networks deal with this issue by fixing the number of these basic functions and allowing the algorithm to learn and estimate their parameters. It is a broad concept and has numerous types with different architectures and applications. Furthermore, neural networks can be adapted to different classification and regression tasks easily [86, 87].

Being compact, they can process new data very fast but a large amount of Computation is required to be trained. In the next subsection, neural network model that is quite useful for smart data analysis will be discussed.

Multilayer Perceptrons (MLP)   Multilayer perceptrons, also referred to as feed forward neural networks (FFNN), are predominant models of neural networks and have numerous practical applications. For understanding the algorithm, we will explain a simple two-layer FFNN model. Suppose we have some F basic functions and need to learn their parameters. The regression or classification task can be formulated as:

$$f\left(x, w^{(1)}, w^{(2)}\right) = \varphi^{(2)}\left(\varphi^{(1)}\left(x^T w^{(1)}\right)^T w^{(2)}\right),$$

where $w^{(1)} = \left(w_1^{(1)}, \ldots, w_M^{(1)}\right)^T$, $\varphi^{(1)} = \left(\varphi_1^{(1)}, \ldots, \varphi_F^{(1)}\right)^T$, $w^{(2)} = \left(w_1^{(2)}, \ldots, w_F^{(2)}\right)^T$,   and

$\varphi^{(2)} = \left(\varphi_1^{(2)}, \ldots, \varphi_F^{(2)}\right)^T$ [15]. A two-layer MLP is visualized in Figure 3.6. There are three important layers: input layer, hidden layer, and output layer. Neurons are the units in input layer representing the data points of input vector x, the hidden layer elements are represented as $\varphi_i^{(1)}$, and those of output layer are denoted by $\varphi_i^{(2)}$ which finally outputs f. The nonlinear functions of different units in a particular layer determine the activities of units in the next layer. Such function is termed as activation function represented as $\varphi(.)$.

The type of data manipulation task decides the nature of activation function in last layer of perceptron. For example, a softmax function is used in the case of

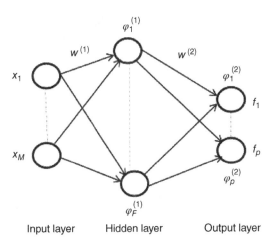

Figure 3.6   A two-layer multilayer perceptron.

multi-class classification [26, 88]. Determining an ideal and balanced set of weights w is an NP-complete problem. Different learning techniques are implemented for training of the model here like adaptive gradient, stochastic gradient, etc. [15, 88]. Moreover, numerous techniques are implemented to overcome the issue of overfitting and improve the generalization of the network such as weight decay, dropout, early stopping, etc. [89].

Scope in IoT Data Analytics With just two layers, this type of FFNN has limited generalization and representation which can be enhanced by incorporating more hidden layers in the model resulting in deep neural networks. Various models of neural networks find diverse applications in real world. In [90], a model based on artificial neural networks (ANN) is deployed to predict the states of IoT elements and authors in [91] use neural networks for analysis of health data. In [92], ANN is used for security purpose in smart homes and the authors of [93] have used neural network model for proposing a design for home energy management system. The real-time applications of neural networks in different IoT application areas like security, agriculture, healthcare, etc., become much more prevalent when we shift to more sophisticated versions of neural networks like recurrent neural networks (RNN), convolutional neural networks (CNN), and other deep learning architectures.

3.2.2.2 Feature Extraction

3.2.2.2.1 *Principal Component Analysis (PCA)* PCA has its foundation in the theorem of orthogonal projection where data points are projected onto an L dimensional linear subspace, known as the principal subspace, having most projected variance [93]. It aims at determining a comprehensive orthonormal set of L linear M-dimensional basis vectors $\{w_j\}$ and the respective linear projections of data points $\{z_{nj}\}$ such that mean reconstruction error (J) is minimized

$$J = \frac{1}{N}\sum_n \left|\widetilde{x}_n - x_n\right|^2,$$ (3.16)

$$\widetilde{x}_n = \sum_{j=1}^{L} z_{nj}w_j + \bar{x},$$ (3.17)

where $\bar{x}$ is the mean of all data points [15, 24].

Scope in IoT Data Analytics PCA works well with high-dimensional data as data size gets highly reduced by applying it but incurs high computational cost. It falls in the category of highly important preprocessing procedures in the world of machine learning. It has a wide applicability in data visualization, compression, whitening, and fault detection. In real-world situations, it is used in neuroscience, monitoring public places [80], IoT cyber security [94], quick investigation of anomalies in IoT network traffic [95], and so on.

3.2.2.2.2 Canonical Correlation Analysis (CCA) CCA is a linear dimensionality reduction technique. It is highly similar to PCA. It is somewhat improved version of PCA in terms of having ability to handle and correlate two or more variables whereas PCA can deal with just one variable [15]. Its main purpose is to determine a corresponding pair of highly cross-correlated linear subspaces. An ideal solution is derived by solving a generalized eigenvector problem [26].

Scope in IoT Data Analytics In [80], the author finds comparison of PCA and CCA in intermittent faults detection and masking failures of closed settings. Its real-time application areas are quite similar to that of PCA.

3.2.3 Conclusion

IoT is set to fulfill its promise of turning this world into a smarter place where every facility can be availed just at one click. This is realizable due to the enormous inclusion of variety of sensor devices in various modules of an IoT model for different real-time applications. Providing the base for IoT, these objects generate and transmit huge amounts of data that need to be stored, organized, processed, and analyzed so as to derive value and insight from it. The analysis of these smart data is very necessary for optimization and enhancement of various services provided by IoT in different application fields. In this chapter, we discussed various supervised and unsupervised machine learning approaches and their highly significant role in smart analysis of IoT data. A detailed taxonomy of various machine learning algorithms together with their strengths, challenges, and shortcomings is discussed. Following this, a review of application areas and use cases for each algorithm is presented. It is quite helpful in having a better understanding of usage of each algorithm and helps in choosing a suitable data analytic algorithm for a particular problem. It is concluded that machine learning has a lot of scope in the world of IoT and is proving highly beneficial for efficient analysis of smart data.

References

1 Weiser, M. (1991). The computer for the twenty-first century. *Scientific American* 265 (3): 94–10.
2 Sundmaeker, H., Guillemin, P., Friess, P., and Woelfflé, S. (2010). *Vision and Challenges for Realising the Internet of Things*. Brussels: European Commission—Information Society and Media.

3 Atzori, L., Iera, A., and Morabito, G. (2015). The internet of things: a survey. *Computational Networks* 54 (15): 2787–2805.

4 Cecchinel, C., Jimenez, M., Mosser, S., and Riveill, M. (2014). An architecture to support the collection of big data in the internet of things. *IEEE World Congress on Services*, Anchorage, AK (27 June to 2 July 2014), 442–449. IEEE.

5 Muhammad, A.I., Oladiran, G.O., and Magdy, A.B. (2016). A review on Internet of Things (Iot): security and privacy requirements and the solution approaches. *Global Journal of Computer Science and Technology* XVI (VII). Version I.

6 Madakam, S., Ramaswamy, R., and Tripathi, S. (2015). Internet of Things (IoT): a literature review. *Journal of Computer and Communications* 3: 164–173.

7 Rob, V.K. (2008). *The Internet of Things: A Critique of Ambient Technology and the All-Seeing Network of RFID*. Institute of Network Cultures.

8 Da Xu, L., He, W., and Li, S. (2014). Internet of things in industries: a survey. *IEEE Transactions on Industrial Informatics* 10 (4): 2233–2243.

9 Saleem, T.J. and Chishti, M.A. (2020). Exploring the applications of machine learning in healthcare. *International Journal of Sensors, Wireless Communications and Control* 10: 1.

10 IBM (2020). Bringing big data to the enterprise. http://www-01.ibm.com/ software/data/bigdata/ (accessed 20 April 2020).

11 Saleem, T.J. and Chishti, M.A. (2019). Data analytics in the Internet of Things: a survey. *Scalable Computing: Practice and Experience* 20 (4): 607–629.

12 Preethi, N. (2014). Performance evaluation of IoT result for machine learning. *Transactions on Engineering and Sciences* 2 (11): 161–166.

13 Marjani, M., Nasaruddin, F., Gani, A. et al. (2017). Big IoT data analytics: architecture, opportunities and open challenges. *IEEE Access* 5: 5247–5261.

14 Chen, M., Mao, S., Zhang, Y., and Leung, V.C. (2014). *Big Data Related Technologies, Challenges and Future Prospects*. Heidelberg: Springer.

15 Mahdavinejad, M.S., Rezvan, M., Barekatain, M. et al. (2018). Machine learning for Internet of Things data analysis: a survey. *Digital Communication Network* 4: 161–175.

16 Tsai, C.W. (2015). Big data analytics: a survey. *Journal of Big Data* 2 (1): 1–32.

17 Saleem, T.J. and Chishti, M.A. (2019). Data mining for the Internet of Things. *ICDM* (Posters).

18 Sheth, A. (2014). Transforming big data into smart data: deriving value via harnessing volume, variety, and velocity using semantic techniques and technologies. *Data Engineering (ICDE), 2014 IEEE 30th International Conference*, Chicago, IL (31 March to 4 April 2014), 2.

19 Gartner Inc. (2019). Gartner's Hype Cycle Special Report for 2019. http://www. gartner.com/technology/research/hype-cycles/ (accessed 27 April 2020).

20 Samuel, A.L. (1959). Some studies in machine learning using the game of checkers. *IBM Journal of R&D* 3 (3): 210–229.

21 Mitchell, T.M. (1997). *Machine Learning*, 7e. New York: McGraw Hill.

22 Safavian, S.R. and Landgrebe, D. (1991). A survey of decision tree classifier methodology. *IEEE Transactions on Systems, Man and Cybernetics* 21 (3): 660–674.

23 Barber, D. (2012). *Bayesian Reasoning and Machine Learning*. Cambridge University Press.

24 Bishop, C.M. (2006). *Pattern Recognition and Machine Learning*. Springer.

25 Murphy, K.P. (2012). *Machine Learning: a Probabilistic Perspective*. MIT Press.

26 Bengio, I.G.Y. and Courville, A. (2016). *Deep Learning*. MIT Press.

27 Alsharif, M.H., Kelechi, A.H., Yahya, K., and Chaudhry, S.A. (2020). Machine learning algorithms for smart data analysis in Internet of Things environment: taxonomies and research trends. *Symmetry* 12: 88.

28 Blockeel, H. (2013). *Machine Learning and Knowledge Discovery in Databases*. Berlin: Springer.

29 Kanj, S., Abdallah, F., Denoeux, T., and Tout, K. (2016). Editing training data for multi-label classification with the k-nearest neighbor rule. *Pattern Analysis and Applications* 19: 145–161.

30 Chomboon, K., Chujai, P., Teerarassamee, P., Kerdprasop, K., and Kerdprasop, N. (2015). An empirical study of distance metrics for k-nearest neighbor algorithm. *Proceedings of the 3rd International Conference on Industrial Application Engineering*, Kitakyushu, Kitakyushu, Japan (28–31 March 2015), 1–6.

31 Prasath, V., Alfeilat, H.A., Lasassmeh, O., and Hassanat, A.B. (2017). Distance and similarity measures effect on the performance of K-nearest neighbor classifier—a review. arXiv:1708.04321

32 Comsa, I., Trestian, R., and Kashyap, R. (2019). Machine learning for Internet of Things. In: *Next-Generation Wireless Networks Meet Advanced Machine Learning Applications*, 57–83. Hershey, PA: IGI Global.

33 Cover, T. and Hart, P. (1967). Nearest neighbor pattern classification. *IEEE Transactions on Information Theory* 13 (1): 21–27.

34 Ghaderi, A., Frounchi, J., and Farnam, A. (2015). Machine learning-based signal processing using physiological signals for stress detection. *Proceedings of the 2015 22nd Iranian Conference on Biomedical Engineering (ICBME)*, Tehran (25–27 November 2015), 93–98.

35 Sharmila, A. and Geethanjali, P. (2016). DWT based detection of epileptic seizure from EEG signals using naive Bayes and k-NN classifiers. *IEEE Access* 4: 7716–7727.

36 Ma, X., Wu, Y.J., Wang, Y. et al. (2013). Mining smart card data for transit riders' travel patterns. *Transportation Research Part C: Emerging Technology* 36: 1–12.

37 Zhu, R., Ji, X., Yu, D. et al. (2020). KNN-based approximate outlier detection algorithm over IoT streaming data. *IEEE Access* 8: 42749–42759.

38 Li, W., Yi, P., Wu, Y. et al. (2014). A new intrusion detection system based on KNN classification algorithm in wireless sensor network. *Journal of Electrical and Computer Engineering* 5: 1–8.

39 Metsis, V., Androutsopoulos, I., and Paliouras, G. (2006). Spam filtering with naive Bayes which naive Bayes? *CEAS*: Third Conference on Email and Anti-Spam, July 27–28, 2006, Mountain View, California USA.

40 Webb, G.I., Boughton, J.R., and Wang, Z. (2005). Not so naive Bayes: aggregating one dependence estimators. *Machine Learning* 58 (1): 5–24.

41 Han, W., Gu, Y., Zhang, Y., and Zheng, L. (2014). Data driven quantitative trust model for the internet of agricultural things. *Proceedings of the 2014 International Conference on the Internet of Things (IOT)*, Cambridge, MA (6–8 October 2014), 31–36.

42 Gumus, F., Sarkar, C.O., Erdem, Z., and Kursun, O. (2014). Online Naïve Bayes classification for network intrusion detection. *IEEE/ACM International Conference on Advances in Social Networks Analysis and Mining (ASONAM 2014)*, Beijing (17–20 August 2014), 670–674.

43 Karakostas, B. (2016). Event prediction in an IoT environment using naïve Bayesian models. *Procedia Computer Science* 83: 11–17.

44 Gnjewar, P.D., Barani, S., Wagh, S.J., and Sonavane, S. (2020). Food monitoring using adaptive naïve Bayes prediction in IoT. *Advances in Intelligent Systems and Computing* 940: 424–434.

45 Somwanshi, H. and Ganjewar, P. (2018). Real time dengue prediction using Naïve Bayes Predictor in the IoT. *International Conference on Inventive Research in Computing Applications (ICIRCA)*, Coimbatore (11–12 July 2018). IEEE, 725–728.

46 Bhmer, W. and Obermayer, K. (2015). Regression with linear factored functions. In: *Machine Learning and Knowledge Discovery in Databases*, Lecture Notes in Computer Science (eds. A. Appice, P.P. Rodrigues, V. Santos Costa, et al.), 119–134. Springer International Publishing.

47 Homann, J.P. and Shafer, K. (2015). *Linear Regression Analysis*. Washington, DC: NASW Press.

48 Montgomery, D.C., Peck, E.A., and Vining, G.G. (2012). *Introduction to Linear Regression Analysis*, 821. Hoboken, NJ: Wiley.

49 Wagle, S., Sathe, T., Vamburkar, G., and Gaikaiwari, A. (2015). *Regression Based Prediction Algorithm for Remote Controlling of IoT Based Applications*. IEEE https://doi.org/10.1109/CoCoNet.2015.7411202.

50 Peng, T., Sellami, S., and Boucelma, O. (2019). IoT data imputation with incremental multiple linear regression. *Open Journal of Internet of Things* 5 (1): 69–79.

51 Shirke, M.A. and Chandane, M.M. (2018). IoT data based predictive modeling for energy usage of appliances in smart home. *International Journal of Engineering & Technology* 7: 931–934.

52 Hasan, M., Islam, M.M., Zarif, M.I.I., and Hashem, M.M.A. (2019). Attack and anomaly detection in IoT sensors in IoT sites using machine learning approaches. *Internet of Things* 7: 100059. ISSN 2452-6605.

53 Balducci, F., Impedovo, D., and Pirlo, G. (2018). Machine learning applications on agricultural datasets for smart farm enhancement. *Machines* 6: 38. https://doi.org/10.3390/machines60300038.

54 Ding, S., Qi, B., and Tan, H. (2011). An overview on theory and algorithm of support vector machines. *Journal of University of Electronic Science and Technology, China* 40: 2–10.

55 Ponte, P. and Melko, R.G. (2017). Kernel methods for interpretable machine learning of order parameters. *Physics Review B* 96: 205146.

56 Alber, M., Zimmert, J., Dogan, U., and Kloft, M. (2017). Distributed optimization of multi-class SVMs. *PLoS ONE* 12: e0178161.

57 Nikam, S.S. (2015). A comparative study of classification techniques in data mining algorithms. *Orient Journal of Computer Science and Technology* 8: 13–19.

58 Nikravesh, A.Y., Ajila, S.A., Lung, C.H., and Ding, W. (2016). Mobile network traffic prediction using MLP, MLPWD, and SVM. *Proceedings of the 2016 IEEE International Congress on Big Data (BigData Congress)*, San Francisco, CA (27 June to 2 July 2016), 402–409.

59 Liu, P., Choo, K.K.R., Wang, L., and Huang, F. (2017). SVM or deep learning? A comparative study on remote sensing image classification. *Soft Computation* 21: 7053–7065.

60 Azim, R., Rahman, W., and Karim, M.F. (2016). Bangla hand-written character recognition using support vector machine. *International Journal of Engineering Works* 3: 36–46.

61 Cang, Z., Mu, L., Wu, K. et al. (2015). A topological approach for protein classification. *Computational and Mathematical Biophysics* 3: 140–162.

62 Nguyen, Q.C., Shin, D., and Kim, J. (2009). Real-time human tracker based on location and motion recognition of user for smart home. *Third International Conference on Multimedia and Ubiquitous Engineering*, Qingdao, China (4–6 June 2009), 243–250.

63 Petersen J., Larimer, N., Kaye, J.A. et al. (2012). SVM to detect the presence of visitors in a smart home environment. *Annual International Conference on IEEE Engineering in Medicine and Biology Society*, San Diego, CA (28 August to 1 September 2012), 5850–5853.

64 Fleury, A., Vacher, M., and Noury, N. (2010). SVM-based multimodal classification of activities of daily living in health smart homes: sensors, algorithms, and first experimental results. *IEEE Transactions on Information Technology in Biomedicine* 14 (2): 274–283.

65 Das, B., Cook, D.J., Krishnan, N.C., and Schmitter-Edgecombe, M. (2016). One-class classification-based real-time activity error detection in smart homes. *IEEE Journal of Selected Topics in Signal Processing* 10 (5): 914–923.

66 Ni, P., Zhang, C., and Ji, Y. (2014). A hybrid method for short-term sensor data forecasting in internet of things. *11th International Conference on Fuzzy Systems and Knowledge Discovery (FSKD)*, Xiamen, China (19–21 August 2014), 369–373.

67 Krzywinski, M. and Altman, N. (2017). *Points of Significance: Classification and Regression Trees*. Berlin: Nature Publishing Group.

68 Loh, W.Y. (2011). Classification and regression trees. *Wiley Interdisciplinary Reviews: Data Mining and Knowledge Discovery* 1 (1): 14–23.

69 Breiman, L. (2017). *Classification and Regression Trees*. Abingdon: Routledge.

70 Trevor, H., Robert, T., and Jerome, F. (2001). *The Elements of Statistical Learning: Data Mining, Inference and Prediction*, vol. 1, 371–406. New York: Springer-Verlag.

71 Belgiu, M. and Dragu, T.L. (2016). Random forest in remote sensing: a review of applications and future directions. *ISPRS Journal of Photogrammetry and Remote Sensing* 114: 24–31.

72 Shotton, J., Sharp, T., Kipman, A. et al. (2013). Real-time human pose recognition in parts from single depth images. *Communications ACM* 56 (1): 116–124.

73 Kaur, P., Kumar, R., and Kumar, M. (2019). A healthcare monitoring system using random forest and Internet of Things (IoT). *Multimedia Tools and Applications* 78: 19905–19916.

74 Hassan, A.R. and Bhuiyan, M.I.H. (2016). Computer-aided sleep staging using complete ensemble empirical mode decomposition with adaptive noise and bootstrap aggregating. *Biomedical Signal Processing and Control* 24: 1–10.

75 Au, T., Chin, M.-L.I., and Ma, G. (2010). Mining rare events data by sampling and boosting: a case study. *International Conference on Information Systems, Technology and Management* (11–13 March 2010), 373–379. Springer.

76 Shinde, A., Sahu, A., Apley, D., and Runger, G. (2014). Preimages for variation patterns from kernel PCA and bagging. *IIE Transactions* 46 (5): 429–456.

77 Coates, A. and Ng, A.Y. (2012). Learning feature representations with K-Means. In: *Neural Networks: Tricks of the Trade, Lecture Notes in Computer Science*, vol. 7700 (eds. G. Montavon, G.B. Orr and K.R. Müller). Berlin: Springer.

78 Hromic, H., Phuoc, D.L, Serrano, M. et al. (2015). Real time analysis of sensor data for the internet of things by means of clustering and event processing. *IEEE International Conference on Communications (ICC)* (8–12 June 2015), 685–691. IEEE.

79 Shukla, M., Kosta, Y., and Chauhan, P. (2015). Analysis and evaluation of outlier detection algorithms in data streams. *International Conference on Computer, Communication and Control (IC4)* (10–12 September 2015), 1–8. IEEE.

80 Monekosso, D.N. and Remagnino, P. (2013). Data reconciliation in a smart home sensor network. *Expert Systems and Applications* 40 (8): 3248–3255.

81 Costa, C. and Santos, M.Y. (2015). Improving cities sustainability through the use of data mining in a context of big city data. *The 2015 International Conference of Data Mining and Knowledge Engineering* (1–3 July 2015), vol 1, IAENG, 320–325.

82 Tao, X. and Ji, C. (2014). Clustering massive small data for IoT. *2nd International Conference on Systems and Informatics (ICSAI)* (15–17 November 2014), 974–978. IEEE.

83 Ester, M., Kriegel, H.P., Sander, J. et al. (1996). A density-based algorithm for discovering clusters in large spatial databases with noise. *KDD'96: Proceedings of the Second International Conference on Knowledge Discovery and Data Mining* 96: 226–231.

84 Çelik, M., Dadas, F., er-Çelik, A.S., and Dokuz, A.S. (2011). Anomaly detection in temperature data using DBSCAN algorithm. *2011 International Symposium on Innovations in Intelligent Systems and Applications (INISTA)* (15–18 June 2011), 91–95. IEEE.

85 Khan, M.A., Khan, A., Khan, M.N., and Anwar, S. (2014). A novel learning method to classify data streams in the internet of things. *Software Engineering Conference (NSEC), 2014 National* (11–12 November 2014), 61–66. IEEE.

86 Glorot, X. and Bengio, Y. (2010). Understanding the difficulty of training deep feed forward neural networks. *Aistats* 9: 249–256.

87 Eberhart, R.C. (2014). *Neural Network PC Tools: A Practical Guide.* Academic Press.

88 He, K., Zhang, X., Ren, S., and Sun, J. (2016). Deep residual learning for image recognition. *Proceedings of the IEEE Conference on Computer Vision and Pattern Recognition*, Las Vegas, NV (27–30 June 2016), 770–778.

89 Kotenko, I., Saenko, I., Skorik, F., and Bushuev, S. (2015). Neural network approach to forecast the state of the internet of things elements. *XVIII International Conference on Soft Computing and Measurements (SCM)*, St. Petersburg, Russia (19–21 May 2015), 133–135. IEEE.

90 Ramalho, F., Neto, A., Santos, K. et al. (2015). Enhancing ehealth smart applications: a fog-enabled approach. *2015 17th International Conference on E-health Networking, Application Services (HealthCom)*, Boston, MA (14–17 October 2015), 323–328. IEEE.

91 Badlani, A. and Bhanot, S. (2011). Smart home system design based on artificial neural networks. *Proceedings of the World Congress on Engineering and Computer Science 2011*, San Francisco (19–21 October 2011), 106–111.

92 Ciabattoni, L., Ippoliti, G., Benini, A. et al. (2013). Design of a home energy management system by online neural networks. *IFAC Proceedings Volumes* 46 (11): 677–682.

93 Abdi, H. and Williams, L.J. (2010). Principal component analysis. *Wiley Interdisciplinary Reviews Computational Statistics* 2 (4): 433–459.

94 Pour, M.S., Bou-Harb, E., Varma, K. et al. (2019). Comprehending the IoT cyber threat landscape: a data dimensionality reduction technique to infer and characterize internet-scale IoT probing campaigns. *Digital Investigation* 28 (Supplement): S40–S49. ISSN 1742-2876.

95 Hoang, D.H. and Nguyen, H.D. (2018). A PCA-based method for IoT network traffic anomaly detection. *20th International Conference on Advanced Communication Technology (ICACT)*, Korea (South) (11–14 February 2018).

4

IoT Data Analytics Using Cloud Computing

Anjum Sheikh[1], Sunil Kumar[2], and Asha Ambhaikar[3]

[1]*Electronics and Communication, Kalinga University, Naya Raipur, Chhattisgarh, India*
[2]*Electrical and Electronics Engineering, Kalinga University, Naya Raipur, Chhattisgarh, India*
[3]*Computer Science and Engineering, Kalinga University, Naya Raipur, Chhattisgarh, India*

4.1 Introduction

IoT has revolutionized the world with its versatility, ubiquity, and fast expansion. Researchers estimate that the next few years will witness such a rapid escalation of the technology in the coming years that the number of IoT end nodes or devices will be more than the global population. The application areas of IoT are numerous like healthcare, environmental monitoring, home automation, smart transportation, agriculture, etc. As the sensors can efficiently gather and transmit data and have the ability to be deployed anywhere, the sensor-enabled IoT nodes or devices can be used for purposes like smart meters for home, garbage collection, remote weather stations, alert for the farmers to milk their cows, and the list of applications is limitless. With the emergence of the concept of smart cities, IoT has been considered to be one of the best solutions for industrial as well as commercial purposes. The different nature of technology has enabled the usability of the sensors in a diverse environment in every part of the world. It has succeeded in enhancing the comforts of its users.

IoT interconnects many technologies, networks, devices, and human beings for attaining a set of accomplished objectives. This technology is closely interrelated with data. All the applications collect, store, process, and transfer data. The data exchanged on the network need to be appropriately analyzed for taking crucial decisions. Any kind of inaccuracy or miscalculations while dealing with data can be disastrous. For example, if we consider an example of a healthcare system, any negligence while transferring the medical parameters can mislead the doctors to arrive at wrong decisions for their patients which can be fatal and may cause death. Substantial growth in the number of IoT end-users and devices will increase the amount of data exchanged on the networks. Taking into account the

Big Data Analytics for Internet of Things, First Edition. Edited by Tausifa Jan Saleem and Mohammad Ahsan Chishti.
© 2021 John Wiley & Sons, Inc. Published 2021 by John Wiley & Sons, Inc.

seriousness of maintaining precise and error-free data, it is essential to implement appropriate techniques to deal with the voluminous amount of data generated by the IoT applications.

The benefits of IoT can be increased if it is possible to store the data and at the same time, be able to analyze the data. Along with this, the storage system should be capable of handling various types of data originating from different devices. It is therefore needed to have a storage system that will provide inexhaustible space and at the same time be able to handle the interoperability, scalability, and heterogeneity of the IoT systems. The answer for IoT data storage initiatives by the developers and organizations lies in cloud computing. The organizations had to set up big data centers and make substantial financial investments to configure networks, hardware, and infrastructure because of the overwhelming data generated by them. Cloud computing has enabled organizations to cut down on the development costs without making additional investments on the hardware and software. It further helps the business enterprises to analyze all characteristics of IoT for searching more accessible methods to modify the existing operations and also develop new applications.

Another computing platform used by organizations for effective and efficient management of data is fog computing. It uses a decentralized architecture where the data are processed by the nodes. Fog computing does not replace Cloud, but it assists the IoT ecosystems in reducing the time consumed by the Cloud in processing and analyzing data. It acts as an intermediate layer among the data sources and Clouds and allows local processing of data. Some more computing frameworks used for processing IoT data are edge computing and distributed computing. Edge computing enables the processing of data at edge devices. In contrast, distributed computing can be used for processing voluminous IoT data by partitioning data into packets, and then allow processing of each packet on different computers.

The companies in the earlier days relied on data monitoring capabilities while working with IoT. But with the advancement of technology, demand for data analytics is growing and the companies are looking forward to adopting it to obtain benefits like empowering their employees, to develop trust and satisfaction among its consumers, or controlling some industrial process automatically. This chapter tries to discuss the Cloud computing framework for IoT data analytics.

This chapter is arranged as follows: Section 4.2 briefly discusses the procedure of IoT analytics, Section 4.3 describes the architecture of Cloud computing platform, IoT analytical platforms using the Cloud have been discussed in Section 4.4, utilization of Machine learning for IoT analytics are described in Sections 4.5 and 4.6 lists out the challenges faced while using the Cloud for analytics, and Section 4.7 ends the chapter with the conclusion.

4.2 IoT Data Analytics

The existence of IoT devices is heavily dependent on data exchanged by them. The IoT users are not interested in raw data but they require a worthwhile translation of their data. This gives rise to the necessity of adopting data analytics by which the raw data can be converted into a productive form which will provide better solutions to the users. The organizations are using IoT analytics to develop new applications and products for the contentment of customers. Analytics can improve the existing operations and infrastructure and make intelligent choices related to investments for infrastructure development in the forthcoming years.

4.2.1 Process of IoT Analytics

The process of IoT data analytics passes through five different steps that enhance the value of data [1]:

- **Generation of Data:** A variety of data is generated by the different kinds of IoT devices. The data can be of various types like room temperature, humidity, electricity consumption for smart home applications and, health parameters like blood pressure, body temperature, pulse rate, etc., for smart healthcare. The diversification and volume of data will depend on the kind of applications used by the people.
- **Collection of Data:** All the generated data have to be accumulated in a depository sometimes referred like a data warehouse. As the volume of IoT data is increasing rapidly, its storage and especially safe storage have become challenging. Most of the organizations lack the infrastructure for gathering data for which they rely on cloud service providers for storage facilities.
- **Analysis:** This stage helps in knowing the value of collected data to deliver better performance in the future. The data analysis team studies the collected information, the past behavior and, responses, processes it using different methods and, to derive better insights.
- **Decisions:** After analyzing the data, the organizations take appropriate decisions to deliver better services for improving their business and relationship with their consumers. The decisions should promote cost-effective solutions to the problems that were detected in their previous systems.
- **Prediction:** The last stage of an analytics process is the prediction of challenges and reasons for failure that may arise in the future. An upgraded or new application after data analytics may be successful in solving the existing problems but the organizations should meticulously study the data to identify the probable complications and should be prepared to face them for preventing the complete failure of their systems.

4.2.2 Types of Analytics

The classification for IoT data analytics is done according to the type of operation to be implemented. IoT data analytics are basically of four types [2]:

- **Descriptive Analytics:** It is the simplest form of analytics and that enables reducing voluminous data into a more valuable piece of information. It uses arithmetic operations on the gathered historical data to convert it into a form that can be easily interpreted. The results for descriptive analytics are available as visualizations in the form of graphs, pie charts, bar graphs, etc.
- **Predictive Analytics:** This type of analytics helps to forecast about happenings in the future to help the organizations to plan accordingly to achieve their business goals. Predictive analytics applies a different type of mathematical, data mining, and machine learning methods on the gathered data to anticipate eventual situations.
- **Prescriptive Analytics:** This type of analytics predicts multiple consequences for an action to strengthen decision-making for achieving targets on time. It can be viewed as many predictive models running simultaneously, one for each possible action. Predictive analytics help business users to select the best option.
- **Diagnostic Analytics:** It uses the gathered data to know the reasons for the occurrence of certain events or behaviors. Diagnostic analytics helps to take a deeper view of the data by using correlation, data discovery, and data mining approach.

4.3 Cloud Computing for IoT

Cloud computing consists of virtually optimized data centers that are capable of providing hardware, software, and information reserves to the users according to their requirements. The organizations working with IoT infrastructure can derive maximum benefits by using the cloud for the storage and analysis of data. Integration of IoT and cloud computing has gained importance with the increase of demand to manage the available data for the creation of valuable services.

Cloud computing is a highly reliable and scaling platform that enables the distribution of resources and costs for a large number of users. The organizations are using the cloud to store their data as it provides unlimited storage, keeps the data secure, universal access for the end-users and their devices, and at the same time eliminates the need to maintain big data centers. The cloud platform is, therefore, a cost-effective solution for the organizations to store and process data as investments on the infrastructure for safe storage of data are minimized. Some important characteristics of cloud are broad network access, rapid elasticity, measured service, on-demand self-service, and resource pooling [3]. A cloud platform is said to have broader access as the cloud resources are accessible all over the network and support the usage of traditional interfaces for accessing the information with the help of commonly used devices like mobile phones, laptops, tablets, etc. Rapid elasticity in the cloud is possible as allotment of storage for the data resources takes place rapidly and automatically due to which the customers tend to feel as there is an unlimited storage space available for their data. The measured service feature supports automatic control and optimization of resources by effective usage of metering facilities for the different kinds of available services like bandwidth allocation and storing and processing of data. As the cloud is capable of providing on-demand services, users can use the computing facilities of the cloud whenever demand for it arises and it does not require interacting with the service providers for the fulfillment of the demands. All the resources are pooled and are assigned dynamic locations due to which the user does not know the exact location of his data.

Cloud computing is a kind of infrastructure where storage as well as the processing of data are being carried outside the device [4]. Using cloud computing for IoT data enables faster information exchange, processing of the collected data to meet certain demands, and arriving at intelligent decisions that can benefit both the users and the business organizations. The authors in [5] have listed some of the benefits of using cloud computing platforms for IoT:

- Cloud uses built-in applications and specially designed portals to provide cost-effective conclusions by enabling connection, management, and tracking of a variety of devices irrespective of their locations. It allows the sharing of information and applications at a faster rate and access to different types of data being continuously collected by the sensing devices.
- Cloud is considered to be one of the most suitable reservoirs available at a lower cost for the large amount of data that is being created and exchanged through the IoT devices. It supports both structured (relational database) as well as unstructured (text, email, audio, video, etc.). The cloud platform continuously provides opportunities for collection and unification of data and sharing it with a third party.
- IoT devices have limited storage and processing capabilities due to which performing analytical operation is challenging. Cloud helps in this regard by providing limitless storage and virtual processing facilities by which predictive algorithms and other analytical algorithms can be used over the data to take some fruitful decisions at reduced investments for the development of an application or an organization.
- With the help of IoT, billions of devices and people are getting connected that have increased the amount of data being exchanged on these networks. An

increase in the number of devices comes with increased possibilities of security risks. The cloud platform facilitates the development of improved applications and services that would be able to handle the probable risks.

4.3.1 Deployment Models for Cloud

Deployment models are used to decide a mechanism for the availability of cloud services to users. The four types of deployment models defined by the National Institute of Standards and Technology (NIST) are Private, Public, Hybrid, and Community [6, 7].

4.3.1.1 Private Cloud

A Private cloud is an infrastructure generally owned and managed by a single organization. The owner of the cloud can control the activities of purchasing, maintenance, and support. One of the advantages of using a private cloud is that the organization can have direct access to all the systems, able to troubleshoot any problems without the intervention of service providers. The owners of the cloud can update their systems to reduce the chances of systems failure. With this type of deployment model, an organization controls the type of applications to be developed, and test multiple versions of an application to provide better services to its customers. Ensuring security is easier as the owner of a private cloud will have full control over the application, systems, and data.

One of the disadvantages of a private cloud is that its implementation needs more investments as the infrastructure has to be competent enough to fulfill all the present as well as future requirements. Another difficulty associated with this kind of deployment model is the availability of qualified and skillful professionals who are capable of developing applications according to the demand of the consumers and at the same time are aware that the software and hardware developed by them should be compatible with the clients.

4.3.1.2 Public Cloud

A Public cloud is a set up owned by an external service provider or an organization that provides cloud service. It can be used by any person or any organization by paying for the services and the limits for the availability of services or resources is determined by a service level agreement (SLA). A primary goal of the public cloud providers is to ensure that their services are accessible by all the kinds of devices through the Internet and should not require any additional software. The organizations that use a public cloud do not need to worry about hardware or software deployments but they have to pay only for the services. Using the public cloud is therefore considered to be an economical option.

Higher scalability of a public cloud allows the organizations to scale their capacity either temporarily or permanently for the development of new applications or for adding new users to an application without any changes in the associated infrastructure.

Data security is a major concern while using a public cloud as the service provider is the owner of data and anyone working with the service provider can access your data. An organization using a public cloud can face problems as the service provider can take the systems offline for maintenance. This situation can be avoided for a short duration by an agreement with the providers regarding convenient time for taking the systems offline but it cannot be avoided for long periods. In a public cloud, the service provider has the responsibility of looking after the components required to implement the service (server, data, application, and storage) while the customer organization has to look after the components required to use the service by the clients.

4.3.1.3 Hybrid Cloud

Hybrid cloud shown in Figure 4.1 interconnects private and public cloud set up to deliver advantages of both the deployment models to the organizations using them. It enables the organizations to move data and applications between private and public clouds according to the operations to be performed. Security of data while storing it in a hybrid cloud can be maintained by classifying data based on its confidentiality. Public cloud can be used to store the data that does not require a powerful secure environment while sensitive data that are critical for the functioning of an organization can be stored in a private cloud. Similarly, the organizations can store their applications in any one of the clouds. For the applications that require the same data, either duplicate copies can be maintained or data can be moved around. Moving of data in the hybrid environment can be complicated due to the bandwidth restrictions. Implementation of a hybrid cloud is complex as the features of both private and public cloud has to be considered. The cost of implementation and maintenance is higher for hybrid clouds as compared to other deployment models.

4.3.1.4 Community Cloud

A Community cloud is a set up used by multiple organizations that are working for a common goal. The organizations share the computing resources, support, and maintenance activities, and the cost required for the infrastructure. Sharing of costs reduces the economic burden on the organizations and hence the community model is considered to be a cost-effective. As multiple organizations are working together, there can be conflicts for ownership and assigning responsibilities. Joint ownership through an agreement and shared responsibility can be beneficial by reducing the burden of management on a particular organization.

4.3.2 Service Models for Cloud Computing

The service models of the cloud (Figure 4.2) determine methods that will be used to provide services to the client. The three types of service models for cloud computing are Software as a Service, Platform as a Service, and Infrastructure as a Service [7].

4.3.2.1 Software as a Service (SaaS)

In SaaS, the applications are available to the end-users as services which are accessible through either web browser working as a thin client or by using a program interface. Online emails, Google apps, and social media platforms are examples of SaaS. The service provider installs and operates the applications in the cloud and the users gain access to it through the clients. The users work with the applications by entering their credentials without installing any additional programs and purchasing software or licenses. The features of the applications used by each customer are stored by the provider which are can be accessed on-demand. The SaaS model provides scalability in the pricing of applications in case there is removal or joining of users. Some of the advantages for the SaaS users are its resistance to software attacks, easy to handle, and frequent upgrades of software for making it more comfortable for the users without any additional costs.

4.3.2.2 Platform as a Service (PaaS)

PaaS is a platform that enables an organization to channelize its efforts to develop and manage its applications. The developers can create new applications and offer them as services through the Internet. It reduces the complexities and extra investments required to maintain and buy the elemental hardware and operating systems. The user is not able to control the infrastructure of the cloud but will be able to regulate the installed applications by utilizing available configurations. The process of delivering and creating cloud applications on the PaaS platform is simplified with the provision of services to the developers like the addition of extra components on demand to update an application and advanced services required for controlling, maintaining, and reporting. Integration with third-party services by using standard interfaces and protocols accelerates the enhancement of applications according to the wishes of the users.

4.3.2.3 Infrastructure as a Service (IaaS)

The IaaS service model is a basic building block of cloud that supplies infrastructure components like storage, network devices, firewall, and some basic resources required for processing data. Amazon web services and Google compute engine are examples of IaaS. The user will be able to control operating systems, applications, and networking components. The basic infrastructure is provided according to the arrival of demands. Installation and deployment of applications are done by a distributed system formed by the interconnection of virtual machines. The customers are charged according to the virtual space utilized by them and in some cases based on their usages like for months, days, or hours. It is preferred by organizations and especially by new entrepreneurs because they do not require spending on the infrastructure but can hire the servers and network resources needed for their business purpose. With the IaaS model, multiple users will be able to share the same infrastructure and allow all the users to access available resources over the Internet. It provides dynamic scaling of resources due to which the users are not worried about the improvement of software or hardware. Any changes made in the infrastructure resources do not affect the system performance.

4.3.3 Data Analytics on Cloud

Cloud is used for data analytics for many IoT applications. A cloud-based system for IoT analytics used for smart cities is given in Figure 4.3 [8].

The array of sensors continuously collects the raw data and sends it to the storage system or data depository in the cloud, i.e. the data generated from the sensors will gather in a depository. The first task of feature selection is used to select only the data for which analysis has to be done. Task 2 uses analytical algorithms, calculates mean, standard deviation, variance, and other such parameters that will be helpful in the classification of data according to the goals of the organization. Machine learning is used by task 2 for training a classification model based on the features identified by task 1. Task 3 receives the classifications and takes decisions for the selection of the best model that would help in meeting the profit-making, economically viable, and consumer contentment goals of an organization.

4.4 Cloud-Based IoT Data Analytics Platform

Data analytics software should be able to handle different types of data, huge volumes of data, and should be able to use some security protocols for the data coming from the edge devices. It should be able to provide processing capabilities for the edge as well as the on-premises devices and should be able to utilize

Figure 4.1 Hybrid Cloud.

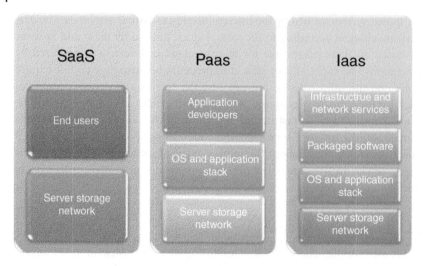

Figure 4.2 Service models.

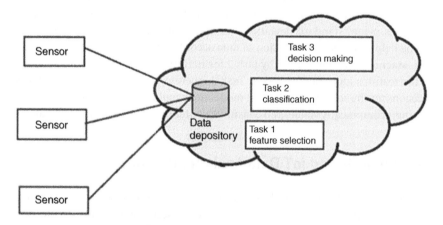

Figure 4.3 IoT analytics on Cloud [8]. *Source:* Modified from John Soladatos, (2017), *Building Blocks for IoT Analytics,* River Publishers.

artificial intelligence and machine learning for better outputs. The growth of any analytical platform depends on its ability to gather, organize, manage, and reconstruct the data to yield favorable outcomes for the fulfillment of business-related visions by the organizations. This section will provide a brief description of some of the commonly used IoT data analytics platforms.

4.4.1 Atos Codex

Atos Codex is an analytical solution developed by Atos, which is recognized as a leading organization for rendering digital services at the international level.[1] Atos Codex provides an end-to-end data analysis solution for the whole information technology value chain. The platform intends to help the organization in maximizing the worth of its data cost-effectively. This platform is available as-a-service from the cloud and can be advantageous for the organizations whose functioning and profits can be enhanced by analyzing the data obtained from its employees, partners, and consumers. For example, a telecommunication provider can use the data analytics tools for developing revenue-generating methods by and upgrade their services by analyzing network activities, the behavior of consumers, and their likes or dislikes. Atos Codex is a business-driven platform that combines the IoT, Machine Learning (ML), and Artificial Intelligence (AI) to provide expert skills and technologies to its customers for developing new data-driven services while maintaining the regulatory measures and specifications for securing smart business services.

4.4.2 AWS IoT

AWS IoT analytics is a service that formulates procedures for the conversion of raw and unstructured data collected by sensors into a sophisticated analytical data that can be used to improve the characteristics of an IoT application.[2] The data analysis by AWS IoT goes through five different stages:

- *Collect:* The process of collection of data is done by the channels. A channel is considered to be an entry point of the analytics system and collects data from multiple sources and various formats.
- *Process:* The data collected by the channels are sent to the pipeline for processing the data before their storage in the data store. This stage performs filtering, enrichment, and transformation of messages while preprocessing the received data. Filtering activity is done by determining threshold values and defines maximum/minimum filters to remove irrelevant data. Mathematical logic is used for the transformation of messages to perform common calculations like conversion of temperatures from Fahrenheit to Celsius. The data can be enriched by the addition of fundamental contexts to IoT data by applying customer-defined lambda functions on the messages.
- *Store:* The messages processed by the pipeline are sent to the data store. The types or number of data stores depends upon the pipeline configuration and the

---

1 Retrieved from https://atos.net/wp-content/uploads/2018/09/Atos-Codex-Utilities.pdf.
2 Retrieved from https://docs.aws.amazon.com/iot/latest/developerguide/iot-dg.pdf.

type of messages that may arrive from various devices or location. The data store is capable of storing raw as well as processed data. Raw data collected by the devices are stored and processed according to the occurrence of requests of data. AWS IoT analytics use optimized data stores for time series data that support fast query response. It helps to manage data retention policies, permissions required to access the data, and sending data to external access points.

- *Analyze:* The data from the data store are used to prepare different kinds of data-sets like SQL dataset or container dataset. SQL query engine to run ad-hoc queries for obtaining quick results for any queries that need immediate attention. Analysis of the performance of devices and their usage, continuous monitoring of device data to facilitate timely maintenance is done by using time series analysis. It also supports the creation of sophisticated models based on machine learning techniques by using Jupyter notebooks. The built-in notebook templates can be used for predictive maintenance, segmentation of devices according to consumer's behavior, and forecasting the production requirements of a product by analyzing the change in the number of consumers in a given time. Statistical classification by using logistic regression, predicting the output of a process to time by using a neural network-based technique called as a long short-term memory, and K means clustering for clustering the devices are some of the techniques used for data analysis.
- *Build:* This stage is used for the improved analysis and visualization of the received analytics which can be used to develop new systems and applications by proper utilization of the available hardware and infrastructure.

4.4.3 IBM Watson IoT

IBM Watson supports three basic services: connection, data analysis, and blockchain. This platform connects with the devices to gather data and transform them into a worthwhile format. The analytical services utilize the gathered raw data, use predictive and cognitive analysis to help the organizations to identify better prospects for boosting their business.[3] Watson IoT uses analytics for IoT solutions by the following methods:

- It allows interaction through languages used commonly by the people and then processes it into a form required by machines or algorithms. The collected data are forwarded securely toward the cloud by using lightweight protocols like MQTT.
- It performs textual analysis to examine the gathered data and tries to generate new information from it by discovering correlations among the datasets.

---

3 Retrieved from https://ibm.com.

- ML is used to automate the process of transforming or conversions of data and at the same time closely supervises the continuously arriving data for obtaining better results.
- This platform enables analysis of unstructured data obtained in the form of videos and images and monitors it to classify them according to specific patterns, locations, or sceneries.

4.4.4 Hitachi Vantara Pentaho, Lumada

IoT data analytics by the Pentaho platforms help the organizations to improve the satisfaction levels of the consumers by combining sensors as well as the machine-generated data with the regular data stores.[4] It utilizes ML techniques and data mining processes to discover unique patterns for industry equipment and devices. Organizations using Pentaho are benefitted due to its capabilities of assembling and examining the structures as well as the unstructured data and use programming languages like R and python. Data analytics performed by Pentaho platforms have been helpful for the organizations in reducing the inefficiencies and high operational costs by using the outcomes of analysis to improve the product quality, attract positive feedback of customers, and increase profits. The organizations have achieved favorable results by using Pentaho due to their abilities that include:

- Automating the process of data collection from all the possible sources using Metadata Injection. The users can blend the obtained data, cleanse, develop, test, and analyze it to create data models in very less time due to metadata injection.
- Custom datasets developed by automation help organizations visualize the data for making reports and taking business decisions. The streamlined data refinery technique helps to reduce the time needed for this process.
- ML orchestrations enable more effective data preparation, developing better predictive models, planning for advanced features, and update the existing models by the inclusion of new data.
- The recorded observations, opinions, and experiences are used directly to modify the IoT services and applications by using embedded analytics.

Lumada uses a multi-data cloud and is considered to be an intelligent IoT data analytic platform that studies the demands of its customers and tries to fulfill their demands cost-effectively. Handling structured and unstructured data, high scalability, and secure data management are some of the key characteristics that have helped the organizations to come up with intelligent business solutions and

---

4 Retrieved from https://www.hitachivantara.com/en-us/pdfd/datasheet/iot-analytics-blueprint-datasheet.pdf.

applications. It is an analytic platform that enables the establishment of business operations to meet some stringent requirements of the customers by analyzing their views related to some products and assets. Lumada will be able to perform data analysis in a very short time by using onboard data sources and pipelining techniques. It supports many services for the formulation of digital solutions to develop new business models with the help of machine learning and deep learning by using batch and stream analytics.[5]

The architecture of the Lumada IoT platform given in Figure 4.5 consists of the following layers:

- Edge uses the data collected by the sensors and assets to perform the task of analyzing and filtering data securely. The results are used for knowing the actual status of the assets in the present scenario and then develop business operations based on it.
- The core is used for connecting assets, gathering data, and supporting identity and access management. Asset Avatars are used by the core for the storage of digital depictions of physical assets.
- The analytics layer uses machine learning, deep learning, batch analytics, and stream analytics for a steady rise in optimization and transformation.
- The studio combines all the obtained analysis, warnings, and notifications for enabling business in dashboards and provides quick and meaningful conclusions for your data.
- Foundry provides support for quick building of service apps used for deploying, composing, and packaging product solutions.
- The data management layer helps in managing the data assets by data collection, blending, and orchestration.

4.4.5 Microsoft Azure IoT

Azure IoT is a cloud-based analytical platform managed by Microsoft that enables speedy processing of huge data collected by the different types of IoT devices by applying AI and ML.[6] Figure 4.6 illustrates the Azure IoT reference architecture. The Azure platforms are highly scalable due to which the enterprises will be able to survey numerous alternatives by adding or connecting any number of devices during analytics. This platform provides end-to-end security solutions by using authentication to preserve the confidentiality of data being exchanged over the networks and by assuring the security of data in the cloud. An Azure platform works on

---

5 Retrieved from https://www.hitachivantara.com/en-us/pdfd/solution-profile/lumada-data-services-solution-brief.pdf.
6 Retrieved from https://www.microsoft.com.

four basic steps in which the data are collected, stored, analyzed, and then appropriate actions are taken based on the results obtained from the analysis.

According to the reference architecture for IoT given in Figure 4.3, an application will have the given subsystems:

- The devices possess the ability to develop secure registration on the cloud and are also able to send and receive data. It uses Device Provisioning System (DPS) to allow registration and connection of a large number of devices to particular endpoints in IoT hub.
- The cloud gateway or IoT hub allows safe reception and storage of data along with facilities of managing the devices. The data from the cloud gateway is sent to the storage and for stream processing.
- The stream processing stage obtains a large stream of data records from the cloud gateway for processing and evaluation. Stream analytics can be used efficiently for complicated evaluations by using stream aggregations, external data sources, and time windowing functions. The outcomes of the evaluation are used by the business integration process to make decisions for execution suitable actions like sending emails or messages, setting alerts, and storing informative messages. The storage is divided into a warm-path and a cold-path. The warm-path storage is used to carry data that require immediate availability for reporting and visualizations whereas cold storage holds data for longer periods and are utilized for batch processing.
- The user interface for IoT applications uses visualizations and analytics to facilitate the management of a large number of devices.

4.4.6 Oracle IoT Cloud Services

Data analytics is an indispensable constituent of Oracle IoT platform (Figure 4.7) and IoT applications. ML and some specially designed algorithms used by the platform enable applications to interpret the received data to produce appropriate recommendations and solutions. Figure 4.4 depicts an Oracle IoT cloud platform in which devices, links, and analytic processors are some of the basic components that interact to carry to perform data analytics.

Oracle IoT can use the data gathered by a large number of smart devices to formulate strategies and opinions for heading toward a fruitful business by using the following steps:

- The first stage is to connect sensors, programmable and non-programmable devices to the Oracle IoT cloud service. The programmable devices are connected using client libraries. These libraries support bidirectional communication between the devices and the cloud. Sensors and the non-programmable devices are connected to the cloud by using gateways.

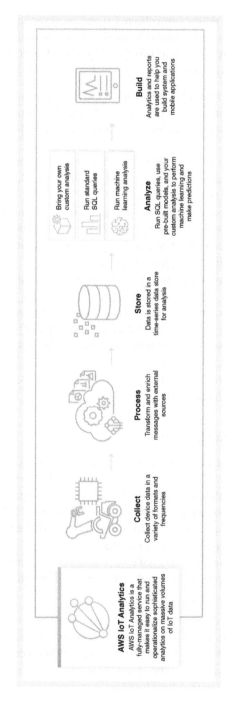

Figure 4.4 Features of AWS IoT [8]. *Source:* Modified from AWS IoT Core: Developer Guide 2020.

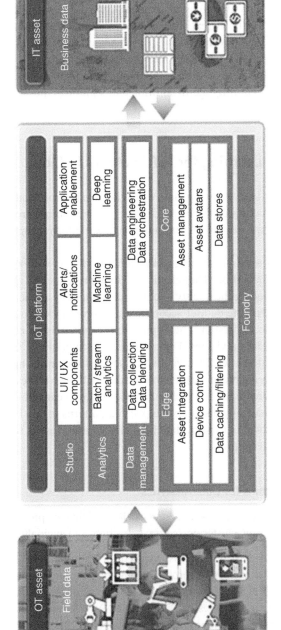

Figure 4.5 Lumada IoT architecture.

- The raw data streams collected from all the kinds of devices have used an input to the analytics processor. A suitable data analysis pattern is selected and applied to the data and it is sent to integrated cloud services. It uses predictive analysis for the incoming data in real time to identify novel solutions for rectifying satisfaction levels of consumers.
- The cloud service is used to integrate the analyzed data with the enterprise applications that issue commands, send queries to any device, and develop some distinguished intelligent solutions to enhance the performance of business enterprises.

4.5 Machine Learning for IoT Analytics in Cloud

IoT sensors continuously sense the data which need to be processed and sent back to the devices for taking decisions. ML and cloud computing are two essential components for data analytics for achieving cost-effective operation of an application [9].

ML plays a great role in analyzing a large amount of data without any dependence on human intervention. The selection of an appropriate ML algorithm plays a great role in enhancing the quality of data analysis for any application. This can be done by organizing the dataset according to different tasks. To select an algorithm, one needs to understand the application, characteristics of the available data, and the data-driven visions. For example, K-means can be used to handle voluminous data with varied data types while support vector machine (SVM) and PCA-based anomaly are preferred for handling data with high noises [10]. Machine learning has been envisaged as a technology that helps to convert data into information [11], automatically discover complicated patterns for making intelligent decisions with the help of existing data [12], and improve results with the availability of data for solving problems.

4.5.1 ML Algorithms for Data Analytics

ML algorithms when used for IoT analytics can be used for data processing tasks like classification, regression, clustering, feature extraction, and anomaly detection [10].

- **Classification:** A classification task is used for an application to assign class labels to specific items from the problem domain. An example of classification is identifying emails as spam and not spams [13]. Some of the ML algorithms used for classification are K-Nearest Neighbors (KNN), SVM, and Naïve Bayes. KNN measures the nearest neighbor for K given data points that define the number of nearest neighbors that should be examined for defining a class of sample data points [14]. Naive Bayes is an algorithm that classifies an input vector z based on the Bayesian theorem by assuming conditional independence over the training dataset [15]. SVM is a kind of ML algorithm that creates the best decision boundary that can separate n-dimensional space into classes for correct classification of the data points in the future. The classification of data points is done by finding the hyper-plane that distinguishes two classes efficiently [16].

- **Regression:** This class of ML algorithm is used for training output variables that are real or continuous. Different types of regression models are identified based on the type of relationship among the independent and dependent variables. Linear regression and support vector regression (SVR) are some of the ML algorithms that can be used for analytics. The linear regression technique is to determine a linear relationship between output and input variables. This technique is used for forecasting and determining cause and effect relation among the variables by using a best-fit straight line [17]. SVR is a regression model based on SVM that is used for the prediction of a continuous variable. Linear regression tries to minimize the error values between the predicted and actual values but SVR tries to fit the errors inside a certain threshold [18].

- **Clustering:** It is an ML technique that is commonly used for statistical analysis in many fields. The algorithm is used to form a group of data points. The groups are formed according to similarities in the data points. The data points in a group will have similarities and there will be many dissimilarities among the data points from two different groups. K-means and Density-based spatial clustering of applications (DBSCAN) with white noise are two types of clustering algorithms used for IoT analytics. In K-means clustering, the data points with similar characteristics are divided in K number of groups called a cluster. The data points are grouped based on the similarity of distance from the center of the cluster, called a centroid. Grouping is implemented by selecting the value of the sum of the square of distances between data and centroid [19]. DBSCAN forms a cluster of data points depending on density and will be able to identify groups with irregular shapes and sizes from big databases that have been degraded due to the presence of noise. A cluster is formed by grouping data points of constant densities and considers the data points with low density as outliers. This algorithm is thus effective for the identification of noise in data [20].

- **Feature Extraction:** It is a process reducing the dimensionality of a raw dataset to form a group of small dimensions to make data processing easier. A large dataset consists of a large number of variables and therefore requires more computing resources. It combines variables to form a smaller dataset that preserves the properties of the original dataset. It thus improves the effectiveness and speed of ML as it reduces the amount of data available for analysis and therefore number of resources required for processing is reduced [21]. Principal component analysis (PCA) is one of the feature extraction methods that is used to

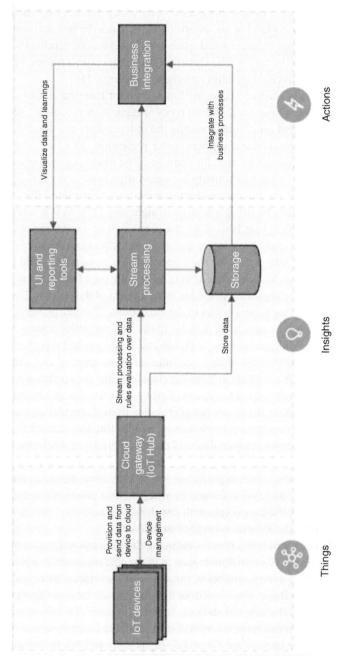

Figure 4.6 Azure IoT reference architecture.

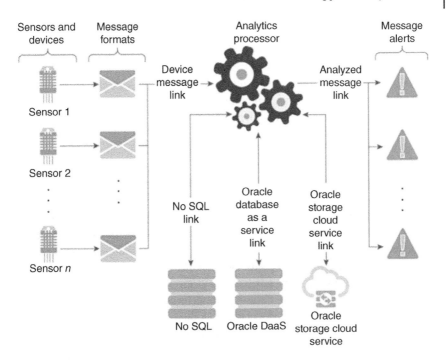

Figure 4.7 Oracle IoT cloud platform.

reduce the dimensionality of the dataset to improve the quality of analytics and minimize loss of information. This technique is implemented by the creation of new variables called principal components that are similar to the original variables. The components that do not affect final output classification are eliminated due to which this technique is effective in maximizing information content and reducing redundancy [22].

- **Anomaly Detection:** This technique is used for identifying patterns that do not match a specified pattern or some items of a dataset. The group of indifferent items may cause errors resulting in the degradation of the performance of algorithms. Using ML algorithms increases the speed of detection of anomalies [23]. One class support vector machine (OCSVM) is an anomaly detection scheme based on SVM which will be able to identify abnormalities in a vast amount of normal data. In this technique, the SVM is trained for only one class of data, called the normal class. It learns the limits of these data points known as decision boundary and detects the points lying outside the boundary to maximize the separation of points and origin. The decision boundary separates the majority of the data with only a few points on the other side of origin which are called outliers [24].

4.5.2 Types of Predictions Supported by ML and Cloud

ML uses self-developed algorithms that are trained over labeled and unlabeled datasets and then used for predicting new patterns of data. While using ML for data analytics, the organizations come across problems like lack of expertise for handling analytics and investments to maintain and upgrade infrastructure to meet the computational requirements. The integration of ML with the cloud can be advantageous for business enterprises as clouds use pay per use model which reduces the need of investing for expensive types of equipments. The cloud supports cheap data storage and helps in bridging the expertise gap faced by organizations as the ML features are implemented by the service providers. The cloud platform provides storage for the massive data of the organizations while ML is used for analytical operations to provide innovative solutions. The ML algorithms discussed in the previous section like classification, clustering, and anomaly detection can be used for predictive analytics for predicting future outcomes using the data based on past experiences. A simple model of predictive analytics using ML algorithms is shown in Figure 4.8 that uses incoming labeled data to develop its algorithm. The machine studies the characteristics of the dataset framing its own prediction rule that will be used to give the required results. Most of the cloud providers like Amazon and AWS help to implement three types of predictions: Binary, Category, and Value.[7]

- **Binary Prediction:** The binary prediction is used for applications where the analytics can be done based on only two types of responses like "yes" and "no." This kind of prediction can be helpful in the evaluation of data for applications where the responses are not complex and can be categorized easily like yes or no. Binary prediction can be used for fraud detection, deciding appropriate time

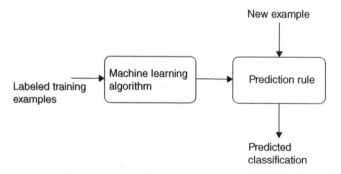

Figure 4.8 Predictive analysis using machine learning [25]. *Source:* Based on Rob Scaphire, *Theoretical Machine Learning*, Princeton Course Archive.

---

7 Retrieved from https://techbeacon.com/enterprise-it/machine-learning-cloud-how-it-can-help-you-right-now.

for upselling of their products, search engines used for recommending audio or videos to the users based on collected data, and learned responses.

- **Category Prediction:** This technique uses the dataset to classify the data points in a specific category. It is used for datasets that contain various types of data. Analysis of the data is done by specifying different categories to enable better understanding and processing of data. It can be used for different types of applications to identify additional significance to the data. Category prediction can be used by the insurance companies to classify the kind of claims provided by them to the clients. Similarly, the e-commerce websites can use it for categorizing their products according to price or brands to promote smart buying options to the customers.

- **Value Prediction:** This technique utilizes learning models to identify some patterns in the given dataset and enables classification according to specific numbers or amounts. Value prediction can be used by a company to predict its sales in the forthcoming weeks and plan its manufacturing process accordingly to meet the demand of customers.

4.6   Challenges for Analytics Using Cloud

Cloud computing is a platform that provides support for storage as well as the processing of data of IoT data. The integration of cloud and IoT has many benefits for the organizations that need data-driven solutions for their progress. Along with the benefits are a few challenges those need to be addressed. This section describes some of the probable challenges that will be faced while performing analytics using the cloud.

- **Storage of Data:** IoT analytics on the cloud will yield fruitful results with appropriate actions taken while dealing with data. Storage, processing, and quality of data are some of the challenges while working with analytics [26]. It has become extremely important for the cloud service providers to divert additional efforts in providing storage space large enough to handle the exponential growth in the IoT data coming toward the cloud. The pay per use model of the cloud allows the organization to pay according to the requirement of storage space. The amount of data being gathered by the organizations is increasing and therefore the investments being made by them on cloud storage will increase correspondingly. So the organizations need to be judicious by eliminating unnecessary data and placing only the required data on the cloud. This step will reduce the burden of the cloud and also the costs incurred by the organizations to hire the cloud facilities. To reduce the amount of storage, the devices should be allowed to collect only the meaningful data and the data that would have any effect on the analytical procedure should be avoided.

- **Quality of Data:** The availability of accurate data at the appropriate time is another important feature to enable decision-making with the help of analytics. IoT devices are energy-constrained due to which the dead batteries or failure of nodes may result in loss of messages. In such circumstances, missing values in the dataset can affect the outcomes due to incorrect analytics. An uninterrupted flow of data toward the cloud is dependent on the kind of application and communication technique. The IoT devices should be able to operate through different transmission mediums like Wi-Fi, Zigbee, and similarly for applications like smart cars using Wi-Fi is not possible, your application should support transmission through the cellular network. Several algorithms have been developed to minimize the energy consumption of IoT nodes. But still, there is a need to focus on energy consumption issues because of the tiny size of the nodes and smaller size batteries. For any application, it is therefore essential to know the reasons for the failure of nodes caused by the discharge of batteries so that preventive measures can be implemented to minimize energy consumption.

- **Processing of Data:** Evaluating the amount of data to be gathered and processed plays a great role in determining the extent of cloud requirements. IoT data are useless if they get deposited on the cloud or servers without being processed. It is therefore essential to be familiar with the analytical skills and tools required for the best utilization of your data. The computing platform should be able to handle the enormous data that may get added at any time to enable the processing of real-time data. The analytical software should be able to work on various operating systems and hardware arrangements so that it will be able to select among the available processing scenarios. Many organizations have been working hard for developing a suitable software solution that will be able to work on various processing platforms.

- **Security of Data:** More devices and more data mean more opportunities for hackers to have unauthorized access for the data. Security plays a considerable role in IoT data analytics due to the involvement of an enormous volume of private data [27]. Awareness of security risks for IoT data at different levels is required by all the organizations. Protection of IoT data is important at the devices and also during their transmission from devices to the cloud. The data when stored in the cloud are at great risk, especially when using the public cloud because information of all the users is accessible by a third-party cloud service provider. Some of the security risks related to data required for analytics are breaching and loss of data. Data breaching is a threat in which the hackers try to exploit the weakness of the users to gain unauthorized access to the data. Loss of data is caused due to changes in data in a way that cannot be reverted to its original state, mishandling of data by the cloud service providers, and unexpected deletion of information by human error or a malicious outsider. Encryption techniques and strong passwords can be used for preventing breaching and data loss, respectively, but as users have no control over the encryption keys, it becomes difficult to protect data in some cases.

- **Charges for Delivering Services:** It is challenging to determine expenses due to the availability of the resources on demand by the users. Estimating and evaluating the expenditure for accessing cloud services becomes difficult in the absence of any standard criterion specified by the service providers. The cloud users should be able to evaluate and examine the entire process of analytics to know about any hidden costs involved while using cloud services. Several organizations are struggling to manage their cloud expenses. It is therefore essential to assure excellent services, availability, and scalability while making the SLA to develop trust among the organizations that use the cloud for storing and processing data.
- **The Flexibility of Business:** The usage of cloud services should be flexible so that the business enterprise would be able to store data in the cloud and move it out later on. Lock-in period for using the cloud should be avoided so that the organizations are free to switch service providers and the computing services should be capable of functioning easily with the onsite IT environment.

4.7 Conclusion

This chapter provides an overview of performing analysis of IoT data on the cloud computing platform. Service models and deployment models of cloud have been described to know the architecture of the cloud. It also provides information about some of the cloud service providers used by the organization. AI-based techniques like machine learning can be effectively used for analytics and some of the ML algorithms that can be used for analytics have been studied. The last section tries to list some of the challenges faced while using cloud computing for IoT analytics. The field of analytics is vast which requires additional investigation of the challenges faced while storing or processing data to derive maximum benefits of the versatile technology.

References

1 Ahmed, E., Yaqoo, I., Hashema, I.A.T. et al. (2017). The role of big data analytics in Internet of Things. *Computer Networks: The International Journal of Computer and Telecommunications Networking* https://doi.org/10.1016/j.comnet.2017.06.013.

2 Mohamed, K.S. (2019). *IoT Cloud Computing, Storage, and Data Analytics, The Era of Internet of Things*, 71–91. Springer.

3 Mell, P. and Grance, T. (2009). *A NIST Notional Definition of Cloud Computing*, version 15 (10 July 2009).

4 Stergioua, C., Psannis, K.E., Kim, B.-G., and Gupta, B. (2016). Secure integration of IoT and cloud computing. *Future Generation Computer Systems* 78: 964–975.

5 Atlam, H.F., Alenezi, A., Alharthi, A., Walters, R.J., and Wills, G.B. (2017). Integration of cloud computing with Internet of Things: challenges and open issues. IEEE *International Conference on Internet of Things (iThings) and IEEE Green Computing and Communications (GreenCom) and IEEE Cyber, Physical and Social Computing (CPSCom) and IEEE Smart Data (SmartData)*, Exeter, United Kingdom (June 2017).

6 Rountree, D. and Castrillo, I. (2014). Cloud deployment models. In: *The Basics of Cloud Computing* (ed. H. Jiang), 35–47. Elsevier.

7 Laszewski, T. and Nauduri, P. (2012). *Migrating to the Cloud: Client/Server Migrations to the Oracle Cloud*, 1–19. Elsevier.

8 Soladatos, J. (2017). *Building Blocks for IoT Analytics*. River Publishers.

9 Wu, C., Buyya, R., and Ramamohanarao, K. (2016). Big data analytics = machine learning + cloud computing. In: *Big Data: Principles and Paradigms* (eds. R. Buyya, R. Calheiros and A. Dastjerdi). Burlington, MA: Morgan Kaufmann.

10 Mahdavinejad, M.S., Rezvan, M., Barekatain, M. et al. (2018). Machine learning for Internet of Things data analysis: a survey. *Digital Communication, and Networks* 4 (3): 161–175.

11 Liu, X., Datta, A., Lim, E.-P. et al. (2015). *Computational Trust Models and Machine Learning*. CRC Press.

12 Harrington, P. (2012). *Machine Learning in Action*. Manning Publications.

13 Soofi, A.A. and Awan, A. (2017). Classification techniques in machine learning: applications and issues. *Journal of Basic & Applied Sciences* 13: 459–465.

14 Cover, T. and Hart, P. (1967). Nearest neighbor pattern classification. *IEEE Transactions on Information Theory* 13: 21–27. https://doi.org/10.1109/TIT.1967.1053964.

15 Zhang, H. (2004). *The Optimality of Naive Bayes*. American Association for Artificial Intelligence.

16 Kecman, V. (2005). Basics of machine learning by support vector machines. *StudFuzz* 179: 49–103.

17 Divya, K.S., Bhargavi, P., and Jyothi, S. (2018). Machine learning algorithms in big data analytics. *International Journal of Computer Sciences and Engineering* 6 (1): 63–70.

18 Smola, A.J. and Scholkopf, B. (2004). A tutorial on support vector regression. *Statistics and Computing* 14: 199–222.

19 Fahim, A.M., Salem, A.M., Torkey, F.A., and Ramadan, M.A. (2006). An efficient enhanced K-means clustering algorithm. *Journal of Zhejiang University Science A* 7 (10): 1626–1633.

20 Khan, M.M.R., Abu Bakr Siddique, M., Arif, R.B., and Oishe, M.R., (2018). ADBSCAN: adaptive density-based spatial clustering of applications with noise for identifying clusters with varying densities. *4th International Conference on Electrical Engineering and Information & Communication Technology (iCEEiCT)*, IEEE, 107–111.

21 Khalid, S., Khalil, T., and Nasreen, S. (2014). A survey of feature selection and feature extraction techniques in machine learning. *Science and Information Conference*, London, UK (August 2014).

22 Jolliffe, I.T. and Cadima, J. (2016). Principal component analysis: a review and recent developments. *Philosophical Transactions of the Royal Society A* 374: 20150202.

23 Hodge, V. and Austin, J. (2004). A survey of outlier detection methodologies. *Artificial Intelligence Review* 22 (2): 85–126.

24 Zhang, R., Zhang, S., Muthuraman, S., and Jiang, J. (2007). One class support vector machine for anomaly detection in the communication network performance data. *5th WSEAS Int. Conference on Applied Electromagnetics, Wireless, and Optical Communications*, Tenerife (14–16 December 2007), 31–37

25 Scaphire, R. *Theoretical Machine Learning*. Princeton Course Archive.

26 Balachandran, B.M. and Prasad, S. (2017). Challenges and benefits of deploying big data analytics in the cloud for business intelligence. *Procedia Computer Science* 112: 1112–1122.

27 Stergiou, C., Psannis, K.E., Kim, B.-G., and Gupta, B. (2016). Secure integration of IoT and cloud computing. *Future Generation Computer System* 78: 964–975.

5

Deep Learning Architectures for IoT Data Analytics

Snowber Mushtaq[1] and Omkar Singh[2]

[1] *Department of Computer Science and Engineering, Islamic University of Science and Technology, Jammu and Kashmir, India*
[2] *Department of Electronics and Communication Engineering, National Institute of Technology, Srinagar, India*

5.1 Introduction

Internet of Things (IoT) has evolved as a modern research area because of the increased connected devices. This allows multiple devices to connect over a network without human intervention. It has evolved due to the merging of multiple technologies that has been deployed in various fields of day-to-day life, including smart, elderly care to provide support to elder and physically challenged people, and also toward the commercial applications. IoT is a concept that allows objects to communicate with each other, operate in co-ordination with additional things to create novel applicability, and attain mutual targets [1]. Every connected device must be considered a thing, in terms of IoT. Things can be any physical sensors, actuators, and an embedded system with a microprocessor. Data collected from sensors deployed in IoT can be used to understand, monitor, and control complex environments around us, promoting greater intellect, more forward decision-making, and more reliable performance [2]. Entities communicate with each other, known as a device-to-device communication. Individual communication can either be short-range or long-range. Short-range communication can be realized practicing wireless technologies such as Wi-Fi, Bluetooth, and ZigBee, and wide-range has achieved using mobile networks such as WiMAX, GSM, GPRS, 3G, 4G, LTE, and 5G. The commercial applications of IoT include medical health, transportation, industrial applications, etc. Artificial Intelligence is presently a part of IoT considering it helps in automated processing, control of devices,

Big Data Analytics for Internet of Things, First Edition. Edited by Tausifa Jan Saleem and Mohammad Ahsan Chishti.
© 2021 John Wiley & Sons, Inc. Published 2021 by John Wiley & Sons, Inc.

and produces promising results. A significant number of connected devices are increasing exponentially, and it is expected to grow around 75 billion by 2025. Thus, data management clarifications are needed for efficient management and transmission of data. ML has provided a way to the management of a large volume of data [3]. The leading success of DL contributed a new direction to tackle problems related to the administration of space. That has the property to categorize a large volume of data sent and received by devices. DL renders support to a large sensor data for effective learning of underlying features in intelligent devices. Explicit merge of IoT and DL has provided new applications of IoT including disease analysis, health monitoring, intelligent control of various appliances, robots, traffic analysis and prediction, and autonomous driving [4]. Big data is the advanced analytical method for better prediction and manipulation of large dimensions of data. DL has been deployed for an end-to-end delivery of reliable noise-free data in IoT.

DL is on inflation from the recent time, and it is there to stay for about a decade of now. The manufacturers are using DL algorithms to generate more income. They are training their workers to learn this experience and contribute to their firm. A lot of startups are beginning up with innovative DL resolutions that can solve challenging puzzles. Also in academia, a lot of investigation is taking place each day, and the way DL is transforming the world entirely is mind-boggling. The DL structures have produced much better than contemporary methods and have achieved the best performances. Thus, to survive in either industry or academia possessing DL skills will most likely play an important role in the advancing years.

Artificial Intelligence is the field of modern computer science that simulates human brain functions, generally problem-solving and learning. It simulates human intellect in machines and is based on the ethic that it can start from simple to complex problem execution. It has been deployed in endless fields including finance, healthcare, etc. Artificial Intelligence has impacted various day-to-day activities providing security, verification, and validity. This learns from experiences and adjusts the inputs to perform intelligent tasks. One adds intelligence by discovery, extracts deeper into data with accuracy, and extracts most valuable information out of data. ML is a subfield of Artificial Intelligence or an application of Artificial Intelligence that makes a machine intelligent without being directly programmed and earns experience or performance over time. This focuses on automated learning without human involvement. Moreover, it can identify patterns, make decisions, and build analytical models automatically. ML algorithms are classified into supervised, unsupervised, semi-supervised, and reinforcement algorithms. DL is a subfield of ML, also known as deep neural networks or deep neural learning. DL has gained popularity nowadays due to the improvement in ML and advancement in chip processing [5]. Hither utilizes a stratified level of artificial neural networks to carry out the processing. DL uses the underlying neural network architecture so-known as deep neural networks. Neural networks are inspired by biological neurons that are the basic structure in the human and animal brain. Neural networks can gain experience by storing it and practicing. It is similar to the human brain where the connections are between the neurons, and the connection strength between the neurons determines the knowledge. Deep Neural architectures have more than one hidden layer. Single-layer neural networks have one hidden layer of neurons and are used to solve linearly separable problems. That is, there is a single hyperplane to divide the input data into two separate classes. Solving the XOR problem is a classical neural network problem where the XOR's input cannot be separated by a single line. Hence, multilayer neural networks are used to solve problems like linearly inseparable ones. DL has more number of layers to extract more depth of features. The next layer extracts more depth features based on the previous layer. DL can handle high dimensional and complex unstructured data like text, video, pictures, etc. This has the ability to learn from different sources of data and to handle a large volume of data [6]. It has more significant learning and prediction ability than ML. They have hierarchical structures for connection between different layers. DL models are more robust than ML in terms of the extraction of features. Some popular DL architectures include restricted Boltzmann machine (RBM), deep belief networks (DBN), autoencoders (AE), convolutional neural networks (CNN), recurrent neural networks (RNN), long short-term memory (LSTM), etc.

Figure 5.1 explains the relationship between DL and Artificial intelligence. Learning algorithms in Artificial Intelligence use computational methods to learn directly from data without relying on a predestined comparison as a standard. The algorithms advance their performance as the number of examples available for learning increases. DL is a specific form of ML. In learning algorithms, we have a set of examples called training set as input data, and we have to predict the output set.

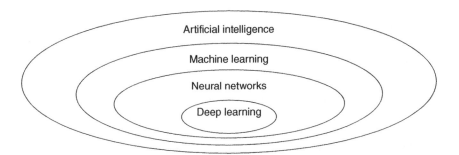

Figure 5.1 Deep neural networks in the context of artificial intelligence.

5.1.1 Types of Learning Algorithms

We mainly have four broad categories of learning algorithms.

5.1.1.1 Supervised Learning

In supervised learning, the training set is provided with a set of labels to predict the target value. Its main application includes regression (target values are continuous values) such as predicting weather and population growth. Additionally, it can also be applied to classification problems like speech recognition, fraud detection, digit recognition, diagnostics, etc., by employing algorithms like support vector machine (SVM), random forest, and others. Learning here is divided into two parts. (i) Training phase: Here the data is provided with a label, and the algorithm learns the relationship between input data and labels and determines the output for the test data. (ii) Testing phase: Here the data without a label is provided as input, and the algorithm finds the relationship between input data and previous labels and determines the output for the test data [7].

5.1.1.2 Unsupervised Learning

In unsupervised learning training, a set is not provided with a set of labels. This is meant for finding useful structures or knowledge in data when no labels are available. It is also used in detecting abnormal server access patterns, detecting irregular access to websites, outlier detection, social network analysis, organizing computing clusters, etc. It solves clustering problems like target marketing, customer segmentation, and recommendation system. It also tries to find patterns to predict future values and also applied for dimensionality reduction problems.

5.1.1.3 Semi-Supervised Learning

It is a combination of supervised and unsupervised learning, that is, both labeled and unlabeled data are used. This is an improvement over unsupervised learning [8].

5.1.1.4 Reinforcement Learning

In reinforcement learning, appropriate steps or a sequence of steps are to be taken for a given situation to maximize the payoff. Its applications include AI gaming, real-time decisions, skill acquisition, and robot navigation. Figure 5.2 explains broad categories of learning.

5.1.2 Steps Involved in Solving a Problem

Framework flow to solve a problem:

- Problem Definition: Understand the problem simply that is being solved and describe it.
- Analyze Data: Understand all the data accessible for solving the problem.
- Prepare Data: Prepare the structure of the dataset from previous data.

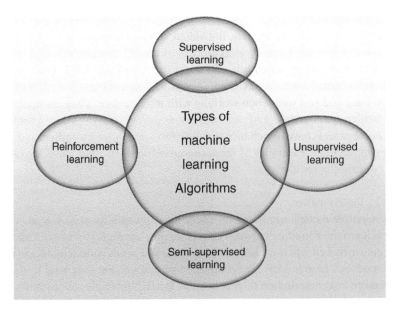

Figure 5.2 Types of learning algorithms.

- Evaluate Algorithm: Develop some baseline precision from which to improve.
- Improve Results: Support results to confirm a more perfect model.
- Present Result: Explain the solution so that it can be understood by users.

5.1.2.1 Basic Terminology
- Representation: understand the problem
- Evaluation: find effectiveness
- Optimization: solve the problem and feature representation.

5.1.2.2 Training Process
The training set consists of N training samples labeled as $\{(x_i, y_i), \ldots (x_m, y_m)\}$. (x_i, y_i) is a training instance. x_i is the ith m-dimensional input vector, y_i is the ith n-dimensional output vector. $x = \{x_1, x_2, .., x_m\}$ is the m-dimensional input matrix and $y = \{y_1, y_2, .., y_n\}$ is the n-dimensional output matrix. Artificial Intelligence techniques build a model from the input and map its corresponding output until the system stabilizes.

5.1.3 Modeling in Data Science

Modeling is done whenever we have to understand any part of the science whether it is physics, chemistry, mathematics, etc. For better understanding and representation of concepts, a model is created and is used to generate an idea and test the theories. Modeling in data science consists of two main categories: (i) Generative and (ii) Discriminative.

5.1.3.1 Generative

In the case of Generative models, individual class distribution is modeled. The Generative models are based on joint probability distribution $p(x, y) = p(y) * p(y|x)$ and conclude based on Bayes rule and select the most likely label. When conditional independence is not convinced, they prove less accurate. They need fewer data to train and test well when working with missing data. They are applied to find hidden parameters and are used for solving classification problems. Generative models have less chance to overfit but are affected by outliers and need more computation. Examples of the Generative model include naïve Bayes and linear discriminative analysis.

5.1.3.2 Discriminative

Discriminative models are based on conditional probability $p(Y|X = x)$. These models learn the boundary between classes, so they are useful for classification. It does not work well when we have missing data, and needs sufficient data to train these models. These models are not affected by outliers, but they tend to over fit and require less computation than generative models. Examples of discriminative models include logistic regression, SVMs, etc.

5.1.4 Why DL and IoT?

DL in IoT is growing rapidly and has been applied in various day-to-day applications. We start our discussion from health applications, including disease prediction and monitoring of health. It has been deployed in homes for smarter signal processing and making our homes smarter. It also has been deployed for smart transportation, which includes traffic analysis and prediction for autonomous driving. Smart Industry is another application of DL in IoT. It has promising results in the field of automated inspection in industries.

- **Smart Healthcare:** DL is omnipresent and is widely used in various applications. It is playing a vital role in the field of Medical science in disease prediction and health monitoring [9]. DL is used to discover patterns from medical data including medical images like MRI, CT-SCAN sources, and provide excellent capabilities to predict diseases and monitor healthcare.
- **Smart Transportation:** DL has been deployed in transportation for traffic flow prediction, traffic monitoring, and autonomous driving. Sensors are embedded in the vehicles, and when mobile devices are installed in the city, it is possible to provide suggestions related to route optimization, parking reservations, accident prevention, and autonomous driving.
- **Smart Industries:** An important application of IoT and DL in the present life is the evolution in the field of industries for easier and more flexible

business processes. The increase in IoT-embedded sensors, such as RFID and NFCE, enables the interaction between IoT sensors embedded in the products. Therefore, those goods can be traced throughout production and transportation processes until they reach the user. The monitoring process and the data generated through this process improve machine throughput. The inspection and assessment of products can be done smartly by employing DL.

- **Social Media:** One of the most common applications of DL is Automatic Tagging Suggestions On a social media platform. Facebook uses face detection and Image recognition to find the face of the character which resembles its database and hence recommends us to mark that person based on DeepFace. Facebook's DL project DeepFace is responsible for the identification of faces. It also provides Alternative Tags to images already uploaded on Facebook.

5.2 DL Architectures

In the following section, we will address various DL architecture and their applications, mainly in the field of IoT. The DL architectures discussed in this paper include the RBM, DBN, AE, CNN, RNN, and LSTM.

5.2.1 Restricted Boltzmann Machine

RBM [10] is a modification of the Boltzmann machine. It forms an undirected bipartite graphical model. It consists of two types of nodes: (i) visible nodes and (ii) hidden nodes. The visible and hidden nodes are connected in the form of a complete bipartite graph. Thus, here we have two types of layers: (i) hidden layer and (ii) visible layer. The visible layer consists of visible nodes; input is fed to these layer nodes. The hidden layer consists of hidden nodes or latent nodes. There is neither connection between visible nodes nor any connection between hidden nodes, so it is known as a RBM. They need expensive computation for performance.

They are based on statistical mechanics and have an impact on energy and temperature, so they are known as energy-based models.

The energy function of a RBM is defined as:

$$E\left(v,h\right) = -b'v - c'v - h'wv$$

where w is the weight between hidden and visible layer and b, c are biases of visible and hidden layers, respectively. V, h are visible and hidden layer nodes.

The visible and hidden layer nodes are conditionally independent.

$$P\left(h_i|v\right)=\prod_i \left(h_i|v\right)$$

$$P\left(v_j|h\right)=\prod_i \left(v_j|h\right)$$

RBM is based on probabilistic graphical models. They are ideal for unsupervised learning and can be used for unsupervised pre-training of conventional neural networks.

Let us assume we have $V_1, V_2, \ldots, V_N$ visible nodes of N-dimensional and H_1, $H_2, \ldots, H_M$ hidden nodes of M-dimensional. We have bias associated with each node of the hidden and visible layer. The bias associated with the visible node V_i is $b_i(v)$ and bias associated with the hidden node H_j is $c_j(h)$, and W_{ij} is the weight associated between the visible node V_i and hidden node H_j. The RBM aims to learn weight W_{ij}. Figure 5.3 explains the architecture of the RBM.

5.2.1.1 Training Boltzmann Machine

Training of Boltzmann machine is done through contrastive divergence (CD) [11] using Gibbs Sampling:

1) The visible state is fed with the training dataset.
2) Positive phase:
   Hidden state nodes are calculated from visible nodes known as positive statistics of edge E_{ij} which is $P(H_j = 1|V)$. The individual activation probabilities for the hidden layer can be calculated as

$$P\left(H_j =1|V\right)=\sigma\left(b_j + \sum_{i=1}^{m} W_{ij}V_i \right)$$

$$=\frac{1}{1+e^{\left(b_j+\sum_{i=1}^{m}W_{ij}V_i\right)}}$$

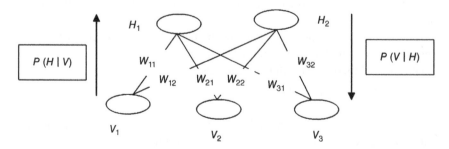

Figure 5.3 Restricted Boltzmann machine basic architecture.

3) Negative phase:

In the negative phase, the input data is reconstructed using the hidden state values known as negative statistics of edge E_{ij}, which is $P(V_i = 1|H)$.

The individual activation probabilities for visible layer nodes can be calculated as:

$$p\left(V_i|H\right) = \sigma\left(a_i + \sum_{j=1}^{n} W_{ij}H_j\right)$$

$$= \frac{1}{1 + e^{\left(a_i + \sum_{j=1}^{n} W_{ij}H_j\right)}}$$

4) Update the weight of the edge:

The updated weight is calculated as:

$$\Delta\left(W_{ij}\right) = W_{ij} + \alpha\left(\text{positive}\left(E_{ij}\right) - \text{negative}\left(E_{ij}\right)\right)$$

and α is the learning rate.

5) Repeat till the threshold is achieved.

5.2.1.2 Applications of RBM

RBM has various applications in the field of classification of data in the form of text, images, etc. It can also be used for reducing a number of variables for obtaining principle variables known as dimensionality reduction and can be used for feature selection and feature extraction. Other applications include processing of unstructured data, images, videos, collaborative filtering, topic modeling, missing value issue, etc. RBM has been applied for supervised and unsupervised learning, and an energy-efficient RBM processor has been proposed for on-chip learning for IoT [12]. Sourav Bhattacharya and Nicholas D. Lane [13] proposed RBM as a DL architecture for robust activity recognition using smart watches, like sleeping patterns and emotional states. They are used for predicting energy consumption in IoT.

The variants of RBMs include (i) discriminative restricted Boltzmann machine (DRBM): they are meant for the selection of parameters, which are analytical for performance. (ii) Hybrid restricted Boltzmann machine (HRBM): it includes features of both generative and discriminative models. (iii) Conditional restricted Boltzmann machine (CRBM). (iv) Robust Boltzmann machine (RoBM): it deals with eliminating noise and influence of corrupted pixels in an image.

5.2.2 Deep Belief Networks (DBN)

DBNs are one of the variants or emerging application [14] of the RBM. They form generative graphical models and are one of the solutions of vanishing gradient problem. Vanishing gradient means loss of error with respect to weights.

This means when we move backward in case of backpropagation, the earlier layers of neuron are smaller as compared to the last layer. RBM has the advantage of fitting the input features, and thus hidden layer output of one layer can be used as input to a visible layer of another layer. DBNs are constructed by stacking more than one RBM. The Greedy approach is used for learning of DBNs. The number of layers stacked depends on the algorithm and thus extract features from already extracted features from the previous layer. Training the network means learning the weights. Here, training is performed in a greedy layer-by-layer unsupervised algorithm [15]. Hinton et al. [16] proposed fine-tuning after training of each layer using backpropagation.

Training in DBN is divided into two phases: In the first phase, each layer has been trained using CD algorithm. The posterior probabilities $p(h_1|v)$ and $p(v|h_1)$ are computed from the first layer. And then they are computed again and again till they converge as computed in the RBM. The h_1 computed from visible layer after convergence is taken as v (input) for the next layer and compute h_2 and same has to be propagated till last layer. In the second phase, whole DBN is fine-tuned. Figure 5.4 explains the architecture of DBNs.

5.2.2.1 Training DBN
1) Train the first layer of RBM using CD algorithm and learn the weights W_1.
2) Freeze the weight vector W_1 of the first layer. Use the H_1 (output of the first layer) from $P(H_1|V)$ as input for the next layer.
3) Freeze the weights of second layer W_2 that defines the second layer features. Use the H_2 (output of the second layer) as input to the next layer.
4) Proceed recursively for the next layers.

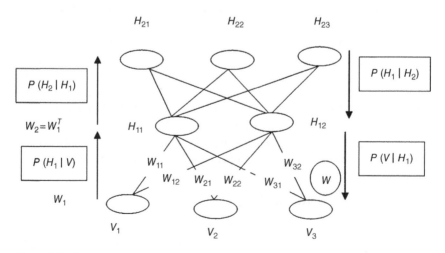

Figure 5.4 Deep belief network architecture with three layers.

5.2.2.2 Applications of DBN

It has been used to avoid overfitting and underfitting problems. It has been used for recognizing and generating images, capturing motion in a video, dimensionality reduction, audio classification, and natural language classification. They can act as feature detectors and has been applied for real-life applications like electroencephalography [17] and drug discovery [18]. Hinton et al. [16] have deployed for the recognition and classification of handwritten characters. Ni Gao et al. [19] have deployed DBN in intrusion detection. It can provide new ideas for further research in Intrusion Detection Systems. They are suitable for hierarchical feature extraction as they use a greedy layer-by-layer training algorithm. It has been used for threat identification in the security and classification of faults in IoT.

5.2.3 Autoencoders

AE are unsupervised learning algorithms. They encode input in a form, from which input can be reconstructed. They convert the input into a code that is a summary of input. Thus, the size of the input is same as that of output. They are also called autoassociator. They are meant for dimensionality reduction and data compression in the areas of information processing. They extract information continuously and convert into smaller representations.

5.2.3.1 Training of AE

AE consist of three main parts: (i) encoder, (ii) code, and (iii) decoder. Figure 5.5 explains the architecture of AE. AE are self-supervised learning algorithms because they generate their labels. The loss function measures how close the reconstructed output is to the input and is given by mean squared error:

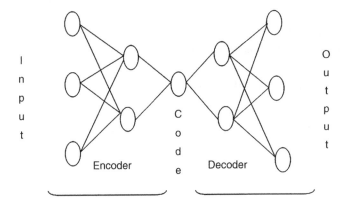

Figure 5.5 Autoencoder architecture.

$$j(x,y) = \|x - y\|^2.$$

AE add a bottleneck to the network. The bottleneck forces the network to create the compress of the input. It works well if the data are correlated and does not work well if the data are not correlated.

Encoder: converts input x into some hidden representation $h(x)$:

$$h(x) = \sigma(b + wx).$$

Decoder: reconstructs the input $\hat{x}$:

$$\hat{x} = \sigma(c + w^*h(x)),$$

where b, c are biases and w is the weight vector.

$w^* = w^T$ and are known as tied weights.

Variants of AE include: (i) denoising autoencoders [20]: these are the type of AE where we try to reconstruct the input from corrupted input data. They try to undo the effects of corruption in data. They improve the accuracy of OCR. (ii) Sparse autoencoder and (iii) variational autoencoder.

Variational autoencoder is a standard type of AE which shares the basic architecture of AE. Here, training is done to avoid overfitting. Encoding is done over a distribution of latent space (z). Low-dimensional view of high-dimensional input has been created.

5.2.3.2 Applications of AE

The applications of AE include anomaly detection, information retrieval, data denoising, etc. They are more powerful than principal component analysis (PCA). Mahmood Yousefi-Azar et al. [21] proposed AE-based feature learning for internet security and for various classifiers for anomaly detection in publically available datasets. Changjie Hu et al. [22] proposed improved version of AE for dimensionality reduction. It has the same number of input and output and is well suited for dimensionality reduction and feature extraction. In IoT, it can be used for emotion recognition and fault diagnosis in machinery.

5.2.4 Convolutional Neural Networks (CNN)

Convolution neural networks or ConvNet or CNN is one of the DL architecture inspired by the connectivity of neurons in human brain. It takes images (one-dimensional or two-dimensional or n-dimensional) as input, converts it into smaller representation without losing information for better prediction. It consists of three primary layers: (i) convolution layer, (ii) pooling layer, and (iii) fully connected layer. Figure 5.6 explains the role of different layers of CNN architecture.

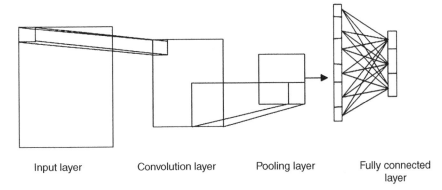

Input layer   Convolution layer   Pooling layer   Fully connected layer

Figure 5.6   Convolution neural network architecture.

5.2.4.1 Layers of CNN

- **Convolution layer:** This is the main layer in a CNN. It acts as a filter to an input; when this filter is applied continuously to the input, we get feature map. Feature map gives us the locations of the important information in an image. CNN can learn filters specific to a particular training dataset. Here, multiplication is performed between input and a set of weights (called filter or kernel). The multiplication between the kernel and input is performed elementwise called dot product. The multiplication is followed by addition. Kernel size is smaller than input size; the same kernel has been applied multiple times to the different portions of an input by systematically overlapping different portions of input from left to right and top to bottom.

  Whenever we are applying convolution, our output shrinks. So to avoid shrinking, we can use padding to get output dimensions the same as an input. Convolutions are of two types: (i) valid and (ii) same. In case of valid, the output we get is smaller than input and no padding is applied. In the other type, the output dimensions are the same as input, and padding is applied here. The three main parameters that control the size of output are: (i) stride, (ii) kernel size, and (iii) padding.

  The step size of the filter when processing the convolution is known as a stride. The same stride size is applied both horizontally and vertically. More the stride values are smaller, the output we get. Kernel size is the filter that we apply to the input. If we want to output the same as input, then we need to include padding. CNN has the property that a part of output depends on only a smaller part of an input that is known as the sparsity of connections. A part of the image is possibly important for other parts of an image known as parameter sharing.

- **Pooling layer:** The function of a pooling layer is to reduce the representation size of an image. It works on feature maps extracted from the previous layer. It gives us the representation of a summary of the input, so that processing and computation are reduced. The common pooling functions include

1) Max pooling: It calculates the maximum value of a portion of a feature map.
2) Average pooling: It calculates the average of a portion of a feature map.
3) Global pooling: This is also known as a global pooling layer. It downsamples whole feature map to a single unit.

- **Fully connected layer:** It is the last layer of CNN. Here, we get the class label for our input. It has a lot of connections forming a complete graph with the output from the previous layer. It gives us the decision of whole processing. Input to this layer is fed in the form of a flattened vector. At this layer, results are available.

Variants of CNN: CNN has become popular in the last few years. It has been applied to various fields and has achieved the best performance. Popular variants of CNN include (i) LeNet, (ii) AlexNet, (iii) ResNet, (iv) VGG, (v) GoogLeNet, (vi) Inception Networks, (vii) Xception, etc.

5.2.4.2 Activation Functions Used in CNN

Activation functions are nonlinear functions that decide output for the next layer. The activation functions used in CNN include:

1) **Sigmoid Function:** It is also known as logistic activation function. It translates input $(-\infty, +\infty)$ to $(0, 1)$ used for binary class problem.

$$\sigma(x) = \frac{1}{1 + e^{-x}}.$$

Figure 5.7 shows the graph of sigmoid function. Soft-max logistic equation is used for multi-class problems. This is used in the last layer of CNN.

2) **Tanh():** This is same as sigmoid activation function but here it translates to $(-1, 1)$. Figure 5.8 shows a graph of Tanh() function.

$$\tanh(x) = \frac{2}{1 + e^{-2x}} - 1.$$

3) **ReLU:** ReLU stands for a rectified linear unit. It gives linear output for positive values and 0 for negative values. It converges faster and is used in the convolution layer of CNN in the middle layers.

$$ReLU(x) = \max(0, x).$$

Figure 5.9 shows a graph of ReLU function. It has the problem of dying ReLU, which means if it gets 0 value, it is difficult for it to get back. Its variants include Leaky ReLU, PReLU, and ELU.

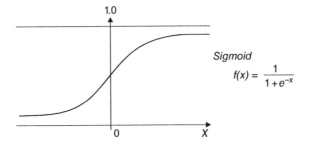

Figure 5.7 The sigmoid function.

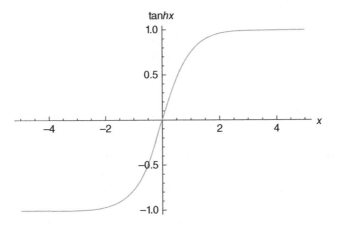

Figure 5.8 Tanh() function.

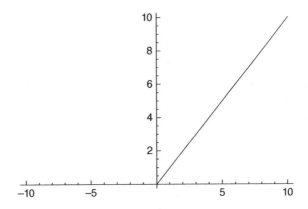

Figure 5.9 ReLU function.

5.2.4.3 Applications of CNN

CNN has proven satisfying in the field of image recognition, image transformations, face recognition, speech recognition, image labeling, recommendation system, handwriting recognition, natural language processing (NLP), behavior recognition, object recognition, etc. Iman Azimi et al. [23] have deployed CNN in the IoT healthcare system. Alberto Pacheco [24] et al. deployed it for the development of Smart Classroom and IoT Computing. A wavelet-based CNN approach has been proposed for automated machinery fault diagnosis [25]. Parkinson's disease patients' prediction can be done using CNNs [26]. CNNs are also used for mapping sequence of ECG samples to a sequence of rhythm classes to develop high diagnostic performance similar to professionals [27]. CNNs have been used for automatic segmentation of MRI scans to predict the risk of osteoarthritis [28]. V. Gulshan et al. [27] have used CNNs to identify retinopathy. A CNN-based approach is introduced for surface integration inspection in [29] to extract patch features and find defected areas via thresholding and segmenting. A wavelet-based CNN approach has been proposed for automated machinery fault diagnosis [25].

5.2.5 Generative Adversarial Network (GANs)

Generative adversarial networks (GANs) are a DL architecture developed in 2014 by Ian Good Fellow. They are used for unsupervised learning and can catch and copy the dissimilarity within a dataset. They are inspiring recent discoveries in DL. GANs are generative models and build new data instances that match provided training data. They can build images that resemble images of individual faces, even though these faces do not relate to any existing person.

5.2.5.1 Training of GANs

It mainly consists of two neural network architectures that contest with one another for learning. The two networks compete with each other to generate some new probability distribution that is indistinguishable to original. GANs generate the imitate data from some original data. It consists of two main parts:

- Discriminator: The role of the discriminator module is to discriminate between original and fraud data.
- Generator: The role of the generator module is to generate the fraud data in such a way that it can deceive the discriminator. That is, data produced here must look like the original.

Figure 5.10 shows the basic framework of GANs for identifying fake or real things. Learning in GANs means deceiving the discriminator, i.e. the discriminator treats fake instances as real. The discriminator weights are kept fixed, the error is backpropagated, and weight adjustment is done at the generator. Figure 5.11 explains the basic architecture of GANs.

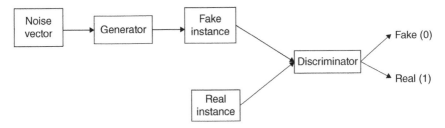

Figure 5.10   The framework of generative adversarial networks (GANs).

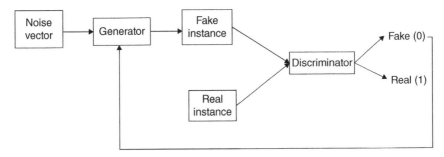

Figure 5.11   Generative adversial networks (GANs).

5.2.5.2   Variants of GANs

The variants of GANs include progressive GANs, conditional GANs, Image to Image GANs, and Cycle GANs.

5.2.5.3   Applications of GANs

GANs have been applied mainly in the field of data generation [30]. They have been successfully applied in the field of computer vision and NLP. The applications in the field of computer vision include image translation, image synthesis, and video generation. Ledig et al. [31] have applied GANs for improvement in the resolution of an image. Li and Wand [32] applied it in the field of texture synthesis. Tulyakov et al. [33] proposed GANs for video generation. They have a great opportunity to be applied in new fields and have been applied for the labeling of unlabeled images [34].

5.2.6   Recurrent Neural Networks (RNN)

CNN discussed previously is a feed-forward neural network and can solve a lot of problems related to unstructured data. But it cannot solve a problem related to time series. To solve problems related to time series, RNN has been developed. RNNs are called recurrent because they accomplish the same task for every component of a sequence, with the output staying dependent on the early estimates. RNNs produce recorded success in many NLP tasks. The most regularly used type of RNNs is LSTMs. They are more qualified at capturing long-term dependencies. Their applications include mainly NLP, speech recognition, handwriting recognition, etc. The nodes of RNN are directly connected to each other to form a loop-like structure. RNNs are designed for time series input data with no fixed sized input. The data is fed in the form of series, a single data item in a series have effect from neighbor data items and affect the others in neighborhood. They have one or more input vectors and generate one or more output vectors.

5.2.6.1 Training of RNN

Training RNN is comparable to training a Neural Network. Here also backpropagation algorithm is used, but with modification. The parameters are shared by all time-steps in the network; the gradient at each output depends on the estimates of the current time step, and previous time steps. To calculate the gradient at each point, backpropagation is needed to be called backpropagation through time (BPTT). Figure 5.12 explains the basic architecture of RNN.

The hidden state at a particular time is a function (F_w) of previous state (S_{t-1}) and current input (X_t).

$$S_t = F_w\left(S_{t-1}, X_t\right),$$
$$S_t = \tanh\left(W_s S_{t-1} + W_x X_t\right),$$

where W_s, W_x are weights.

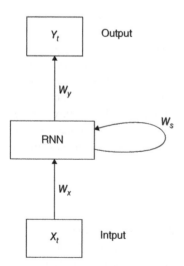

Figure 5.12 Recurrent neural network basic framework.

5.2.6.2 Applications of RNN

The RNN is trained using BPTT algorithm [35]. It is difficult to learn long-term dependencies in the input sequence due to the decrease in error gradient as it propagates backward known as vanishing gradient problem. This leads to larger weights which in turn lead to unstable training. B. Chandra, Rajesh Kumar Sharma [36] has proposed the improved version of RNN which reduced vanishing gradient problem and has promising results for image classification. It learns complex patterns in the IoT.

5.2.7 Long Short-Term Memory (LSTM)

LSTM is a special type of RNN proposed by German scholar Sepp Hochreiter and Juergen Schmidhuber to solve vanishing gradient problem [37]. LSTMs are a distinct kind of RNN, competent in learning long-term dependencies and retaining information for a lengthened period of time. The LSTM model is assembled in the form of a chain structure. But, the perpetual module has a diverse arrangement. Instead of a single neural network like a standard RNN, it has four layers with a different style of communication. It stores information in the hidden layer and particularly discards information of no use in the hidden layer. Thus, it has the decision power of what to store and what to discard. It shares the basic architecture of RNN but has more compound structure inside.

5.2.7.1 Training of LSTM

LSTM consists of a chain of repeating modules called memory cells which consists of four gates. These four gates decide what information to keep or discard, handle dependencies, and handle the storage of information [38]. The four gates are:

- Input gate (i)
- Output gate (o)
- Forget gate (f)
- Memory gate (c).

The information is passed based on the weights and these weights are adjusted during the learning process. Each memory cell correlates with a time step [39]. It removes information through the forget gate, it specifically controls information transmission using the sigmoid function. Thus, outputs values between 0 and 1. One determines to keep the information and zero to discard the information. They maintain a constant error throughout; therefore, they are able to learn over multiple layers and time steps. They sustain information for a more prolonged time. Each cell has its individual decision power related to information. Forget gate removes irrelevant information, thus helps in intensification in the process. Input gate adds information, output gate selects necessary information for the output. Figure 5.13 explains the architecture of LSTM.

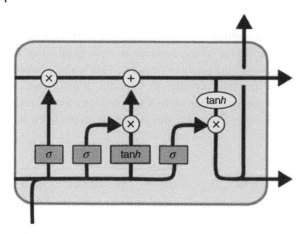

Figure 5.13   LSTM architecture.

5.2.7.2 Applications of LSTM

LSTM has mainly been developed for handling sequential data, NLP [40]. They have been developed for intrusion detection for learning patterns [41]. It has also been proposed for the classification of voltage dips and automatic extraction of features [42]. LSTM has become one of the most popular techniques in DL; research is going on in this field. It has been also proposed to control dynamic systems [43]. It has proven better for the classification and identification of fault with 99.80% accuracy [44]. The other applications include text generation, handwriting recognition [45], handwriting generation, music generation, language translation, and image captioning. In industries, it has been used for the analysis of equipments of IoT [46] and also for time series data generation in IoT, anomaly detection, etc. Time series data are highly complex, dynamic data, and has higher dimensionality, so it requires IoT of resources. LSTM allows forgetting unnecessary data and thus reduces the computation power needed. It can keep track of dependencies between data.

5.3 Conclusion

A thorough study of how DL carries new opportunities to the IoT has been carried out. Many IoT applications have been enabled with DL technology. DL architectures are very powerful to resolve large-scale data analysis problems. Investigation on how to build DL models and their applications for IoT has been carried out. DL extends an innovative perspective to solve the traditional queries of IoT. Both IoT and DL have seen outburst growth of research in the last few years. DL has gained popularity in the areas of computer vision, speech recognition, NLP, video summarization, and they contribute more reliable accuracy than humans. They have already been used in IoT in data fitting. IoT is the integration of devices and various day-to-day applications are dependent on it. DL offers an excellent opportunity to join hardware and software, and develop real-time applications. DL models tune themselves and make changes according to new needs and have already performed well with real-time time-series data. Because of its unsupervised learning nature, it has achieved great success while deployed in IoT and internet search engines. In the future, it can be incorporated into various aspects of day-to-day life to make our tasks easier.

References

1 Saleem, T.J. and Chishti, M.A. (2019) Data mining for the Internet of Things data analytics. *ICDM* (Posters), New York; (17 July to 21 July 2019).

2 Saleem, T.J. and Chishti, M.A. (2019). Data analytics in the internet of things: a survey. *Scalable Computing: Practice and Experience* 20: 4.

3 Thompson, W.L. and Talley, M.F. (2019). Deep learning for IoT communications: invited presentation. *2019 53rd Annual Conference on Information Sciences and Systems (CISS)*, Baltimore, MD (20–22 March 2019), 1–4.

4 Ma, X., Yao, T., Hu, M. et al. (2019). A survey on deep learning empowered IoT applications. *IEEE Access* 7: 181721–181732.

5 Choudhury, S. and Bhowal, A. (2015). Comparative analysis of machine learning algorithms along with classifiers for network intrusion detection. *2015 International Conference on Smart Technologies and Management for Computing, Communication, Controls, Energy and Materials (ICSTM)*, Chennai (6–8 May 2015).

6 Zhang, L., Wang, S., and Liu, B. (2017). *Deep Learning for Sentiment Analysis: A Survey*. National Science Foundation (NSF), and by Huawei Technologies Co. Ltd.

7 Kubat, M. (2017). *An Introduction to Machine Learning*. Cham: Springer.

8 Mohammed, M., Khan, M.B., and Bashier, E.B.M. (2016). *Machine Learning: Algorithms and Applications*. Boca Raton, FL: CRC Press.

9 Saleem, T.J. and Chishti, M.A. (2020). Exploring the applications of machine learning in healthcare. *International Journal of Sensors, Wireless Communications and Control* 10 (4): 458–472.

10 Salakhutdinov, R., Mnih, A., and Hinton, G. (2007). Restricted Boltzmann machines for collaborative filtering. *Proceedings of the 24th International Conference on Machine Learning*, Corvalis, OR (June 2007), 791–748.

11 O'Leary, H.M., Mayor, J.M., Poon, C.-S., Kaufmann, W.E., and Sahin, M. (2017). Classification of respiratory disturbances in Rett syndrome patients using

restricted Boltzmann machine. *Engineering in Medicine and Biology Society (EMBC) 2017 39th Annual International Conference of the IEEE*, Seogwipo (11–15 July 2017), 442–445.

12 Tsai, C., Yu, W., Wong, W.H., and Lee, C. (2017). A 41.3/26.7 pJ per neuron weight RBM processor supporting on-chip learning/inference for IoT applications. *IEEE Journal of Solid-State Circuits* 52 (10): 2601–2612.

13 Bhattacharya, S. and Lane, N.D. (2016). From smart to deep: robust activity recognition on smartwatches using deep learning. *2016 IEEE International Conference on Pervasive Computing and Communication Workshops (PerCom Workshops)*, Sydney, NSW (14–18 March 2016), 1–6.

14 Pan, Z., Yu, W., Yi, X. et al. (2019). Recent progress on generative adversarial networks (GANs): a survey. *IEEE Access* 7: 36322–36333.

15 Bengio, Y. (2009). Learning deep architectures for AI. *Foundations and Trends in Machine Learning* 2 (1): 1–27. https://doi.org/10.1561/2200000006.

16 Hinton, G.E., Osindero, S., and The, Y.W. (2006). A fast learning algorithm for deep belief nets. *Neural Computation* 18 (7): 1527–1554.

17 Movahedi, F., Coyle, J.L., and Sejdic, E. (2018). Deep belief networks for electroencephalography: a review of recent contributions and future outlooks. *IEEE Journal of Biomedical and Health Informatics* 22 (3): 642–652. https://doi.org/10.1109/jbhi.2017.2727218.

18 Ghasemi, P.-S. and Mehri, f. (2018). The role of different sampling methods in improving biological activity prediction using deep belief network. *Journal of Computational Chemistry* 22 (3): 642–652. https://doi.org/10.1109/JBHI.2017.2727218.

19 Gao, N., Gao, L., Gao, Q., and Wang, H. (2014). An intrusion detection model based on deep belief networks. *2014 Second International Conference on Advanced Cloud and Big Data*, Huangshan (20–22 November 2014), 247–252.

20 Yousefi-Azar, M., Varadharajan, V., Hamey, L., and Tupakula, U. (2017). Autoencoder-based feature learning for cyber security applications. *2017 International Joint Conference on Neural Networks (IJCNN)*, Anchorage, AK (14–19 May 2017), 3854–3861.

21 Hu, C., Hou, X., and Lu, Y. (2014). Improving the architecture of an autoencoder for dimension reduction. *2014 IEEE 11th Intl Conf on Ubiquitous Intelligence and Computing and 2014 IEEE 11th Intl Conf on Autonomic and Trusted Computing and 2014 IEEE 14th Intl Conf on Scalable Computing and Communications and Its Associated Workshops*, Bali (9–12 December 2014), 855–858.

22 Vincent, P., Larochelle, H., Bengio, Y., and Manzagol, P.A. (2008). Extracting and composing robust features with denoising autoencoders. *Proceedings of the 25th International Conference on Machine Learning* (July 2008), 1096–1103.

23 Azimi, J.T.-M., Anzanpour, A., Rahmani, A.M., Soininen, A.M., and Liljeberg, P. (2018). Empowering healthcare IoT systems with hierarchical edge-based deep learning. *2018 IEEE/ACM International Conference on Connected Health:*

Applications, Systems and Engineering Technologies (CHASE), Washington, DC (26–28 September 2018), 63–68.

24 Pacheco, P.C., Flores, E., Trujillo, E., and Marquez, P. (2018). A smart classroom based on deep learning and osmotic IoT computing. *2018 Congreso Internacional de Innovación y Tendencias en Ingeniería (CONIITI)*, Bogota (3–5 October 2018), 1–5.

25 Wang, J., Zhuang, J., Duan, L., and Cheng, W. (2016). A multi-scale convolution neural network for featureless fault diagnosis. *Proc. Int. Symp. Flexible Autom. (ISFA)*, Cleveland, OH (1–3 August 2016).

26 Hammerla, N.Y., Halloran, S., and Plötz, T. (2016). Deep, convolutional, and recurrent models for human activity recognition using wearables. *Proc. 25th Int. Joint Conf. Artif. Intell. (IJCAI)* (July 2016).

27 Gulshan, V., Peng, L., Coram, M. et al. (2016). Development and validation of a deep learning algorithm for detection of diabetic retinopathy in retinal fundus photographs. *Proceedings of the JAMA* 316 (22): 2402–2410.

28 Prasoon, A., Petersen, K., Igel, C. et al. (2013). Deep feature learning for knee cartilage segmentation using a triplanar convolutional neural network. *Proc. Int. Conf. Med. Image Comput. Comput.-Assist. Intervent. (MICCAI)*, Nagoya, Japan (22–26 September 2013).

29 Park, J.-K., Kwon, B.-K., Park, J.-H., and Kang, D.-J. (2016). Machine learning based imaging system for surface defect inspection. *International Journal of Precision Engineering and Manufacturing – Green Technology* 3 (3): 303–310.

30 Kim, S.K., McMahon, P.L., and Olukotun, K. (2010). A large-scale architecture for restricted Boltzmann machines. *2010 18th IEEE Annual International Symposium on Field-Programmable Custom Computing Machines*, Charlotte, NC (2–4 May 2010), 201–208.

31 Ledig, C. Theis, L., Huszar, F., and Caballero, J. (2017). Photo-realistic single image super-resolution using a generative adversarial network. *Proc. IEEE Conf. Comput. Vis. Pattern Recognit. (CVPR)*, Honolulu, HI (21–26 July 2017), 105–114.

32 Li, C. and Wand, M. (2016). Precomputed real-time texture synthesis with markovian generative adversarial networks. In: *Computer Vision – ECCV 2016. ECCV 2016*, Lecture Notes in Computer Science, vol. 9907 (eds. B. Leibe, J. Matas, N. Sebe and M. Welling). Cham: Springer https://doi. org/10.1007/978-3-319-46487-9_43.

33 Huang, R., Zhang, S., Li, T., and He, R. (2017). Beyond face rotation: global and local perception GAN for photorealistic and identity preserving frontal view synthesis. *Proc. IEEE Int. Conf. Comput. Vis. (ICCV)*, Venice, Italy (22–29 October 2017), 2439–2448.

34 Zhang, X.-Y., Yin, F., Zhang, Y.-M. et al. (2018). Drawing and recognizing Chinese characters with recurrent neural network. *IEEE Transactions on Pattern Analysis & Machine Intelligence* 40 (4): 849–862.

35 Rumelhart, D.E., Hinton, G.E., and Williams, R.J. (1986). Learning representations by back-propagating errors. *Nature* 323 (6088): 533–536. https://doi.org/10.1038/323533a0. [Online].

36 Chandra, B. and Sharma, R.K. (2017). On improving recurrent neural network for image classification. *2017 International Joint Conference on Neural Networks (IJCNN)*, Anchorage, AK, Anchorage, AK, (14–19 May 2017), 1904–1907.

37 Hochreiter, S. and Schmidhuber, J. (1997). Long short-term memory. *Neural Computation* 9 (8): 1735–1780.

38 Lyu, Q. and Zhu, J. (2014). Revisit long short-term memory: an optimization perspective. In Advances in *Neural Information Processing Systems Workshop on Deep Learning and Representation Learning* (December 2014), pp. 1–9.

39 Anani, W. and Samarabandu, J. (2018). Comparison of recurrent neural network algorithms for intrusion detection based on predicting packet sequences. *2018 IEEE Canadian Conference on Electrical & Computer Engineering (CCECE)*, Quebec City, QC (13–16 May 2018), 1–4.

40 Cho, K., Van Merrienboer, B., Gulcehre, C. et al. (2014). *Learning Phrase Representations Using RNN Encoder-Decoder for Statistical Machine Translation* (June 2014).

41 Tuor, A., Kaplan, S., Hutchinson, B. et al. (2017). Deep learning for unsupervised insider threat detection in structured cybersecurity data streams. arXiv preprint, arXiv:1710.00811.

42 Balouji, E., Gu, I.Y.H., Bollen, M.H.J., Bagheri, A., and Nazari, M. (2018). A LSTM-based deep learning method with application to voltage dip classification. *2018 18th International Conference on Harmonics and Quality of Power (ICHQP)*, Ljubljana (13–16 May 2018), 1–5.

43 Wang, Y. (2017). A new concept using LSTM Neural Networks for dynamic system identification. *2017 American Control Conference (ACC)*, Seattle, WA (24–26 May 2017), 5324–5329.

44 Wang, W., Qiu, X., Chen, C., Lin, B., and Zhang, H. (2018). Application research on long short-term memory network in fault diagnosis. *2018 International Conference on Machine Learning and Cybernetics (ICMLC)*, Chengdu (15–18 July 2018), 360–365.

45 Creswell, T., White, V., Dumoulin, K. et al. (2018). Generative adversarial networks: an overview. *IEEE Signal Processing Magazine* 35 (1): 53–65.

46 Zhang, W., Guo, W., Liu, X. et al. (2018). LSTM-based Analysis of Industrial IoT Equipment. *IEEE Access* 6: 1. https://doi.org/10.1109/ACCESS.2018.2825538.

6

Adding Personal Touches to IoT

A User-Centric IoT Architecture

Sarabjeet Kaur Kochhar

Department of Computer Science, Indraprastha College for Women, University of Delhi, New Delhi, India

6.1   Introduction

Internet of Things or IoT refers to an interconnected network of computing devices, seamlessly connected together, sharing information without the need of any human interference. Smart TVs, smart speakers, toys, wearables, smart meters, smart security systems, and smart devices used to monitor traffic and weather conditions, etc., are only some examples of the IoT devices that are ubiquitous these days [1]. Since these devices are connected to a broader network, they can be used to control a variety of tasks and carry out automated tasks remotely depending upon a certain set of conditions being met, even with relatively modest hardware capabilities of their own. This is the reason why IoT devices are deemed crucial for automation systems. The occurrence of every single event on these practically innumerable smart devices and sensors leads to the generation of a large amount of data [2, 3]. There is no doubt therefore that the amount of data generated by the IoT devices alone is expected to grow to several zeta bytes in the next five years.

This gigantic amount of data is diverse and gets collected at various velocities. Such a huge volume of data, also known as Big Data [3], is an asset for organizations around the world and brings with it ample opportunities to look for important trends over time. Analyzing these data, with the help of emerging technologies such as Machine Learning (ML), KDD, Data Mining, and Artificial Intelligence (AI) unravels tremendous benefits for both the parent organizations of these smart devices and the end-users [4, 5].

Big Data Analytics for Internet of Things, First Edition. Edited by Tausifa Jan Saleem and Mohammad Ahsan Chishti.

The integration of Big Data Analytics (BDA) with the IoT applications has led to the successful development of many complex systems such as smart cities, smart homes, smart security systems, etc. [6]. These systems have been able to deliver the promised physical comfort and ease of doing things to the end-user. However, it is important to distinguish between connected systems that deliver the designated functionality and truly smart systems. A smart system should possess the intelligence of being able to automatically learn, adapt, and respond back relevantly to the needs of an end-user. Also, many connected systems fail to climb the ladder of popularity and are merely rejected by the users because they seem irrelevant to the user, i.e. they fail to strike a chord with the end-user that would guarantee them a place in their life.

Personalization can be defined as the ability to incorporate the needs, preferences, constraints, and sentiments of the users. It is a powerful instrument that has the potential of shaping the quality of future IoT products and services to keep pace with constantly evolving customer needs [7–9]. Key technologies like ML and sentiment analysis can be used for BDA to provide novel insights about the user's personal habits, daily routines, preferred actions, moods and sentiments, etc. Using these personal insights, IoT systems can establish a personal bond with the users, taking into consideration their little wants and adapting themselves to comfort them. Importantly, only the products or services which have the power to capture a user's interests can influence their choices, the way they spend their money and time. The power of granting pleasurable experience to the users ensures their engagement and makes them come back for more. The organizations gain by increased profits, customer base, and brand value. Thus, personalizing IoT systems presents a win-win situation for both the businesses and the end-users.

This chapter investigates how personalization is assuming an important, irreplaceable role in the development of IoT systems being deployed across multiple domains and the lives of associated varied strata of users such as the business owners, marketing professionals, business analysts, data analysts, designers, and the end-user. The work takes stock of the current scenario and establishes through use cases and examples that personalization is already being exploited for huge benefits but the concept itself is being given a rather ad-hoc treatment. This is evident as personalization finds no mention in the IoT architecture itself. It is left to dangle on as a last-minute job in most of the IoT systems developed so far.

Due to its ability to deliver numerous benefits for all associated communities and potential to shape the quality of future IoT systems, this work wishes to argue for recognition of personalization as a basic building block of IoT systems and grant it its due place in the IoT architecture. The idea is not to demand a static architecture that features personalization but to think of personalization as an integral part while conceiving and designing the IoT systems. To show the feasibility of realizing such a vision, a user-centric IoT architecture is proposed in the following text.

The chapter is structured as follows: Section 6.2 introduces the reader to the salient concepts of Big Data, IoT, and BDA that have come to dominate the current decade. It also probes how ML, KDD, and Data Mining have emerged as key technologies to analyze and empower the Big Data generated from the IoT systems. Section 6.3 sets the stage for personalizing the IoT systems by elaborating its numerous benefits with the help of use-case scenarios from multiple domains. Section 6.4 reviews some existing works on personalization from varied domains. Section 6.5 lays down the template of a generic, user-sensitized IoT architecture that tweaks the data layer of the conventional IoT architecture to store contextual, behavioral, and sentimental facets of the user. It also demonstrates the use of ML techniques and sentiment analysis to build a characterization engine and a sentiment analyzer that work together to deliver actionable knowledge about user insights and preferences. Concerns regarding privacy and presenting users only with the limited knowledge that is in line with their interests are documented and directions for future research are laid down in Section 6.6. Section 6.7 concludes the chapter.

6.2 Enabling Technologies for BDA of IoT Systems

The maturity of database technology and the development of enormous processing power have led to the collection of massive amounts of data. The gigantic mountains of data are being continually fed by endless data sources, which are present almost everywhere around us, such as transactions at the online retail mega marts, e-wallet transactions, point of sales transactions, stock ticks, meteorological sensors, biomedical data sequences, consumer behavior data, employee profile data, web clickstream data, textual data and media exchanges on social media platforms, etc.

This ubiquitous data hides gleaming gems of knowledge in its bowels. It is itself an asset that can reveal insights for an organization to hone its capabilities and gain a competitive advantage. Using patterns mined from this ubiquitous data, technologies such as *Knowledge Discovery in Databases (KDD)*, *Data Mining*, and *ML* have demonstrated the power to completely transform the world scenario by revealing novel item sales patterns and customer tendencies in businesses; providing valuable inputs for stock markets; design of highly successful electoral campaigns; characterization of moods and preferences of patients for psychological treatments; spatiotemporal analysis of global pandemics and clustering the genetic databases for prediction of lifestyle diseases, evolutionary research, and drug design, to name a few [10].

The term *Big Data* has evolved mainly due to the gigantic *volume* of the collected data. Some other salient characteristics underline the disposition of this kind of data. An important attribute that defines big data is the *variety* of sources from where the data are collected. Accordingly, the collected data are generally available in no unique standard format. It can be either completely structured such as the data stored in company databases or ERP modules; semi-structured such as the XML or comma-separated files or completely unstructured such as the emails and web pages, etc. The *velocity*, i.e. the rate of collection of data, is another attribute that characterizes the big data. The data may be collected as batches in the static databases, may usher in real time, or be streamed in 24×7. Together the volume, variety, and velocity are known to constitute the three Vs of big data.

BDA is a branch for the computational study of big data to reveal patterns, trends, and associations. BDA has an impressive list of applications to its credit such as sales, marketing, customer profiling, business intelligence, understanding of business change, planning and forecasting, fraud detection, market sentiment analysis, and healthcare, to name a few.

Almost two decades ago, the concept of the *IoT* was formally introduced by Kevin Ashton. IoT is an inter-network of interoperable devices that are connected using a common network. These devices can collect information and exchange data with each other without human intervention. This seamless connection of devices has been made possible by the affordable and reliable sensors, the cloud computing platforms to which these sensors are connected via the underlying network protocols, and the ML and artificial intelligence technologies that offer insights from the large amounts of gathered data. A host of devices such as thermostats, electronic appliances, smart lights, smart speakers, alarm clocks, vending machines, transport sensors, smart TVs, toys and wearables, etc., have been used to implement numerous successful IoT applications for personal, commercial, industrial, and government use. Smart homes, traffic monitoring, weather monitoring, proactive healthcare systems, automated waste management, and smart cities are examples of some successful IoT applications.

These applications generate huge amounts of data from all walks of life such as the environmental data, geographical data, astronomical data, personal data, logistics data, etc. The mere volume of these data can surpass the infrastructural capabilities of the major IT organizations. It is therefore pertinent to complement the IoT with BDA to manage these data for their optimal and tactical usage and also to mine gleaming nuggets of knowledge. Figure 6.1 sums up the above discussion by depicting key technologies that can enable deriving user insights and personalization of IoT systems.

One of the major projected aims of the IoT applications was to make the human–computer interaction more personalized and humane. Yet, most of the developments of the IoT applications have been primarily technology-driven and the design of applications has been oriented for physical comfort and ease of

Figure 6.1 Interplay of key technologies for deriving user insights and personalization of IoT systems.

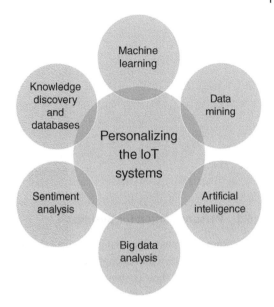

doing things. However, the importance of user preferences, constraints, feelings, and subjective experiences should not be undermined. Tapping into the preferences, moods, and sentiments of the consumer are important inputs that can be used to humanize the overall design of IoT applications and take the human–computer interaction to the next level.

6.3 Personalizing the IoT

Personalization is a powerful tool for organizations to establish a personal bond with the end-user. To meet the end-user's needs and increase user satisfaction, an organization may adapt its IoT services or products according to the insights into the user's personal habits, tastes, mood, and preferences.

Any IoT application is based on four building blocks. Figure 6.2 shows these components viz. the interconnected sensors and devices, Internet and connectivity to the cloud, data analytics, and user interface.

The interaction with users is thus one of the major building blocks of this quadrangle that facilitates IoT applications. The idea of laying enough emphasis on incorporating users' preferences into the design for IoT applications is therefore not overemphasized. Applying a personal touch to the IoT application can yield numerous benefits. The following subsections substantiate this claim by discussing the advantages of personalizing IoT products and services across multiple domains.

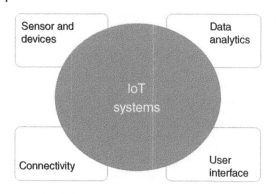

Figure 6.2 The four building blocks of an IoT system.

6.3.1 Personalization for Business

It is common knowledge that it is the customers who make or break a brand. Businesses have been increasingly devoting resources to study their customers. They are making use of their customers' insights to offer them customized services [11]. Resultantly, the customers are not just satisfied, but happy and engaged with the product or service that they are using. This helps businesses retain their customer bases, increase the value of their brands, and stay ahead of their competitors.

For instance, in addition to maintaining a shopping list for the owner, the virtual marts are maintaining a list of items liked or purchased by them in the past. Setting up a smart grocery list already populated with the frequently purchased items makes the shopping experience faster and easier. Reminder for things that a customer may have missed putting in a basket and automatically recommending the items with better offers also enhance the overall user experience.

6.3.2 Personalization for Marketing

Based on the usage patterns and individualized customer experiences, the companies are segmenting their consumers and targeting them accordingly. Personalized advertising campaigns and promotional marketing strategies are designed to target specific audiences that increase revenue and customer base. The rationale is that the set of the population which has responded positively to similar products in the past is more likely to fall for an interesting promotion of a similar product, rather than targeting a new customer who may not even be inclined toward the category of products being offered.

Also, companies are increasingly being benefited by sharing personal data about customer interests with other organizations and lending them space to place personalized ads.

6.3.3 Personalization for Product Improvement and Service Optimization

User's personal insights have the power of broadening the scope of commercial and industrial IoT products and services. Personalization of these services can considerably improve their functionality, thereby transforming how various underlying organizations operate.

For instance, fitness-oriented wearables are quickly being converted into health-oriented devices aimed at helping patients overcome various ailments. They also allow the patient to view how they are progressing. These connections can substantially improve the quality of care to the patient while reducing the cost of healthcare delivery for the provider.

Fitness apps are being personalized for improved healthcare. Smart wearables are tracking the health of patients. They can log the vital statistics such as blood pressure, heart rate, and raise alarm and call on default emergency numbers, in case of need. These new health monitors enable doctors to closely track their patients' vital signs. The detailed logs can help the doctor diagnose and corroborate the patients' activity patterns with their health problems. The patients are also benefited as they can view their routine and progress using many visualization tools. These visualization tools can be developed to graphically display the logged data. Depending upon these data, diet and fitness plans may be customized to suit particular users, helping them manage their diseases. The users may be ranked on a global board, competitively judged against other users, and offered incentives for making progress. A literature review on healthcare services from three aspects in an IoT platform: devices, persons, and timeline can be found in Qi et al. [12].

Ride-hailing companies such as Ola and Uber are using data generated from various sensors attached to their vehicles to monitor the status and location of their cab, the status of a ride (canceled/ongoing/completed), the average driving speed of the driver, and use these statistics to identify issues before they become real problems. Personalizing this analysis to match the driver, cab, passenger, and ride area can help to further improve the performance and efficiency. For instance, assurance of a consistent environment by ensuring the same driver, vehicle, and settings in the cab for regular, long-distance customers can make them feel more welcome and keep them glued to the company's services. Similarly, putting drivers en route on their accustomed route can improve their driving efficiency and simplify their daily chores of refueling and locating service stations for cleaning and maintenance tasks, etc.

Another example is improving the search results returned by a web service by using personalization for automatic filtering of search results. When a user searches the web for some information, he is usually overwhelmed by a large number of search results. It takes a lot of time, effort, and patience to manually go through the irrelevant results to finally be able to zero in on what one wants to find. Based upon parameters such as the location of the user, his previous searches, the kinds of sites that he visits, or the comments that he posts on the web, search engines filter and order the results differently for different users to make the results more relevant for the user. For web publishers, relevant content and ads mean more clicks and ultimately more money.

6.3.4 Personalization for Automated Recommendations

Automated recommendations have been in place for almost two decades now. This involves automatically generating suggestions for similar products or services based on a blend of products or services used by the customer in the past, his preferences as ascertained from his search keywords or his reviews or ratings, past transaction records, surfing details on the product site, etc. For instance, Amazon uses a ML-based service called *Personalize* that helps its developers generate individualized product and content recommendations for its customers. Similarly, numerous Music apps are offering personalized playlists based upon the genre of music, artist, and language liked by the user. Interestingly, recently some studies have focused on correlating the user insights such as their mood, geographical location, and time of the day, etc., to further personalize the recommendation of music titles [13].

However, it is pertinent to note here that a random recommendation for products or services will just lead to irritated users. The recommendations need to be suited to the customer being targeted, strategically timed, and maybe analyzed later for results and fine-tuning future recommendations.

6.3.5 Personalization for Improved User Experience

The end-user is undoubtedly the foundation of any organization's success. Customization of an application or device in accordance with the end user's preferences helps to establish a personal connection, thereby making the users feel valued and important. Resultantly, the underlying service becomes more relevant and enjoyable by the user.

Many web-based video streaming services like Netflix and Amazon Prime personalize the viewing experience differently for different users even under the umbrella of a single login. This not only enables different viewing experiences for each member of the family but also safeguards the privacy and interest of all users. These streaming giants use small nothings like greeting the user with their names and storing the cue of their last played video to spell convenience and increase the quality of the user's experience with the web service. The list of titles viewed in the past by the users is used to recommend flicks of similar genre, language, period, and cast. Cross recommendations are also done. The users are offered recommendations for the content that has been watched and liked by other users, which is similar to the end-user's type. Recommendations such as these make a user's task of searching the titles quicker and easier.

Similarly, enrolling for new services and setting up new devices can be made an enjoyable process for the user. Multiple services such as a video streaming, music streaming, and news streaming service can be integrated to use the preferential settings of the end-user to offer a common look and feel, similar settings for data usage, screen resolution, subtitle font, language settings, payment renewal options, etc. Seamlessly integrating new devices based on the preferences set on older devices and automatically syncing them would save a lot of user effort and time.

6.4 Related Work

In this section, some existing albeit independent works related to the personalization of the IoT systems are reviewed. It is noteworthy that no work on the introduction of personalization in the IoT architecture was found.

Yang et al. [11] highlight the need for personalizing IoT products and underline that this requirement has not been well addressed. They focus on enabling interaction and information sharing for better customization of products. To this end, they propose a model that makes collective use of social networks, IoT and cloud services. A case study for an RFID-enabled system for customized and personalized products is also presented in the paper.

Vallee et al. [14] discuss the challenges of personalization in the health and well-being-oriented IoT services. The problems of sleep quality improvement and recommender systems that aim to help improve it are taken up as a case study. The paper also lists some potential challenges for the task of personalization. The collection of explicit and implicit information about the users is pointed out as one of the hurdles. Lack of interoperability among IoT devices, owing to the use of different technologies, communication protocols, heterogeneous data types and formats, etc., is enumerated as another hurdle for personalization.

A personalized approach for decision-making in remote health monitoring IoT applications has been proposed and validated in [15]. The paper recognizes the problem of missing values as a deterrent to the quality of IoT data, especially when the vital signs are being missed. An approach to impute missing data in real-time health monitoring systems is proposed, taking into account a wide range of parameters and estimate biases. The approach is validated by a human trial in which 20 pregnant women were remotely monitored for 7 months for estimation of the maternal health status by utilizing maternal heart rate.

The problem of drawing out Personalized News Recommendations is investigated in [16]. Twitter user profile vectors are constructed and matched to a set of candidate news items to recommend a set of news to the user. Changes in the user profiles over time are also discovered to form different personalized recommendations for different periods, for example, over a week vs. a weekend.

Russell et al. [17] propose a methodology for personalizing human identification by integrating unobtrusive sensors and microcontrollers. A smart IoT environment is set up using Bluetooth connections. These intuitively connected to the sensors that were used in the wireless sensor network.

Personalization of the Virtual world environments to improve user experiences is investigated by Eno et al. [18]. User profiles are built based on the information contained in their virtual user models called *Avatars*. These user models are extended to incorporate content preferences and offer recommendations. The observed locations are used for personalization by inferring the user interests and identifying locations that a user might be likely to visit.

Kaminskas et al. [13] devise a personalized music recommender system that suits a place of interest (POI). Emotional tags are assigned to both music tracks and POIs and then analyzed. Similarity metrics are used to match a music track to a POI. Jaccard similarity measure is shown to produce the matching preferred by most of the users.

Freyne et al. [19] use an understanding of user reasoning patterns and user needs, uncovered from their ratings to personalize the recommendation of recipes for a user. A standard collaborative filtering algorithm is used to predict a target recipe for users and a content-based algorithm provides ratings based on each ingredient used in the recipe. M5P, a decision tree algorithm, is used to predict scores based on the recipe content and metadata.

6.5 User Sensitized IoT Architecture

The use cases discussed in Section 6.3 establish that personalization derives the cut-throat competition to build a personal connection with the end-user, comforting him and giving him a pleasurable experience, thus ensuring his engagement and making him come back for more. The organizations gain by being able to charm and hence retain the existing customers, design personalized campaigns to target a selected consumer base thereby increasing profits, customer base, and brand value. Thus, personalization is not just a necessity for the IoT systems today but is also a powerful instrument that has the potential of shaping the quality of future IoT products and services to keep pace with constantly evolving customer needs.

The above adjudication prompts us to reflect upon an important revelation that many connected services that facilitate connected experiences fall short merely because they do not provide a real connection to the user and hence stop short of scoring relevant value for the end-user. This brings us to the verge of asking an important question, that if the things that connect us are not inherently personalized, then are they really good enough for the user?

As evident from the discussion in the previous section, most of the IoT services are reaping huge benefits by employing personalization. But still, personalization finds no mention in the IoT architecture. This work wishes to canvass for recognition of personalization as a basic building block of IoT systems. To show the feasibility of realizing such a vision, a user-centric IoT architecture is proposed in the following text.

The idea behind a user-centric IoT architecture is not to propose a static architecture but to propose a generic template with no expectation of a hard-coded definition of the constituent layers and their functionality. The changing needs of users and the environment, where IoT products and services are deployed, dictate that the definition and the layers of the IoT architecture evolve constantly to support functionality, scalability, and availability. Also, the IoT architecture underlying a particular service must be sustainable and maintainable. Without addressing these concerns, the resultant architecture would be a failure. Therefore, the user-centric IoT architecture largely employs the layers of any basic conventional IoT architecture, tweaks the composition of these layers wherever required, and proposes the insertion of a new layer – the personalization layer that dovetails into the existing layers.

The user-centric IoT architecture (as shown in Figure 6.3) is composed of the following four layers: the perception layer, the personalization layer, the tweaked data layer, and the platform layer. The perception layer, also known as the IoT device layer or the IoT layer, is the client-side layer. This layer's main responsibility is to collect useful information and data from the environment. This comprises heterogeneous devices, sensors, and real-world objects that are embedded in IoT products. It thus supports the products and services being used by the end-user. The platform layer is the bottom-most layer. It is fundamental for supporting the Internet and other infrastructural requirements. The data layer resides on the server-side and houses data storage, management, and analytical tools for managing the big data captured and streamed in by the perception layer. The personalization layer sits between the perception layer and the data layer. The perception layer and the platform layers perform their jobs as in any convention IoT architecture. The details of tweaking the data layer and personalization layer are detailed below.

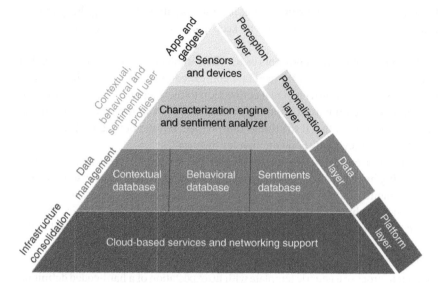

Figure 6.3 Layers of user-centric IoT architecture.

6.6 The Tweaked Data Layer

The proposed user-centric IoT architecture segments the data layer used by the conventional IoT architecture (Figure 6.4) into a context-aware database, behavioral database, and a sentiment database. The context-aware database stores the external details of events associated with the IoT products and services being used by the end-user. These include the details of time, location, and device, and the product or service being used, etc. The objective of the context-aware database is to be able to resurrect the context in which the user is deploying a particular IoT device or service.

The behavioral database on the other hand logs in the details required to study how the users behaves or wishes to conduct themself in a particular situation or respond to an external stimulus. This is possible by logging in details about what activities or actions are being performed by the user like page views, clicking on hyperlinks, email sign-ups, registration, etc. The motivation to store these behavioral details is to be able to understand why an end-user performs the logged actions. It may also be meaningful to store details that indicate how some activity has been performed. For instance, the sequence of hyperlinks clicked for product views before making a final selection. These can reveal important insights about the user's decision-making process, which in turn may help in customizing the IoT systems.

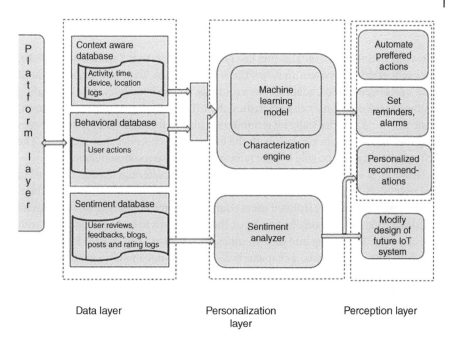

Data layer   Personalization   Perception layer
layer

Figure 6.4 The user-centric IoT architecture.

The third segment of the data management layer is the sentiment database. The objective of this segment is to be able to capture the emotional makeup of an end-user when he deploys a particular product or a service. Tapping sentiment data and being able to relate it with user actions may be a key leveraging aspect both for the end-user and the IoT system. Every organization wishes to understand how an end-user feels about their product. This understanding of user emotions and sentiments holds a key to determine how the IoT system being deployed captures the user's interest. As noted earlier, the products or services which have the power to capture users' interests can influence their choices, the way they spends their money and time. This knowledge can also guide the design of future IoT systems. The sentiment database stores the explicit requests made by the users or the preferences set by them for using the IoT service at hand. Their feedback and reviews about various products and services are also stored in this segment of the database.

Thus, context-aware, behavioral, and sentimental segments of the data layer together enable the corroboration of data in the physical, behavioral, and sentimental contexts in which the IoT product or service is being used.

6.7 The Personalization Layer

The objective of the personalization layer is to transform an IoT system into a smart, user-sensitized IoT system that has the intelligence to learn and reason about a user's preferred actions, adapt to user's routines, and respond relevantly. Depending on the knowledge of user characteristics, the IoT system may automate some mundane tasks and automatically set reminders or alarms for some other routine chores that the end-users perform daily with the IoT product or service they are using. The IoT system may also be able to offer relevant alternative recommendations.

However, it is pertinent to note here that these little offerings need to be suited to the user being targeted because random customizations and ill-suited recommendations may just lead to irritated users. Hence, personal connect is what makes all the difference. The personalization layer, therefore, aims to build a connection with the end-users by tapping into their routines, mannerisms, and moods. It consists of the following components: a characterization engine and a sentiment analyzer.

6.7.1 The Characterization Engine

The characterization engine aims to personalize the IoT system at hand by building a user profile by observing user routines, actions, mannerisms, and behavioral characteristics. It helps the IoT system to learn and analyze the user's preferred actions and adapt to them.

The characterization engine accepts data about the events, user activities, and actions along with the metadata such as their location, timelines, devices used, sites accessed, etc., from the context-aware database and the behavioral databases. Using these data, it builds a ML model such as a rule-based classifier that helps in characterizing a user. This model is then used to predict the user's preferred actions, i.e. the actions the end-user would have liked to perform in a given set of circumstances.

Once the preferred actions are known, the IoT system can be integrated to automatically perform some of these actions such as switching lights and air conditioning on and off at a particular time of day. Alarms may be set or alerts may be sounded for the other preferred tasks such as running, cooking, studying, sleeping, reordering groceries, etc. Relevant recommendations may also be made to the users depending upon their predicted actions. For instance, playlists of songs available on the Internet for working out, new recipes for cooking, and relevant items while ordering groceries may be popped up just in time for the user to perform routine chores.

The ML model for the characterization engine may be chosen depending upon the suitability of the underlying ML algorithm to the IoT application and data at hand. In the proposed user-centric IoT architecture, we make use of a rule-based classifier. A rule-based classifier is a classification algorithm that extracts knowledge from data in the form of If–Then rules. The rule-based systems have been widely used as an effective way to store knowledge, draw logical inferences, and support decision-making in expert systems and decision support systems. Rule-based classification systems are competitive with other classification algorithms and in many cases are even better than them.

In the personalization layer, the rule-based classifier, which makes up the core of the characterization engine, performs the following tasks:

- **Rule induction:** The process of learning rules directly from the data is called rule induction or rule learning. Since classification falls under the umbrella of supervised learning, a set of training examples is required for rule induction. The training data consists of instances already assigned predefined classes. For the IoT application, this training data can be readily derived from the log of the user's daily tasks and the class labels may be taken as the routine actions performed by the current user. A set of If–Then rules is extracted from the data using a data mining classifier. The IF part of the rule is called rule antecedent or precondition and consists of one or more conditions, logically ANDed together. The THEN part is called rule consequent and leads to the prediction of a class label. Any multi-class, multi-label classification algorithm may be used as a classifier, based on any variation of sequential algorithms or decision tree classifiers that can handle the non-exclusive rules.
- **Rule ordering and ranking:** The rules may be ordered or ranked according to their usefulness for accurate prediction. The redundant or unnecessary rules may be removed thereby increasing the efficiency of the classifier.
- **Class prediction:** The set of discovered rules represents a rule knowledge base that can help characterize the end-users and chalk out their profile. These rules can be employed to assign predefined classes for future instances. Given a particular context or a series of actions performed by a user, this would imply that the IoT system is automatically able to predict what action the user is likely to take in the given circumstances. The IoT system may also be able to correlate the user insights such as their geographical location and time of the day, etc., to offer personalized recommendations and or alternatives to widen his horizon.

Every user request and preference setting is taken as a direct indicator of the user's comfort level and requirement and thus directly added or updated as a rule to his emotional profile. The rule-based classifier can also be easily enhanced and complemented by adding new rules from domain experts based on their domain knowledge.

6.7.2 The Sentiment Analyzer

The second component of the personalization layer, the sentiment analyzer, augments emotional facets extracted from the sentiment database to the user profile constructed by the characterization engine.

Blogging sites and social networking sites such as Facebook, Twitter, Google Plus, etc., are the most popular platforms these days where people express their emotions, opinions about products, and share views about their daily lives. These online communities serve as tools for consumers to get information about new gadgets/services and also get influenced by others. Typically, consumers depend upon the feedback and reviews of other users or their discussions on the online portals before making their make-buy decisions. Apart from advertising, this offers opportunities to organizations to connect with their customers, get a feel of their needs, and benchmarks.

Sentiment Analysis, also known as opinion mining, is a systematic, computer-based analysis of written text or speech excerpts, used for extracting the opinion/attitude of the author or speaker. The term "*sentiment analysis*" itself has been derived from the art of identifying the underlying feelings from the subjective experiences, which have been articulated in text. It is a branch of study that has been associated with many tasks such as sentiment extraction, sentiment classification, subjectivity classification, summarization of opinions or opinion spam detection, among others. It aims to analyze people's sentiments, attitudes, opinions, emotions, etc., toward elements such as products, individuals, topics, organizations, and services. Sentiment analysis has been successfully deployed in BDA for gaining meaningful insights from structured and unstructured information sources such as social media platforms, blogs and text documents, etc.

Among other things, the sentiment database of the user-centric IoT architecture contains users' opinions about the products and services that they have used, viewed, searched, or browsed in the form of their feedback and reviews on the blogging sites or social networking sites, etc. Analyzing these opinions can reveal important insights about the attitude of the end-users, their likes, dislikes, and tendencies. The sentiment analyzer uses techniques from the branch of sentiment analysis to detect the underlying sentiment expressed by the users, from their opinions. A polarity such as positive, negative, or neutral is assigned to the opinion that helps the IoT system to understand how a user feels about a particular product or service. It has already been noted that it is of great essence to any organization to be able to identify how the end-user feels about its product or a related product. These sentiments can be used to modify the design of a product/service for greater user satisfaction, gaining a competitive edge over other products and services and therefore bigger brand names and greater revenues.

Two techniques have been broadly identified to perform sentiment analysis. ML, the first technique, involves building models from a labeled training dataset and evaluates them on a test set to gauge the accuracy of the constructed classifier. Naïve Bayes (NB), N-grams model, support vector machine (SVM), and Part-of-Speech tagging are some popular ML algorithms employed for sentiment analysis.

The second technique viz. *Lexicon analysis* is an unsupervised learning technique that uses lexicon, the vocabulary of a language, or a subject by employing a sentiment dictionary. The lexicon is viewed as a bag of positive and negative opinion words. A popular sentiment dictionary is *SentiWordNet*, which is derived from the WordNet database where each term is associated with numerical scores indicating positive and negative sentiment information. A scoring function is used for scoring sentences as positive, negative, or neutral according to the words of positive or negative polarity used in the sentence, as indicated by the sentiment dictionary. Unlike the ML approach, this approach does not require a large corpus of data for training but the quality of sentiment detection depends on the quality of the lexical resources. Lexicon-based approaches also face challenges in detecting slang, sarcasm, and negation. Since the focus of lexicon-based techniques is limited to scoring a sentence based only on the polarity of words, they also cannot detect the sentiment of the context in which the sentence has been said.

The kind of technique used for sentiment analysis may also be dependent on the data characteristics. For example, NB is the preferred choice of ML techniques when the feature set is small, but SVM outperforms NB when the feature size is large. Sometimes even a combination of both the aforementioned techniques may be used for sentiment analysis. For instance, for microblogging websites, a combination of n-grams and lexicon features has been shown to give a better performance than the POS tagging.

The above discussion brings us to the conclusion that the sentiment analyzer of the personalization layer may be based on the technique depending upon the data at hand for the IoT product or service being rendered.

6.8 Concerns and Future Directions

Though the benefits of personalization are numerous and widespread, it also gives rise to some serious concerns such as privacy violations and the filter bubble.

The organizations need information about user interests to personalize services and gadgets. They typically collect information about the social, demographic, geographic, behavioral, and cultural aspects of the user apart from their web address, browser, search history, favorites, etc. Merely from the knowledge of the type of browser, browser settings, and the device being used to access the web, the organizations can track the users' private information like their age, gender, income, and demographics.

The problem is that the users are not generally made aware of what information is being collected, how it is collected, and how it is being used, let alone be asked for their consent before collection. This raises serious concerns about user privacy [20]. The users' data and search histories, etc., are a valuable asset that is being traded off with the other organizations. The private information about the users has also been used to gain a competitive edge in the cut-throat competition. New ventures such as online shopping, streaming content, Maps, etc., have been launched and steered toward profit with the help of private data about user's preferences, without the faintest clue to the innocent users.

Another major disadvantage of personalization has been documented as the filter bubble [21]. The term is used to indicate that personalization, in the long run, leads to the confinement of users only to a bubble of their familiar, well-liked events, items, news feeds, music and movie recommendations, etc. This is because the web-based services only show them the results that are in line with their interests and browsing history. They are never presented with the information that challenges their views and belief systems. Moreover, they do not even know what has been hidden from them and why.

The future belongs to the development of personalized systems that are transparent to the end-users, educate them, and take their consent before designing algorithms that restrict them or compromise their horizons. It is also pertinent to factor in solutions for privacy preservation while working on systems that use private information about individuals. This also promises to be a great future direction for research while working with IoT, BDA, and Personalization.

6.9 Conclusions

The chapter begins by introducing the reader to the most sought-after and ubiquitous terms of the current decade: Big Data, IoT, and BDA. It probes the emergence and interplay of the key technologies such as ML, KDD, Data Mining, and Artificial Intelligence to analyze and empower the Big Data generated from the IoT systems. The motivation for personalizing the IoT systems and the numerous benefits that follow are elaborated with the help of use-case scenarios from multiple domains. Some works related to personalization are reviewed. A generic, user-sensitized IoT architecture is proposed that segments the data layer of the conventional IoT architecture to store contextual, behavioral, and sentimental facets of the user. It also incorporates ML techniques and sentiment analysis to derive knowledge about user insights and preferences. The architecture puts the user-specific mined knowledge to action by using it to automate some user-preferred tasks, remind about other routine chores, and analyze user actions and sentiments for proposing long-term changes to the IoT products and services. Concerns regarding privacy and presenting users only with the limited knowledge that is in line with their interests are also discussed.

References

1 Jayavardhana, G., Rajkumar, B., Slaven, M., and Marimuthu, P. (2013). Internet of Things (IoT): a vision, architectural elements, and future directions. *Future Generation Computer Systems* 29 (7): 1645–1660. ISSN 0167-739X, doi:https://doi.org/10.1016/j.future.2013.01.010.

2 Lee, I. and Lee, K. (2015). The Internet of Things (IoT): applications, investments, and challenges for enterprises. *Business Horizons* 58 (4): 431–440. ISSN 0007-6813, doi:https://doi.org/10.1016/j.bushor.2015.03.008.

3 Chen, M., Mao, S., and Liu, Y. (2014). Big data: a survey. *Mobile Network and Applications* 19: 171–209. https://doi.org/10.1007/s11036-013-0489-0.

4 Zikopoulos, A., Eaton, C., and IBM (2011). *Understanding Big Data: Analytics for Enterprise Class Hadoop and Streaming Data*, 1e. McGraw-Hill Osborne Media. ISBN: ISBN:978-0-07-179053-6.

5 Gandomi, A. and Haider, M. (2015). Beyond the hype: big data concepts, methods, and analytics. *International Journal of Information Management* 35 (2): 137–144. ISSN 0268-4012, doi:https://doi.org/10.1016/j.ijinfomgt.2014.10.007.

6 Russom P. (2011). Big Data Analytics, The Data Warehousing Institute (TDWI) Best Practices. *Tech. Rep.*

7 Anshari, M., Almunawar, M.N., Lim, S.A., and Al-Mudimigh, A. (2019). Customer relationship management and big data enabled: personalization & customization of services. *Applied Computing and Informatics* 15 (2): 94–101. ISSN 2210-8327, doi:https://doi.org/10.1016/j.aci.2018.05.004.

8 Liu, D.Y., Bartimote-Aufflick, K., and Pardo, A. (2017). Data-driven personalization of student learning support in higher education. *Learning Analytics: Fundaments, Applications, and Trends* 94: 143–169. ISBN: 978-3-319-52976-9.

9 Lopes, C., Cabral, B., and Bernardino, J. (2016). Personalization using big data analytics platforms. In: *Proceedings of the Ninth International C* Conference on Computer Science & Software Engineering (C3S2E'16)*, 131–132. New York;, NY: Association for Computing Machinery https://doi.org/10.1145/2948992.2949000.

10 Akter, S. and Wamba, S.F. (2016). Big data analytics in E-commerce: a systematic review and agenda for future research. *Electron Markets* 26: 173–194. https://doi.org/10.1007/s12525-016-0219-0.

11 Yang, C., Lan, S., Shen, W. et al. (2017). Towards product customization and personalization in IoT-enabled cloud manufacturing. *Cluster Computing* 20: 1717–1730. https://doi.org/10.1007/s10586-017-0767-x.

12 Qi, J., Yang, P., Newcombe, L. et al. (2020). An overview of data fusion techniques for the Internet of Things-enabled physical activity recognition and measure. *Information Fusion* 55: 269–280. ISSN 1566-2535, doi:https://doi.org/10.1016/j.inffus.2019.09.002.

13 Kaminskas, M. and Ricci, F. (2011). Location-adapted music recommendation using tags. *Proceedings of 19th International Conference on User Modeling, Adaption and Personalization* (UMAP'11), Girona (11–15 July 2011), 183–194. Springer-Verlag, Berlin, Heidelberg.

14 Vallee T., Sedki K., Despres S. et al. (2016). *On Personalization in IoT, International Conference on Computational Science and Computational Intelligence (CSCI)*, Las Vegas, NV (15–17 December 2016), 186–191. doi:https://doi.org/10.1109/CSCI.2016.0042

15 Azimi, I., Pahikkala, T., Rahmani, A.M. et al. (2019). Missing data resilient decision-making for healthcare IoT through personalization: a case study on maternal health. *Future Generation Computer Systems* 96: 297–308. ISSN 0167-739X, doi:https://doi.org/10.1016/j.future.2019.02.015.

16 Abel, F., Gao, Q., Houben, G., and Tao, K. (2011). Analyzing user modeling on twitter for personalized news recommendations. *Proceedings of 19th International Conference on User Modeling, Adaption and Personalization* (UMAP'11), Girona (11–15 July 2011), 183–194. Springer-Verlag, Berlin, Heidelberg.

17 Russell, L., Goubran, R., and Kwamena, F. (2015). Personalization using sensors for preliminary human detection in an IoT environment. *International Conference on Distributed Computing in Sensor Systems*, Fortaleza, 236–241. doi:https://doi.org/10.1109/DCOSS.2015.40.

18 Eno, J., Stafford, G., Gauch, S., and Thompson, C.W. (2011). Hybrid user preference models for second life and OpenSimulator virtual worlds. *Proceedings of 19th International Conference on User Modeling, Adaption and Personalization* (UMAP'11), Girona (11–15 July 2011), 183–194. Springer-Verlag, Berlin, Heidelberg.

19 Freyne, J., Berkovsky, S., and Smith, G. (2011). Recipe recommendation: accuracy and reasoning. *Proceedings of 19th International Conference on User Modeling, Adaption and Personalization* (UMAP'11), Girona (11–15 July 2011), 183–194. Springer-Verlag, Berlin, Heidelberg.

20 Habegger, B., Hasan, O., Brunie, L. et al. (2014). Personalization vs. privacy in big data analysis. *International Journal of Big Data*: 25–35.

21 Pariser, E. (2011). *The Filter Bubble: What the Internet Is Hiding from You.* London: Viking/Penguin Press. ISBN: ISBN:978-1-59420-300-8.

7

Smart Cities and the Internet of Things

Hemant Garg[1], Sushil Gupta[2], and Basant Garg[3]

[1]*The PSCADB LTD, Chandigarh, India*
[2]*Department of Bio-Sciences, Lovely Professional University, Punjab, India*
[3]*Ministry of Commerce and Industry, Government of India, Udyog Bhawan, New Delhi, India*

7.1 Introduction

Since the advent of the Internet in the early 1990s, a lot of advances have been made in its backbone infrastructure it being digital technology, as well as its supportive and symbiotic systems. Consequently, the Internet has developed from a simple network of rudimentary computing terminals to a complex interconnected system of devices, sensors, storage devices, artificial intelligence processors, and appliances. The same development has resulted in the Internet becoming the Internet of Things (IoT), which is currently a complex network of digital devices, processors, sensors, storage devices, transport systems, and software that learns like neural networks. The same is able to collect data discriminately, process it, learn from it, store it, and make changes that amount to intelligent responses.

Smart cities are one of the products of the IoT where a city's transport, administration, water and power management, security, and human resources infrastructures are subjected to a network of computing devices supported by sensors, controllers, and storage devices that seek to collect and manage data in order ensure resources are well utilized [1]. The data necessary for such crucial processes are collected from people, vehicles, the weather, natural resources, and other people-intensive devices using sensors such as cameras, motion sensors, valves, gates, chemical sensors, keyboards, touch pads, and other biometric devices.

This chapter investigates the constantly changing relationship between smart cities and the IoT. It begins with an introduction of the development of the IoT

Big Data Analytics for Internet of Things, First Edition. Edited by Tausifa Jan Saleem and Mohammad Ahsan Chishti.

before a general treatise on the integration of internet-based development into modern city's developments. With examples, it will explain how the IoT has polished the development of cities. Finally, it will look at how smart cities are developing and make suggestions on the future development of smart cities.

7.2 Development of Smart Cities and the IoT

The IoT began as a humble soda drink freezer back in 1982 at the Carnegie Mellon University [3]. The machine had been connected to a rudimentary form of internet network to report the temperature of its cold drinks while also offering information related to its inventory. Using sensors, the freezer would alert its owners of the need to replenish stock and any technical issues preventing it from offering cold drinks as intended.

Afterward, in the early 1990s, several computing scientists wrote interesting papers on the issues of internet connection and computing devices that acted as independent entities [3]. In 1991, one such scientist called Mark Weiser authored a paper he referred to as "the computer of the 21ST century." Additionally, more work was done by similar professionals at specialized conferences on the development of computing [1]. Some examples of such conferences include PERCOM and Ubicomp, where scientists suggested the first ideals and ideas that would form the foundations of the internet of all things.

Between the early to mid-1990s, several computing giants made entry into the fast developing world of the internet of all things. Microsoft created a product called "At Work" while Novell, another computing giant of the period came up with NEST. However, the first true advance that computer scientists and software engineers recognize as the first modern step toward architecture that truly represents the IoT was made by Bill Joy. In 1999, he forwarded an idea for a computing architecture incorporating computing devices with specialized sensors communicating with other similar computing devices over the Internet [8]. He called his system a Device 2 Device (D2D) system, and that became the first true demonstration of the internet of all things.

Between the early 2000s and early 2011s, a lot of technological development took place further defining the internet of all things. Radio frequency identification (RFID) and digital watermarking came up in the inventory control and retail market sectors. Similarly, barcodes, near field communication (NFC), and advanced forms of digital watermarking also came up [8]. Combined with a sharp increase in the development of digital technology, computing power, and storage as well as sensor devices, the internet of all things became household and indeed city-wide phenomena.

Smart cities can be traced to former US president Bill Clinton's suggestion in 2005 when he asked CISCO to incorporate its vast technological developments into more sustainable cities. The tech giant took up the challenge and injected more than 25 million US dollars into the task [1]. Consequently, it came up with the Connected Urban Development Program. This creation was applied to cities such as San Francisco, Seoul, and Amsterdam on a pilot project basis. It involved the use of strategically placed sensors to reduce traffic congestion, reduce passenger stalling in underground transport systems, search for and identify points of inefficiency in the cities' administrative systems, as well as weather monitoring in the wake of increased pollution.

However, the agreement between the Bill Clinton Foundation and CISCO ended in 2010 leaving the tech giant with a relatively successful creation. Interestingly, other tech giants such as IBM were increasingly interested in the smart cities. Unlike CISCO whose smart city philosophy was biased in terms of hardware development, IBM relied on its strengths in business software development to create specialized information management and analytics software for existing hardware [3]. Therefore, IBM's creation would manage all information collected by sensors such as cameras, traffic load sensors on roads, gateway sensors in the underground transport system, and terminals in the administrative centers. It would also subject the same information to specialized algorithms for the sake of analytics in order to identify points of inefficiency and offer rationalized solutions.

Between 2010 and 2017, a lot of development has taken place in the creation of smart cities. Smart cities have become more reliant on the internet of all things combining pedestrian-based wearable and mobile digital technology, with both vehicular sensor systems and city architecture; both digital and non-digital. The end result is a system that collects a lot of relevant information on important characteristics of the city such as sewerage and water management, security, human resource management, transport management, power distribution and management, city asset tracking and management, levy and tax collection and disbursement, and a lot more [4].

7.3 The Combination of the IoT with Development of City Architecture to Form Smart Cities

Since the idea of the smart city was conceptualized by former US president Bill Clinton back in 2005, the main parameter propelling its development has been efficiency with resources. Ideally, most cities consist of a slowly developing set of infrastructures that a fast-growing population depends upon. Naturally, the results are traffic jams, water shortages, waste management issues, security loopholes, city administration problems, levy and rate collection issues, human resource management problems, and other similar hurdles. However, companies such as CISCO, IBM, and entrepreneurs interested in the issue have created a system that combines the benefits of the internet of all things with the ideal that the internet system and advanced digital communications could create efficiency in the modern city [8].

7.3.1 Unification of the IoT

One of the first aspects of smart cities that demonstrate the unification of the internet of all things and smart city ideals is traffic control. Traditionally, traffic control has relied on existing urban plans that rely on complex algorithms that neither respond to changes in traffic nor learn about patterns of transport. These traditional systems usually rely on pressure pads on certain streets to sense the traffic load along certain roads, or cameras that simply inform human controllers in control rooms to change traffic lights [5]. The most advanced digitalization used programs to change traffic lights based on pre-calculated traffic patterns. However, the inclusion of swerving traffic cameras at numerous traffic points in and around the city adds the system information input. Additionally, specialized pressure pads could be used to direct certain vehicular traffic away from the city with large trucks and school buses being targets. Combining these with advanced neural network-based systems, the traffic control system becomes a highly adaptable one which not only learns, but adapts. It also benefits from continuous optical and programming developments.

7.3.2 Security of Smart Cities

The second aspect of smart cities that demonstrate the unification of the internet of all things with smart city ideals is security. Traditionally, security has depended on a combination of fixed numbers of cameras with police officers who rely on computing terminals in their cars. However, the current problems of terrorism and increased criminal activity mean that the system could benefit from a complete overhaul [6]. Using an existing and continuously updated database of faces entering and leaving these cities in conjunction with cameras with facial recognition, law enforcement would have an easier time. The system would also be connected to an online neural network and artificial intelligence-based system that relayed all information related to security to wearable technology such as palm-top computers used by police officers. The recent advent of drones also presents an opportunity whereby hard-to-reach regions of the city would undergo continuous surveillance from drones equipped with advanced camera systems incorporating facial recognition and license plate reading capabilities. Combined with accurate GPS-based location systems, the city's security would improve immensely.

7.3.3 Management of Water and Related Amenities

Water distribution and waste water management are important aspects of the modern city dwelling. Many cities have several million people necessitating a permanent source of water. The water has to be analyzed, treated, and distributed using safe and approved methods. One method that the internet of all things provides is use of well monitoring systems that measure the amount of water left in sources such as dams and lakes [2]. Additionally, using specialized computerized sensors, any changes in the chemistry of such important sources of water are detected in order to pre-empt changes in the treatment regime necessary to make it drinkable and usable by human beings. Such systems must be composed of accurate sensors, valve mechanisms, laboratory communication systems, chemical dozers, filtration systems, and even storage, all of which must be able to adapt to changes on both sides of the water distribution system. Similarly, all waste water must be collected, analyzed for heavy metals, industrial waste, and agricultural waste before the appropriate waste water treatment processes are initiated. Using the internet of all things and specialized automatic waste water treatment facilities, the process could be done rather accurately with little human intervention other than monitoring.

7.3.4 Power Distribution and Management

Power distribution and management is one of the most critical infrastructures of any modern city. Whether hydroelectric, nuclear, or geothermal, the process of power generation and distribution must always be monitored remotely. Coupled with the need to control the amount of power being generated according to load and demand, the internet of all things seems like a viable option of reducing any down-time, inefficiency, and detecting technical problems [5]. Additionally, using machine learning and artificial intelligence, the system could learn to self-regulate and even on-board any secondary power system the city may have in place. Such operations are made possible using special sensors on the power grid, neural network-based AI systems, and monitoring stations for the few human elements tasked with monitoring, maintenance, and software development.

7.3.5 Revenue Collection and Administration

City revenue collection and allocation also forms an important layer on the development of the smart city. All modern smart cities are dependent on revenues collected in order to fund all development activities and run their day-to-day operations [2]. Such revenue collection processes may be manual or automated meaning a hybridized computerized system may be useful. Using the internet of all things, the system could learn to adjust revenue collected based on specialized algorithms that detect parameters such as low traffic in parking lots and inflation in the economic system. However, such system requires marginally additional human monitoring based on the economic incentive for fraud and theft. However, modern technologies in encryption and blockchain systems may limit need for human elements in the self-regulating, learning-based, efficient system.

7.3.6 Management of City Assets and Human Resources

City asset and human resource management also comprise efficiency-generation strategies of most modern smart cities. The city's administrative and field-based assets need to be tracked at all times and controlled to prevent misallocation and inefficiency in use. From procurement, special sensors are installed to monitor important aspects of the asset's use such as times of use, location and application, user and service or maintenance records, as well as operational state [1]. Similarly, the city's human resources in the form of employees may be subjected to a system reliant on the internet of all things to increase productivity, reduce redundancy, and identify laxity. Using biometric sensors coupled with computer terminal monitoring software and CCTV, the system will detect areas that may exhibit inefficiency and laxity while increasing productivity. It may even learn to detect certain human resource management aspects such as stress, depression, demotivation and industrial action such as sabotage and go-slows.

7.3.7 Environmental Pollution Management

Pollution and environmental management are important in the development of sustainable smart cities. Traditional city and urban development practices have placed little or no emphasis on environmental protection and pollution detection [5]. However, the internet of all things may assist a modern smart city through specialized remote sensing equipment designed to detect issues such volcanic ash for early warning systems, leaking industrial gases and material in the air or water system. Additionally, strategically placed underground sensors may also be used to detect soil leaching and illegal dumping of military, agricultural, or industrial waste. All the information is collected, analyzed, processed, and fed to the monitoring database or a reporting system that liaises with human elements in civil and government offices [4]. The system may be designed to learn to identify faults introduced into the system through hurricanes and volcanic systems, and inadvertent human activity. Additionally, it may also be used in heavily congested cities such as Beijing to monitor their quality of air and collaborate with that responsible for the control and management of traffic in order to reduce the amount of traffic in the city.

7.4 How Future Smart Cities Can Improve Their Utilization of the Internet of All Things, with Examples

Wearable tech and active lifestyle software prompts to increase human health and reduce costs of treatment which contribute to pollution. Using wearable technology that is coupled to smart devices such as smart phones and smart watches, some smart cities have resorted to advocating for an active lifestyle among their inhabitants [4]. The wearable technology offers information related to heart rate, metabolic rates, and cardiovascular performance to encourage activity. Combined with the need to reduce the number of vehicles in the cities, such tech demonstrates a noble application of the IoT in reducing sedentary lifestyles and pollution. It communicates with the smart phone using NFC via a wrist band or special watch with sensors. Additionally, the information can be sent remotely to a user's laptop computer and analyzed using special software as a method of tracking performance. One example of a smart city using this approach is London.

Driver-less transport systems that incorporate electric propulsion have come on to the scene to increase efficiency while reducing pollution. Automotive technology is one of the fastest developing demonstrations of the uses of the internet of all things. Most modern cars have a myriad of proximity and activity sensors that have propelled the development of autonomous cars. Such cars reduce inefficiency especially when combined with electric propulsion [6]. Los Angeles is one such city where Tesla and Google are both developing autonomous electric cars that could incorporate the internet of all things to streamline smart cities' transportation.

Automatic solar farms with cutting-edge photovoltaic cells tech have been developed to track sunlight while reducing carbon footprint. Combined with urban development programs such as NEST which detects when someone is at absent from home to reduce power consumption, such internet of all things-reliant developments promise to revolutionize power generation and management in future smart cities. Both Chicago and Tokyo have developed specialized systems that depend on the internet of all things and Wi-Fi technology to detect when a home owner is not at home in order to reduce power consumption within online appliances such as the smart television, fridge, and air conditioning systems.

Water resources management systems that sense weather patterns such as heat strokes and flooding to control pool systems, showers, toilet flush dynamics, and dam mechanics to conserve as much water as possible. Smart cities of the future will have to contend with the reducing availability of fresh water, therefore having to manage existing resources more carefully [7]. One approach to achieving this objective is using fresh water monitoring, storage and distribution systems that rely on the internet of all things to ensure better use of these critical resources. California experiences some of the worst heat stroke in Western America, meaning people have to rely on pools and other cooling methods. Unfortunately, the region is also adversely affected by drought during most months of the years [8]. However, systems are under development that uses sensors in the toilet flushing system, shower heads, and pool control system to control the amount of water used up by these outlets. Additionally, the system learns to control water usage based on the weather information available and the household users' needs and behavior.

Pollution control systems incorporate a wide array of sensors and processors even on lampposts and buildings to monitor effects of human and vehicle activity on pollution levels. Smart cities of the future will have to use better systems to detect and manage air and water pollution within and outside their boundaries [7]. One such method which uses the internet of all things in conjunction with specialized sensors and sister infrastructures such as the traffic control system is Chicago's Array of Things. It depends on strategically placed sensors that detect rising levels of air pollution and prompts city dwellers to car-share or take further action by sending the information remotely [2]. It also informs the traffic control system which in turn reduces the number of cars in the city. Dublin also runs a somewhat similar system although it does not have a link to the traffic control system. It is referred to as City Watch and its main reaction is to use special AI cues to inform the city dwellers calling them to action. It has learned what language to use in conjunction with social media habits and how dynamic they are making it truly autonomous yet able to learn and communicate.

7.5 Conclusion

The IoT is a collection of hardware, software, and some human elements that all share information with the aim of developing a semiautonomous or autonomous system. The system collects and shares information with an ability to learn and sometimes make decisions. Smart cities are one of the many applications of the internet of all things where specially designed systems collect information using unique sensors, process it, utilize or share it before storage. The main aim of such applications of the internet of all things is to ensure the cities create increased efficiency with the little resources they have while boosting productivity and even comfort.

Smart cities are one of the products of the IoT where a city's transport, administration, water and power management, security, and human resources infrastructures are subjected to a network of computing devices supported by sensors, controllers, and storage devices that seek to collect and manage data in order ensure resources are well utilized. The data necessary for such crucial processes is collected from people, vehicles, the weather, natural resources, and other people-intensive devices using sensors such as cameras, motion sensors, valves, gates, chemical sensors, keyboards, touch pads, and other biometric devices.

However interesting the combination of modern smart cities and IoT is, some considerations will have to be made. The increasing threat of cyber-terrorism continues to place important infrastructure such as power grids, water supply systems, and transport at risk. Additionally, there is a large group of ethicists who believe that the creation of artificial systems that control such important aspects of human life is questionable. All the same, research and development is ongoing and the benefits are already visible for all to see and experience.

References

1 Dustdar, S., Nastić, S., and Šćekić, O. (2017). *Smart Cities: The Internet of Things, People and Systems*. Cham, Switzerland: Springer.

2 Geng, H. (2016). *Internet of Things and Data Analytics Handbook*. Somerset: Wiley.

3 Georgescu, M. and Popescul, D. (2016). The importance of Internet of Things security for smart cities. *Smart Cities Technologies* 8 (12): 23–33. https://doi.org/10.5772/65206.

4 Karnouskos, S. (2014). The cloud of things empowered smart grid cities. *Internet of Things Based on Smart Objects*: 129–142. https://doi.org/10.1007/978-3-319-00491-4_7.

5 Montori, F., Bedogni, L., and Bononi, L. (2017). A collaborative Internet of Things architecture for smart cities and environmental monitoring. *IEEE Internet of Things Journal*: 1–1. https://doi.org/10.1109/jiot.2017.2720855.

6 Sakamoto, M. and Nakajima, T. (2016). Some experiences with developing intelligent Internet-of-Things. *2016 IEEE International Smart Cities Conference (ISC2)* 5 (2): 121–138. https://doi.org/10.1109/isc2.2016.7580866.

7 Sterling, B. (2014). *The Epic Struggle of the Internet of Things*. New York;: Strelka Press.

8 Szénási, S. (2017). Highway safety of smart cities. *The Internet of Things*: 243–272. https://doi.org/10.1201/9781315156026-15.

8

A Roadmap for Application of IoT-Generated Big Data in Environmental Sustainability

Ankur Kashyap

Bennett University, Greater Noida, Uttar Pradesh, India

8.1 Background and Motivation

Sustainability is caring for the future generation while achieving the present economic goals. The world fraternity is now more concerned about the environment with or without stakeholder's willingness, for example, introduction of Environmental Impact Assessment (EIA) legislation in the western part of the globe [1]. Environmental sustainability may be understood as responsible interaction with the environment to avoid depletion and degradation of natural resources for long-term environmental quality. These days, customers are also more concern about their consumption habits and prefer products that are manufactured by producers who care for the environment. Since customers are taking into account the sustainability aspect in their purchase decisions, big brands are showcasing their environmentally friendly manufacturing practices they follow while manufacturing. However, the question is whether the same concept could become so powerful so as there could be the inclusion of new technologies like IoT and Big Data for achieving environmental sustainability. There are studies which explored the concept with a qualitative approach. Labrinidis and Jagadish [2] conducted a panel discussion concerning sustainability, especially in developing countries with limited resources. It is with the above background that this study would cater to the following research questions.

- What is the role of Big Data in sustainability?
- What is the present status of the application of IoT in the domain of environmental sustainability?

Big Data Analytics for Internet of Things, First Edition. Edited by Tausifa Jan Saleem and Mohammad Ahsan Chishti.
© 2021 John Wiley & Sons, Inc. Published 2021 by John Wiley & Sons, Inc.

- How could higher sustainability be achieved using Big Data?
- What are the potential barriers in the system that hinder the application of IoT in the domain of environmental sustainability?

In continuation of IoT, Big Data is generated from different smart devices that are connected with each other. Large amounts of data which cannot be stored locally on computers and are stored in the cloud come under the category of Big Data. Its size is more than 1000s of GBs (Giga Bytes). It is the unprocessed, raw, and complex data which could not be processed using generally available software. Initially, Big Data is characterized by its large volume, variety of sources, and velocity in terms of speed of its generation [3]. Later, researchers also added more "V"s like variability and value which relates Big Data with inconsistency and inaccuracy, respectively. Sources of Big Data can be broadly divided into three categories: directed, automated, and volunteered [4]. Now, we have access to large or overabundant data which are in contrast to the situation a decade ago when data were scarce and costly. In this study, we are concerned with the Big Data generated from IoT.

8.2 Execution of the Study

Catering to the above-mentioned research question, a brief literature review was conducted to identify the applications of IoT in the field of environment protection and is shown in Table 8.1. Further, the study followed a qualitative approach and organized a focus group discussion with five experts from academics and one expert from industry. The output of the Focused Group Discussion (FGD) was used to carry on the study and to structure the roadmap. Additionally, consent was taken to use their views to identify the significant barriers present in the implementation of proposed roadmap in the Indian scenario.

8.2.1 Role of Big Data in Sustainability

IoT could be described as the interconnection through the Internet of computing devices embedded in everyday objects, enabling them to send and receive data. This system having the ability to send data without human intervention makes it a smart choice. This feature enables IoT to be used in a variety of domains with even broader aspects. This study explores the possibility of application of IoT and Big Data for achieving environmental sustainability.

Some prominent sectors and/or processes where IoT and Big Data could find application in the present and near future are:

- Waste management and pollution monitoring.
- Reducing energy usage.
- Optimizing transportation.

Table 8.1 Recent studies on environmental sustainability through technology.

S. No.	Reference	Objective of the study
1.	Li, Wang, Xu, & Zhou 2011 [5]; Lu, Xie, Chen, & Han 2015 [6]	Explore the characteristics, areas, and challenges of application of Big Data of IoT in environmental protection
2.	Xu & Sun 2011 [7]	Study the changes of digital environmental protection to smart environmental protection
3.	Ganchev et al. 2014 [8]	Study provides the classification of the IoT platforms and proposes a top-level generic IoT architecture for smart cities
4.	Xiaojun, Xianpeng, & Peng 2015 [9]	Real-time air pollution monitoring and forecasting system
5.	Junior, Busso, Gobbo, & Carreão 2018 [10]	Use of technology in process safety and environmental protection in Industry 4.0.
6.	Majewicz & Maślankowski 2018 [11]	Big Data analysis of the environmental protection awareness on the Internet
7.	Bosch & Olsson 2016 [12]	Analyze the role of Big Data in a smart system
8.	Bonilla, Silva, Terra da Silva, Franco Gonçalves, & Sacomano 2018 [13]	Challenges of Industry 4.0 and sustainability implications
9.	Wang, Zhang, Zhang, Zhang, & Li 2013 [14]	Study the IoT-based appliance control system for smart homes
10.	Ji & Anwen 2010 [15]	Study the application of IoT in the emergency management system in China
11.	Haitao, Ji, Xiaojuan, & Jiaorao 2013 [16]	Application of IoT in environmental monitoring
12.	Huang & Li 2013 [17]	Creation of new teaching mode of digital environmental protection

- Reducing wastage of resources.
- Wildlife conservation.
- Weather forecasting and climatic study focused on global warming.
- Predicting natural disasters.
- Crisis management in the event of a major disaster.
- To predict the availability of usable water around the world.

8.2.2 Present Status and Future Possibilities of IoT in Environmental Sustainability

There could be different uses of Big Data like in enterprise management, in online social networks, in medical applications, in collective intelligence, for smart grid, in financial companies or supply chain management [18].

One of the crucial aspects of environmental sustainability is the management of waste produced by people living in the city. According to a report of Times of India, India's urban population of 429 million citizens produces a whopping 62 million tonnes of garbage every year. Out of this, 5.6 million tonnes is plastic waste, 0.17 million tonnes is the biomedical waste, 7.90 million tonnes is hazardous waste, and 15 lakh tonnes is e-waste. A staggering figure of forty-three million tonnes of Solid Waste is collected annually, out of which only 11.9 million, that is 22–28%, is treated, while about 31 million tonnes of waste is left untreated and dumped at the landfill sites. There are no scientific and systematic ways of disposal of this municipal waste. A waste management strategy that emphasizes segregation of garbage at source, recycling, and reusing it should be implemented. The concept of a smart city has emerged in conjunction with technological advancements like Big Data and IoT [8] where there is a proper system for garbage disposal.

Untreated waste material is a significant source of land, air, and water pollution within cities. Another thing which is needed to be understood is that the present system of waste disposal is also elevating the pollution level. In the present system of waste disposal, one important component is open dumping areas that are situated at the outskirt of the city and are very harmful to the environment. With the help of IoT system consisting of digital machines and objects, the solid waste management system could be made more effective [19]. Similar practices are being followed in some developed cities of the western countries. For instance, in Japan, the sewage system of the city is modified by replacing the traditional iron-made utility hole covers by a special one to install water-level measuring devices. Hence, the existing system does not require major construction work, keeping implementation costs down. There are some prototypes available for monitoring air pollution designed to be installed on lamp post which integrates with smartphones and operates at a meager cost [20, 21]. These innovations are boon to countries like India having small resources. They encourage users to implement them for achieving environmental sustainability. Previously, the air pollution monitoring is done via computerized tomography technique which generates a two-dimensional map of pollutant concentration. This system is taken over by wireless sensor network, which is a low cost, easy to set up, and provides a real-time pollutant data [22].

Energy savings could also be facilitated by integrating IoT with the existing system. There are various autonomous applications which could be used at offices, business, manufacturing units, and home for achieving higher energy efficiency [23]. There are innovations and researches have done which connect the various components of Indoor Environmental Quality (IEQ) using smart technology [24–26]. IEQ is a holistic concept, including Indoor Air Quality (IAQ), Indoor Lighting Quality (ILQ), and acoustic comfort, besides thermal comfort. But these systems require particular hardware, and unfortunately, it requires high fixed cost.

Humans have invented several modes of transportation for commuting from one place to another. Today in the modern world, most of the transportation systems use the nonrenewable sources of energy. Especially, in India, most of the people commute through road, causing more fuel consumption. Moreover, apart from public transportation, distribution of products is a major activity of any business which causes damage to the environment. The transport of raw materials and finished goods consumes a large quantity of fuel.

Transportation is one of the sectors which uses IoT at a full scale. In transportation, IoT provides improved communication, control, and data distribution. The applications include personal vehicles, commercial vehicles, trains, UAVs, and other equipment. IoT can reap business benefits by making intelligent decisions at real time. Referring to the public transportation companies like Uber and Ola are using the technology to optimize the route for commuting based on the inputs gained by the multiple sources and devices. The business owners in advanced countries have optimized their supply chain management with the help of IoT. Transporters could track the position of their vehicle with the information about the load they are carrying. They could also apply geo-fencing, fleet management, inventory control, etc.

Other sectors where IoT usage results in big change are the weather and climate monitoring system. One of the most extensive applications of IoT concerning the environment is through installing a variety of sensors and actuators that measure levels of light, humidity, temperature, gas, electrical resistivity, acoustics, air pressure, movement, speed, and weather-related phenomenon. But, a high quality of the network is the prerequisite for such systems [27]. Previously, this system was not very accurate and relied on the data provided by the satellites, which have their limitations. Nowadays, modern sensors are installed in the lower atmosphere for collecting more and precise information. Big Data from various sensors is analyzed and results in much more accurate weather forecasting.

Due to the diversified geography, natural disasters are frequent in the Indian part of the world. India's major portion comes into the earthquake zone, whereas some parts are flood or drought affected. Other than these natural disasters, due to the vast coastline of India, frequent cyclones and storms are also a cause of concern. A system for detecting the signs of a natural disaster works through sensors that are embedded in relevant infrastructure to monitor the condition continuously. The data that have been collected are analyzed to detect any sign of a disaster. Like in case of danger of flooding caused by incessant rains, monitoring river water levels with sensors help in predicting floods. Japan has developed a tsunami warning system based on IoT and Big Data. It has robust predictive analytics and algorithms that can increasingly model future scenarios with surprising accuracy. The systems also use inputs from smartphones that reduce the cost of the system. A similar type of system is also developed in California for predicting the cyclones. The utility of IoT could be very beneficial for prevention as well as rescue relief operations that are carried after a natural calamity.

8.3 Proposed Roadmap

This section of the paper deals with the third research question of the study, i.e., how could higher sustainability be achieved using Big Data? To answer this research question, the author have proposed a roadmap which could be followed to achieve higher environmental sustainability using IoT and Big Data. The roadmap is given in Figure 8.1, which depicts four components, i.e., Input, Process, Result, and Impact. The first step is the identification of the specific area of application by the experts. The input component represents the activities that could be followed in the present system to create changes in the process. It is given in the second component of the proposed chart. The last two components represent the outcome of the process followed and impact on the environment. The figure is self-explanatory and different color code is used to separate the components. The "Input" part is represented with green color, the "Process" part is represented by yellow color, the "Result" component is represented by blue color, and the "Impact" part is shown by red color. One of the real life examples is of Precision Farming System (PFS), where the data from sensors regarding the crop yield, moisture levels, and terrain topography locate the position for putting fertilizers, thus saving costs and is more sustainable [28]. Here, the input is PFS system which brings the changes in the traditional method of the farming process resulting in less wastage of labor and material and thus, has a positive impact on the agricultural production. Similarly, the US-based companies take help of Deep Thunder, a product of IBM, in predicting the locations where the public is going to face outages due to weather conditions so that they can take necessary steps beforehand to prevent it or can send their staff to reduce the outage time by fixing the problem right on time [29].

The British Telecom (BT), one of the world's leading communications services companies having consumers in the United Kingdom and more than 170 countries worldwide, collaborated with the Carbon Trust to study the total carbon footprint of its business. It was found that two-thirds of its all emissions were produced not by the company itself but by its 17,000 supplier establishments [30]. In this way, knowing the facts helps in making a concrete plan instead of working on guesses.

The case studies of Kilimo Salama (meaning "safe agriculture" in Swahili) in Tanzania, Zimbabwe, Kenya, Rwanda, and Nigeria showed the use of IoT in

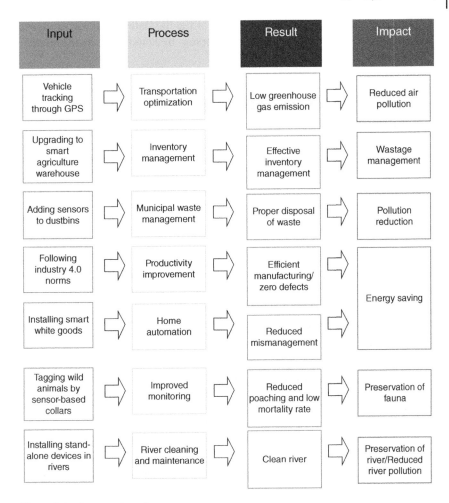

Figure 8.1 Proposed roadmap.

agriculture where the crops of small farmers were monitored by sensors for heavy rainfall or winds and their monetary losses were automatically transferred through mobile payment system M-Pesa [28]. In 2014, Cisco signed partnership agreements on Big Data management for environmental purposes with three local governments in Denmark, the municipalities of Copenhagen, Albertslund, and Frederikssund to provide technical solutions to improve service to citizens and meet the environmental targets set for 2025 [30].

As the prices for sensors, an integral component of IoT applications have declined by about 80–90% over the past decade, the use of IoT is on widespread.

But, the clear objectives and appropriate use of the Big Data are the keywords for gaining utmost from the technology, keeping in mind the environmental sustainability.

8.4 Identification and Prioritizing the Barriers in the Process

This section of the study will be dealing with the identification of the barriers that possibly will be confronted while implementing the proposed roadmap. Three major roadblocks were identified which are relevant for Indian conditions.

8.4.1 Internet Infrastructure

IoT works on the internet platform and requires high speed and reliability. Although we have seen a development in internet infrastructure in India, still there is a lot more to be done. There is a problem of connectivity, and the erratic behavior of the network causes a problem for IoT. In production or business processes, all the activities are needed to be connected via the Internet to gain portability. But, considering the present state of infrastructure, the manufacturer is not willing to invest in IoT.

8.4.2 High Hardware and Software Cost

Another barrier in the implementation of IoT in environmental sustainability domain is the high hardware cost. As the compatible hardware is not manufactured in India and most of the equipments are imported, their cost is very high. This acts as a blockage in the progression of IoT in the Indian situation. Another issued related to hardware is the requirement of storage space which could be local or the cloud environment. The storage also needs specific hardware and software which attracts high cost.

8.4.3 Less Qualified Workforce

For analyzing the Big Data generated from IoT, specific data analyzing capabilities are required. To make use of the IoT technology to its fullest, the data generated by various sensors or separate modules are needed to analyze and interpret the results. The prime requisite for most of the IoT applications is an intelligent analytical mechanism that can carry out tasks like classification, clustering, association rule mining, or time series analysis [31]. Therefore, for this task, either a separate workforce is required or the existing human resource should be well qualified enough to handle the job. This situation could be a potential barrier for integrating IoT with the current system.

8.5 Conclusion and Discussion

The developments in the technology could be used to protect the environment from the damages made by the technology. One such technique is IoT which could be implemented strategically for the betterment of environmental sustainability with the help of various sensors. With new technologies such as micro-electro-mechanical systems (MEMS), it is becoming possible to place such sensors in any object (even in humans). This conceptual roadmap helps in gaining that aim and is suitable for developing countries.

There is a need for the business community as well the government and citizens to be more responsible towards Mother Nature. The production units should be working in a manner that causes the least damage to the environment. It could be said that if the industries follow the norms, they would be less polluting since most of the micro, small, and medium enterprises are not following the right practices as given in government guidelines. It is valid to some extent, but in the near future because of the stringent government control, these manufacturing units with less financial resources are expected to work within the specified limits. The large MNC manufacturing units are needed to set an example and create benchmarks in the industry by using the latest technologies like IoT and Big Data for environmental protection.

References

1 Glasson, J. and Therivel, R. (2013). *Introduction to Environmental Impact Assessment*. Abingdon: Routledge.
2 Labrinidis, A. and Jagadish, H.V. (2012). Challenges and opportunities with Big Data. *Proceedings of the VLDB Endowment* 5: 2032–2033.
3 Laney, D. (2001). 3D data management: Controlling data volume, velocity and variety. *META Group Research Note* 6 (70): 1.
4 Kitchin, R. (2013). Big Data and human geography: opportunities, challenges and risks. *Dialogues in Human Geography* 3: 262–267.
5 Li, S., Wang, H., Xu, T., and Zhou, G. (2011). Application study on the internet of things in environment protection field. In: *Informatics in Control, Automation and Robotics*, Springer, Lecture Notes in Electrical Engineering, vol. 133. Berlin, Heidelberg: Springer.

6 Lu, S.Q., Xie, G., Chen, Z. and Han, X. (2015) The management of the application of Big Data in internet of thing in environmental protection in China. *IEEE First International Conference on Big Data Computing Service and Applications*, Redwood City, CA (March 2015), pp. 218–222. IEEE.

7 Xu, M. and Sun, H. (2011). From digital environmental protection "to" smart environmental protection. *The Administration and Technique of Environmental Monitoring* 4: 112–121.

8 Ganchev, I., Ji Z. and O'Droma, M. (2014) A generic IoT architecture for smart cities, *Irish Signals & Systems Conference 2014 and 2014 China-Ireland International Conference on Information and Communications Technologies (ISSC 2014/CIICT 2014). 25th IET*, 196–199.

9 Xiaojun, C., Xianpeng, L. and Peng, X. (2015) IOT-based air pollution monitoring and forecasting system. *2015 International Conference on Computer and Computational Sciences (ICCCS)* (January 2015), pp. 257–260.

10 Junior, J.A.G., Busso, C.M., Gobbo, S.C.O., and Carreão, H. (2018). Making the links among environmental protection, process safety, and industry 4.0. *Process Safety and Environmental Protection* 117: 372–382.

11 Majewicz, D. and Maślankowski, J. (2018). Big Data analysis of the environmental protection awareness on the Internet: a case study. In: *Interdisciplinary Approaches for Sustainable Development Goals*, 31–41. Springer.

12 Bosch, J. and Olsson, H.H. (2016) Data-driven continuous evolution of smart systems. *Proceedings of the 11th IEEE/ACM International Symposium on Software Engineering for Adaptive and Self-Managing Systems*, pp. 28–34.

13 Bonilla, S., Silva, H., Terra da Silva, M. et al. (2018). Industry 4.0 and sustainability implications: a scenario-based analysis of the impacts and challenges. *Sustainability* 10: 10.

14 Wang, M., Zhang, G., Zhang, C. et al. (2013) An IoT-based appliance control system for smart homes. *Proceedings of the 2013 Fourth International Conference on Intelligent Control and Information Processing (ICICIP)* (June 2013), pp. 744–747.

15 Ji, Z. and Anwen, Q. (2010) The application of internet of things (IoT) in the emergency management system in China. *2010 IEEE International Conference on Technologies for Homeland Security (HST)*, pp. 139–142.

16 Haitao, M., Ji, Z., Xiaojuan, Y., and Jiaorao, X. (2013). Application of IoT in environmental monitoring [J]. *Environmental Science and Management* 01.

17 Huang, F. and Li, X. (2013). Primary research on the structure of digital environmental protection platform. *Experience Science Technology* 79: 231–254.

18 Chen, M., Mao, S., and Liu, Y. (2014). Big Data: a survey. *Mobile Networks and Applications* 19: 171–209.

19 Abba, S. and Beauty, P. (2019). Smart framework for environmental pollution monitoring and control system using IoT-based technology. *Sensors and Transducers* 2: 29.

20 Kularatna, N. and Sudantha, B.H. (2008). An environmental air pollution monitoring system based on the IEEE 1451 standard for low-cost requirements. *IEEE Sensors Journal* 8: 415–422.

21 Muthukumar, S., Mary, W., Jayanthi, S. et al. (2018). IoT based air pollution monitoring and control system. *2018 International Conference on Inventive Research in Computing Applications (CIRCA)*, pp. 1286–1288. IEEE.

22 Sirsikar, S. and Karemore, P. (2015). Review paper on air pollution monitoring system. *International Journal of Advanced Research in Computer and Communication Engineering* 4: 1.

23 Abba, S. and Nyam, I.M. (2018). Design and evaluation of a low-cost and flexible data acquisition system using sensor network for smart homes. *Sensors & Transducers* 227 (11): 73–81.

24 Kelly, S., Suryadevara, N., and Mukhopadhyay, S.C. (2013). Towards the implementation of IoT for environmental condition monitoring in homes. *Sensors Journal, IEEE* 13: 10.

25 Lohani, D. and Acharya, D. (2016) SmartVent: a context aware iot system to measure indoor air quality and ventilation rate, *17th IEEE International Conference on Mobile Data Management (MDM), Porto*, pp. 64–69.

26 Salamone, F., Belussi, L., Danza, L. et al. (2017). Design and development of a wearable wireless system to control indoor air quality and indoor lighting quality. *Sensors* 17: 5.

27 Lee, S.K., Bae, M., and Kim, H. (2017). Future of IoT networks: a survey. *Applied Sciences* 7: 1072.

28 Antony, A.P., Leith, K., Jolley, C. et al. (2020). A review of practice and implementation of the Internet of Things (IoT) for smallholder agriculture. *Sustainability* 12 (9): 3750.

29 Mukred, M.A.A. and Jianguo, Z. (2017). Use of Big Data to improve environmental sustainability in developing countries, Int. *Journal of Business and Management* 12: 11.

30 Cepal, N.U. (2014) Big Data and open data as sustainability tools. A working paper prepared by the Economic Commission for Latin America and the Caribbean. Economic Commission for Latin America and the Caribbean.

31 Saleem, T.J. and Chishti, M.A. Data analytics in the internet of things: a survey. *Scalable Computing: Practice and Experience* 20: 4.

9

Application of High-Performance Computing in Synchrophasor Data Management and Analysis for Power Grids

C.M. Thasnimol and R. Rajathy

Department of Electrical and Electronics Engineering (EEE), Pondicherry Engineering College, Puducherry, India

9.1 Introduction

Demand for electrical energy is increasing day by day. The conventional methods of power generation and distribution are not efficient enough to resolve the thirst for electrical energy. Conservation and preservation of the environmental resources are the main reasons for shifting from centralized generation to distributed generation system. The increased exploitation of distributed generation systems results in a two-way flow of energy. It causes electric congestion, heating, and failure in both supply channel and transmission network, which will lead to blackouts eventually. There is no adequate information and communication system in the grid to examine disruptions in the network. Supervisory control and data acquisition (SCADA) systems are reliable solutions for this in terms of giving a static view of the power system. However, dynamic event monitoring is not possible through this system because of its low reporting rate. SCADA system analyses only two-three sample data per second.

Complete power generation system is mapped by the wide area monitoring system (WAMS). Global positioning system (GPS) enabled the phasor measurement unit (PMU) offer synchronized values of voltage, power, frequency, and fluctuation rate. All the data analyzed in PMU are recorded in Phasor Data Concentrator (PDC). It collects measurements from various PMUs and further consolidates the data with proper time configuration. The final compressed data are transferred to super PDC or the control station.

Big Data Analytics for Internet of Things, First Edition. Edited by Tausifa Jan Saleem and Mohammad Ahsan Chishti.
© 2021 John Wiley & Sons, Inc. Published 2021 by John Wiley & Sons, Inc.

Thirty measurements are recorded in PMU per second. Typically, ninety-four bytes is the standard size of PMU data. A PMU could generate the data scaling to 80GB in a year. A system which carries 1000 PMUs will make the data up to 80TB in one year. Currently, India has installed around 1500 PMUs in various parts of the country. Regulating energy distribution with the help of this enormous PMU data will fetch a positive impact on the reduction of energy loss. Analysis of old data is also essential as it has a more significant role in the post-production fault analysis, feature correlation, and feature forecast.

A collection of colossal PMU data will not draw any conclusion to the causes of errors that occur in the system. This considerable volume of data needs to be analyzed in a real-time environment for analysis like situational awareness, error mapping, and anomaly exposure. Traditional systems are not capable of controlling such substantial data flow. Therefore, novel databases and analysis tools are highly essential and come handy for operators to locate possible faults with the help of PMU data analysis.

A power grid that depends on data-driven solutions helps in quick decision support. However, model-based applications are better in performance because of integrated techniques of big data analysis. The study intends to discuss various problems associated with big data analysis with particular reference to PMU data handling and tries to introduce some of the modern techniques and tools to resolve those pitfalls. The first part of the articles explains WAMS in general and characteristics of big data. The concept of synchrophasor measurements are described in the following section, and the study exhibits some of the significant issues related to PMU data processing. The last part of the article illuminates advanced data processing and storage platform for various PMU-oriented systems.

9.2 Applications of Synchrophasor Data

There are real-time and offline applications of synchrophasor data. Power grid applications which have to be finished within a time frame of two to three minutes come under the real-time section of the synchrophasor data application. Wide area monitoring, voltage stability analysis, oscillation detection, islanding, mode meter, and state estimation are examples of a real-time category of synchrophasor data application. In the real-time application, the data source should be live and near to the system for better performance. Generally, data intended for processing in the real-time applications will be stored in the buffered memory of the databases.

On the other hand, offline data analytics of synchrophasor application is not time-bound. Historical analysis of data, power plant model validation, frequency response analysis, post-event analysis, and baselining fault location are some of the examples of offline synchrophasor data applications. Offline synchrophasor applications use historical data for analytics of the system. Offline applications use well-sorted high-quality historical information. The sorting process involves filtering of time alignment, removal of bad data, and missing data filling. Figure 9.1 shows a few imperative applications related to synchrophasor data and are discussed in detail in the following section.

9.2.1 Voltage Stability Analysis

Power generation and transmission systems produce maximum power and transmit it through the distribution channel because of ever-increasing demand. Therefore, the rate of system failure and transmission faults is always higher than before. Latest error forecast systems and quick fault detection tools should be employed to make the power system more accurate and reliable. Voltage stability index generated from synchrophasor measurement can predict the distribution lines that are operating under unsafe conditions [1]. Optimization techniques could be developed and applied to maintain the voltage stability margin of those defective lines. The stability of voltage can be speculated from Thevenin equivalent impedance that is generated from PMU measurement data [2]. Synchronization errors and cyberattacks may cause the failure of the system due to the wrong

Figure 9.1 Applications of synchrophasor data.

prediction. A secure recursive least square algorithm developed by Zhao, Wang, Chen, and Zhang [3] uses a stable estimator and projection statistics for forecasting voltage volatility. This method of removing gross error in voltage stability projection will improve the performance of the system.

9.2.2 Transient Stability

Integration of multiple power generation and supply systems and support of modern technologies to reduce the power loss and voltage stability control can resolve the power scarcity of the society. A meager fault in any part of the system will harm the smooth supply of the power and affect the stability of the system [4]. Usually, power transfer volume on the interface is monitored to ensure the balance of the system. Checking on the generator phase angle difference obtained from synchrophasor measurements will also tell us the stability of the system. The potential generator buses are defined initially using transient stability simulations [4].

Further, the difference in the generator phase angle is calculated that would depict the power transfer on the transfer interface and the transient stability of the system. A technique to calculate the transient stability margin is proposed by Wu and others [4]. The mode of disturbance technique merging of the transient energy function is used to forecast the transient instability. Mode of disturbance (MOD) is the collection of generators which will fall out of synchronism after a fault. Fault location, transient energy, and details about MODs are available in a system directory. With the help of PMU and MODs data, actual kinetic power will be marked after clearing the fault, and remedial actions are taken.

Transient instability of the system occurs commonly with an unexpected commotion in a composite power network. For example, overloaded power supply line will cause rotor angle instability that will result in finally transient instability of the system. This needs to be rectified through emergency network reconfiguration. Finding proper corrective measures to tackle the disturbance in minimal time is a difficult task, especially in a complex power system. The network restoration method is used [5] to decrease the search space of control action. The network restoration algorithm uses the strategy of the lowest mean square system state error.

A transient stability assessment method based on synchrophasor data and Core Vector Machine (CVM) consists of two phases: offline and online [6]. Feature selection is the first phase that signifies the status of the system. Modification in the CVM using designated features from the synchrophasor data generated from time series simulation will be in the early part. Coordination of real-time PMU data and trained CVM will find out the transient stability of the system. The transient stability of the power system depends on the inertia of its components.

With the high penetration of low-inertia renewable energy sources, the total inertia of the power system became low, which causes oscillation of synchronous generator following a disturbance. Distributed storage method will improve the stabilization process of these oscillations by using real-time measurements from PMU [7].

9.2.3 Out of Step Splitting Protection

Phase angle information will help in executing out of step splitting protection, and it will be without any complex mathematical operations. A power system may have stable and unstable oscillations. The fault in a power system may cause an unstable oscillation (out of step oscillations). The protective relays installed in a system should not fail in a stable oscillation but should report quickly to an unstable oscillation. Short circuit, synchronous, and out of step oscillations are discriminated with the rate of change of positive sequence impedance. Out of step center from PMU measurement is decided after identifying the presence of out of step oscillations [8], and the entire power system is divided into islanding sectors to prevent blackouts.

9.2.4 Multiple Event Detection

Rafferty et al. [9] introduced synchrophasor database-oriented multiple event detection. Events like loss of generation, loss of loads, and islanding detection are identified using frequency measurement. PMU generates a substantial amount of data every second. Extraction of information from PMU data can be done by using data analytics techniques. Principal component analysis (PCA) helps to reduce the dimensionality of PMU data [9]. Moving window approach can assist in measuring the time-varying nature of power system variables and Kernel principal component analysis (KPCA) [10] detects irregularities from the enormous streaming micro PMU data. A novel method that requires minimal data is used to discriminate various events. Hidden Structure Semi-Supervised Machine (HS3M) is also applied [11] for event detection, using unlabeled or labeled data.

9.2.5 State Estimation

Wang and Yaz [6] introduced a power grid synchronization based on real-time state estimation. The extended Kalman filter was used to get an actual state of the system. Dynamic State Estimation (DSE), otherwise known as Forecast Aided State Estimation, uses historical data and real-time measurement data for estimating the system state. System parameters and system state will change in the case of sudden disturbances like line outage and generator outages, etc. Historical data analysis for state estimation will result in erroneous state estimation results under these circumferences. Aminifar et al. [12] propose mixed-integer linear programming-based DSE to tackle such situation. In the event of unexpected trouble in the system, the predicted values will be discarded from the estimation process to increase the quality of estimation. However, during normal conditions, the predictions will be included to improve the accuracy of the state estimation.

9.2.6 Fault Detection

Rule-based data-driven analytics finds out the actual position and nature of the error from synchrophasor measurements [13]. This process does not need any model or topology of the system. Pignati et al. [14] also introduced an algorithm for identifying the location, type, and the fault current from synchrophasor measurements. Pignati et al. [14] discussed a method to find out the location and type of fault using time of arrival (ToA) of the electromechanical wave. Computation of ToA is from phasor angle output of PMU using artificial neural network (ANN). ANN will also help in identifying the type of error. ToA and speed of electromechanical waves from fault point to PMU along with the topology of the network are used to understand the faulty line. Once the fault line is identified by applying the binary search method, the exact location of the fault is determined.

9.2.7 Loss of Main (LOM) Detection

When an implanted generator at the distribution side fails, it should be cut off instantly from the use. Otherwise, the restoration operations will cause severe hazards to safety personnel and out of synchronism reclosure will cause damage to the grid. Ding et al. [15] introduced LOM detection using synchrophasor measurements. Peak ratio analysis of the rate of change of frequency is employed to find out the connectivity of the network. Laverty et al. [16] developed a system to monitor the phase angle of the embedded generator continuously based on a reference cite. The integrated generator's loss of synchronism represents the existence of LOM. Guo et al. [17] explored PCA to detect islanding.

9.2.8 Topology Update Detection

The finding out of topological changes in a system is essential for state estimation and network monitoring. Unidentified topological changes will cause unnecessary stress on the power system that will lead to shutdowns. Any variation in topology will result in the voltage phasors with unique variation, which can be identified with the help of sparsely located PMUs in that area. Ponce and Bindel [18] mentioned about a fingerprint linear state estimator based on the topological detection method. It needs only a minimal number of PMUs.

9.2.9 Oscillation Detection

Early identification of oscillations will avoid instability of the system. The oscillations are of two types, namely sustained or unsustained. Unattended sustained oscillations will damage equipment, causes power quality issues, and even blackouts. The peak of the coherence spectrum of synchrophasor measurement data can help us in finding oscillations. Zhou [19] discussed a cross coherence method using multiple channels of PMU data. Multiple channels of PMU data support to avoid false alarming. This method is suitable for both low and high signal-to-noise ratio systems. The mechanism of oscillation will decide different countermeasures as the reason for each oscillation will differ from one another. A data-driven approach was proposed by Wang and Turitsyn [20] to identify the source and mechanism of sustained oscillations from the synchrophasor data.

9.3 Utility Big Data Issues Related to PMU-Driven Applications

Effective utilization of PMU data is highly necessary for the better performance of the system. Raw PMU data will not be in a position to draw any conclusion about the power system events. Storage of massive PMU data is a primary concern that every power system administration faces. Proper data mining techniques should be implemented to draw a logical conclusion about the problems in the system. Data retrieved through the data mining process are not in an understandable format for the operator. Information security is a potential risk in any system. Mostly, utilities are related to these problems. Many researches are being carried out in this area. The following sections discuss some of the significant issues related to power system big data.

9.3.1 Heterogeneous Measurement Integration

Relational DataBase Management System (RDBMS) is commonly used in WAMS and other power grid monitoring systems for recording and preserving the streaming data. Nevertheless, the RDBMS is not suitable for processing heterogeneous power system data. For managing this high-velocity PMU data, high-capacity servers and hard disks are essential. It will naturally increase the cost of the system. Complex SQL languages and multi-table joint operations are required for integrating heterogeneous power grid measurement data. It will severely affect the read and write performance. Database scaling is another critical hurdle faced by conventional RDBMS for streaming data. Type of PMU data will change according to the site of installation. It can be found out only by looking at the configuration file.

The expansion of the two-dimensional table format of RDBMS is difficult. The option of allocating fields for all types of data in the database will increase redundancy. Guerrero et al. [21] introduced a different data source integration method called Metadata Mining based on the metadata in relational databases.

9.3.2 Variety and Interoperability

The PMU technology is still in the developing phase though it was introduced three decades ago. Many of the utilities are shifting from the SCADA-based monitoring system to PMU-based system because of its advantages. At present, the PMUs in utilities are either installed in different periods or manufactured by different vendors. PMU-enabled IED can also perform various functions in addition to generating synchrophasor data. PMU data types and structure may vary upon vendors. Utilities are forced to integrate communication modules, data concentrators, and visualization tools with PMU data processing. The addition and interoperability of these heterogeneous data are crucial concerns for the power utility. The word interoperability signifies the ability of different utility applications and information technology systems to communicate, share data, and exploit most out of the utility data for the reliable operation of the grid [22]. Hardware-independent interoperability is the essential feature of any computing and data storage platforms when applied to the smart grid.

9.3.3 Volume and Velocity

Synchrophasor data is one of the types of time series data. The storage and processing of time series data are primary concerns of data scientists, especially when the volume of information is significant. Requirements for processing the real-time data also make the analysis complex. The conservative method of centralized control center approach for storage and process of massive streaming PMU data is not efficient because of significant latency issues that are crucial for time-critical smart grid applications. High computing power and ultrafast communication technologies are essential for a centralized system approach. Another problem is the lack of a unified approach to store the time series data in a standard format which can be used by different applications. Systematic storage is the preconditioner for an efficient data analysis application.

9.3.4 Data Quality and Security

Cybersecurity is one of the main issues faced by the smart grid. Most of the smart grid PMU applications have a high dependency on communication and sensing technologies. Therefore, they are prone to vulnerable cyberattacks. Cyberattacks will lead to wrong decisions and may lead to the collapse of the entire power system. Paudel et al. [23] pointed out various types of cyberattacks concerning attacking locations, their impacts, and the existing strategies for alleviating their effects. A detailed survey of cybersecurity of PMU network was conducted by Beasley et al. [24]. Wang et al. [25] developed a density-based spatial clustering method for cyberattack detection and data recovery. An overview of cybersecurity concerning hybrid-state estimation is provided by Basumallik et al. [26]. Mao et al. [27] employed PCA for the real-time detection of attacks and their locations. Paudel et al. [23] demonstrated how the attackers are manipulating the PMU data with the knowledge of the state estimation process. They also show how the attackers are getting successful in bypassing the existing security measures adopted by the utility. Certificate-based authentication was proposed by Farooq et al. [28] to secure the PMU communication network from man-in-the-middle attacks. It is implemented in real-time using python-based terminals.

9.3.5 Utilization and Analytics

The raw data will keep on accumulating in the storage system without giving any useful information to the system operators. Measurement noises, outliers, and missing data will affect PMU measurements and will adversely impact the performance of PMU-driven applications. PMU measurements taken at different events will differ in signal features. The signal features demanded by the steady-state applications are different from that of dynamic power system applications. The oscillatory signals in the PMU measurements originated under dynamic power system events will be treated as bad data for steady-state applications. Therefore, application-specific data analytics methods should be applied to dig the trends and patterns hidden in the PMU data.

Data mining is the process of extraction of useful information from extensive databases and processes. Presently, grids will be smart only if they can extract valuable information from the massive data generated from all the intelligent meters, sensors and various types of forecasting data from weather, load, and generation forecasts. A mathematical model can be created using this extracted information. Using real-time data and the mathematical model of the system, the current state of the system will be estimated, and it will help to determine the possible actions that should be performed to solve potential problems. The time-critical power system applications like fault detection, service restoration, self-healing, and energy management require quick and efficient analysis of real-time and historical data. Various data mining techniques could be used to understand the customer load pattern, event detection, price forecasting, etc., from the raw data available from monitoring devices like AMI, SCADA, PMU, and other forecasting data. Table 9.1 lists different data mining techniques discussed in the literature.

Table 9.1 Data mining techniques.

Technique	Application
Normalized wavelet energy function [29]	Event detection
Multivariate trend filtering scheme [30]	Estimation for microgrid
Fuzzy cmeans (FCM) [30]	Locating the false data attacks
Principal component analysis and second-order difference method [31]	Real-time event detection
Low-rank matrix completion [32]	Missing PMU data recovery
Moving Window principal component analysis [9]	Multiple event detection
SVM [33]	Event detection
Fourier-based automatic ringdown analysis [34]	Extracting dominant oscillatory modal content
Dynamic programming-based SDT (DPSDT) [35]	Event detection
Random matrix theory [36]	Anomaly detection and location
Euclidean distance-based anomaly detection schemes [37]	Covert cyber deception fault detection
Ensemble classifier bootstrap aggregation [38]	Intrusion detection
Data mining of code repositories (DAMICORE) [39]	Islanding detection of synchronous distributed generators
OPTICS [40]	To segment data and finds the outliers in the segmented data
CNN [41], RNN [41]	Power system transient disturbance classification
Swinging door trending [42]	PMU data compression
Principal component analysis [30, 43, 44]	Dimensionality reduction of PMU data

9.3.6 Visualization of Data

Understanding the weaknesses and potential forecasting of complications are the main objectives of contingency analysis (CA). CA results should be presented in an easily understandable way to assist the system operators in perceiving the security status of the system quickly and intuitively because power system protection is highly time-critical. Advanced visualization methods are required to efficiently present the overall security status and level and location of contingencies. The critical challenge of visualization framework is to integrate information originated

Table 9.2 Visualization techniques.

Visualization technique	Description
GridCloud [49]	Open source platform for real-time data acquisition sharing and monitoring
FNET/GridEye [47]	Visualization of system stress
Tableau [50, 51]	SG data visualization
Animation loops [52]	Combine periodic snapshots of the grid into time-lapse videos defined across geographic areas
Sparklines [52]	Summarize trends in time-varying PMU data as word-sized line plots
Google Earth [53, 54]	SG data visualization
GIS [55]	To display real-time operational data

from different data analytics tools and external sources and concisely present them in a single display. The operators must get a quick insight into the current state of the power grid and for acting accordingly.

Stefan et al. [45] presented an overview of visualization techniques for smart metering. A panoramic visualization scheme for smart distribution network (SDN) is presented by Du et al. [46], which can visualize risk warning, fault self-healing, etc. FNET/GridEye is a low-cost GPS-synchronized frequency measurement network. FNET/GridEye servers hosted at the University of Tennessee and Virginia Tech can visualize the system stress as animations of frequency and angle perturbations [47]. Correlating PMU data with PMU location and system topology stress is visualized as variations in frequency and phase angle [48] with the help of FNET/GridEye. Table 9.2 shows some of the visualization techniques discussed in the literature.

9.4 Big Data Analytics Platforms for PMU Data Processing

A strong and powerful computing platform is required for getting intuitions about the real happenings in the power system from the signals originated from thousands of smart meters or PMU. Many open-source projects are happening around the world to realize this high computing platform requirement. One of such projects is Hadoop stalk which is a collection of open-source tools managed by Apache including Hadoop MapReduce, Hadoop Distributed File System (HDFS), YARN, Cassandra, Storm, Spark, and several other tools. Some of the tools within

Hadoop stalk can handle only batch processing, and some others can handle both batch and stream processing. Batch processing is for processing static historical data, while stream processing is for streaming data-like sensor data. MapReduce is a technique for handling batch processing. Apache Storm and Spark are for streaming data processing.

9.4.1 Hadoop

Apache Software Foundation developed Hadoop, which is an open-source platform that includes both MapReduce function and HDFS. MapReduce consists of Map and Reduce operation. Map operation splits an extensive dataset into many datasets of smaller size, and each one is assigned to a node or computer. At the Reduce stage, the results obtained from the cluster nodes are collected and aggregated into the final output. High scalability, fault tolerance, and computational parallelization are the most important features of Hadoop. Currently, the Hadoop framework is mostly used for offline data analytics applications. The high computational time required by input and output files limits its use for online analytics. Hadoop MapReduce belongs to batch processing. Hence, the system will wait until the batch reaches a predefined size. It will result in a high processing delay. Therefore, the Hadoop platform is not suitable for handling streaming PMU data. But it is most efficient for the batch processing of PMU data.

The oscillatory events in the power system and the presence of bad data may deteriorate the PMU measurement and will affect the satisfactory performance of PMU-based steady-state applications. Therefore, PMU data should be appropriately filtered before giving to these applications. A MapReduce framework implemented with Hadoop is employed for a parallel fluctuation analysis, utilizing a cluster of computer nodes [56]. The large PMU dataset is split into various small sets, and each will be assigned to a mapping process. After doing fluctuation analysis on these small datasets by the mapping process, the results will be combined and compared with a threshold value during the reduce phase. This framework is employed for detecting transient events from the collected PMU data. The measurement data from the installed PMU is collected by OpenPDC software and stored in a data historian that can save up to 100 megabytes. All the real-time applications on the PMU data are performed in this phase. Once the data exceeds 100 megabytes, a data file will be created and stored in HDFS. The Hadoop framework will further fetch this data file for offline data mining applications. Gene expression programming [57] was introduced for optimizing Hadoop performance by automatic tuning of Hadoop's configuration parameters.

A Hadoop-based cloud computing platform was developed [58] to perform data mining and data processing of the massive WAMS data. Matthews and Leger [59] proposed a MapReduce framework for real-time anomaly detection from the PMU data. The MapReduce algorithm was implemented in Phoenix++, which is an open-source multicore implementation of the MapReduce framework. The time-sliced PMU data in CSV file format is fed into the anomaly detection algorithm, which is implemented in Phoenix++. The system operators are intimated with an alarm if any anomaly is detected. Regardless of anomaly detection, the PMU data will be further stored in a database for offline processing.

Online monitoring of transmission line parameters is very important as it will help the system operators to take appropriate actions before the malfunctioning of the grid. Hadoop-based distributed transmission line parameter estimation from the massive PMU data is proposed by Sun et al. [53]. Trachian [60] used windowed sub-second event detection in the OpenPDC platform for real-time event detection from streaming PMU data. The instance-based learning approach is used to train Hadoop to detect the presence of a specific event.

9.4.2 Apache Spark

Apache Spark is a data processing engine with inbuilt modules for streaming, machine learning, and batch processing. Resilient Distributed Data Sets (RDDS) is a list-like data structure, physically partitioned, with each partition residing in a different node. Spark is very efficient with distributed and parallel processing capabilities. In comparison to Hadoop that relies on disk-based data processing, data computation speed in Spark is about 100–150 times higher. It can process both batch and streaming data proficiently. Spark is compatible with many programming languages like Python, Scala, SQL, and R. It can run over a variety of platforms like Hadoop and Mesos, or it can run standalone or in the cloud.

Aggregation and processing of enormous data for time-critical applications demand low latency and high-performance parallel processing. The power system data are usually stored in small redundancy block in distributed file systems such as HDFS, and Apache Hive which suffered from considerable latency due to high disk turnaround time. Latency requirements demand direct streaming of data to real-time applications. The existing implementations of IEEE C37.118.22011 synchrophasor communication protocol like PyPMU, S3DK Synchrophasor Application Development Framework (SADF), Matlab library, etc., are not capable of streaming data directly to Hadoop, Spark, and other high-performance computing platforms. Menon et al. [61] designed a streaming interface in Apache Spark. The interface was in Scala language and intended for receiving synchrophasor data directly to Spark applications for real-time processing and archiving.

Zhou et al. [19] proposed a distributed data analytics platform for the FNET/GridEye synchrophasor measurement system. For real-time applications, PMU data are processed in the data center's buffer memory itself, and FNET/GridEye system and OpenPDC are acting as data servers. Near real-time applications utilize the data stored in data historian, which is realized through OpenHistorian 2.0. Post-event and statistical analyses are performed in the analytics cluster, which is achieved through different data analytics platforms such as Apache Spark, R, and Pandas. The analysis algorithms are parallelized across multiple nodes for distributing the computational loads. An intrusion detection algorithm for synchrophasor data is designed by Vimalkumar and Radhika [62] on the Apache Spark platform. A stream computing platform based on Infosphere has been developed by IBM [63] for real-time voltage stability monitoring of the power system.

9.4.3 Apache HBase

Apache HBase™ is a Hadoop database. It is an opensource, distributed, scalable, and non-relational database that stores data as key-value pairs. Data are distributed horizontally between clusters of commodity hardware. It gives real-time read and write functionality for a massive volume of data. A database management system based on Hbase is proposed by Wang et al. [64], in which the data storage is organized according to the data access patterns of respective applications.

A cloud computing platform is proposed by Li et al. [65] for a panoramic synchronous measurement system that integrates the measurement and fault recorders in high voltage side with the measurement system in the low voltage side. These heterogeneous measurement data are stored in non-relational database HBase. Hadoop architecture is utilized for different processing of these massive data. After decoding the measurement data according to the corresponding configuration file, the information-like analogue data, digital data, and the status of switches are extracted and temporarily stored in a memory. After standardization, the elimination of duplicate data and filling of missing data are completed in HBase. Data preprocessing is done with MapReduce, together with various data analytics tools like Hive, Mahout, and Hadoop streaming.

9.4.4 Apache Storm

Apache Storm is an open-source distributed computing platform for real-time processing of streaming datasets. It is scalable, fault-tolerant, and has multi-language support. Low latency and high encryption efficiency are advantages of Storm. Zhang et al. [66] introduced parallel data encryption and cloud storage platform for WAMS data. The encrypted information is stored in Hadoop's HDFS file system. HBase is used for storing the index and keys of the WAMS data based on the source and the time of generation.

9.4.5 Cloud-Based Platforms

Zhang et al. [67] presented a cloud-hosted Hadoop platform for analyzing massive historical PMU data. This platform consists of three layers, namely, service layer, cluster layer, and application layer. Amazon EC2 and Amazon S3 constitute the first layer (service layer). Amazon EC2 is hosting the on-demand virtual servers demanded by Hadoop, and Amazon S3 provides scalable storage options and also hosts a static website for visualizing the data analysis results sent to S3. The cluster layer consists of Hadoop clusters running on virtual nodes hosted by Amazon EC2. The application layer hosts all the applications and analytics on PMU data. They have also presented a method for the detection of frequency excursion from the PMU data.

Mo et al. [68] proposed a SaaS platform for synchrophasor application. Through virtualization, the physical servers in the cloud are partitioned into virtual machines. Virtual machines are allocated to a particular application based on the processing requirement of the synchrophasor application. The end-user is not required to install the specific synchrophasor application on their local computers but can analyze by accessing the apps through a webserver. Pegoraro et al. [69] proposed an adaptive state estimator for the distribution system. They have utilized a cloud-based IoT paradigm.

The accuracy of the measurements is different under dynamic conditions than steady-state conditions. The weighting parameters of the state estimator algorithm are modified according to the detection of a dynamic state in the distribution system. The PMU data are transferred to virtual PMU (vPMU) at a high reporting rate. The cloud-hosted vPMU performs local processing and identifies the state of the distribution system, which is directly observed by that PMU. vPMU varies measurement reporting rate to state estimator applications according to different states of the system. The state estimator changes the weighting function of the WLS algorithm according to the reporting rate of vPMU.

The Linear State Estimator (LSE) that utilizes only PMU measurements are subject to errors due to false data injection. It will degrade the reliability of the system, and hence False Data Detection (FDD) techniques should be incorporated into the LSE algorithm. It will increase the complexity of the state estimation process, and, in turn, will increase its latency. A cloud-based and parallelized LSEFDD algorithm was proposed by Chakati et al. [70]. They used graphical processing unit (GPU) also to fasten the state estimation with an increasing number of PMUs.

Shand et al. [71] proposed a tool for power system model validation employing plug and play PMU. To identify the exact geographical and electrical positions of the PMU, the authors have used GPS and cloud platforms. It helps to detect the presence of bad data which arises due to an incorrect model of the system. Yang et al. [72] proposed a PMU fog architecture for enhancing the QoS requirement of the WAMS communication system. By utilizing the computational capability of PMU devices, onsite preprocessing of data is carried out to detect and mark anomaly data. The market data is given higher priority for transmission to the control center to mitigate the latency issues in the critical applications.

9.5 Conclusions

Research and developmental activities in electrical engineering focus on cross-disciplinary collaborations where advanced communication technologies, machine learning, artificial intelligence, signal processing, etc., could be integrated to develop a robust smart grid monitoring system. More emphasizes are given for a cyber-physical intelligent grid system. It is high time for an electrical engineering research community to develop suitable data storage and computing platform for better management of the power generation and distribution system. More studies are required to create new appropriate analysis techniques for various power system applications which could resolve real-time issues.

References

1 Li, H., Bose, A., and Venkatasubramanian, V.M. (2016). Wide-area voltage monitoring and optimization. *IEEE Transactions on Smart Grid* 7 (2): 785–793. https://doi.org/10.1109/TSG.2015.2467215.

2 Lee, Y. and Han, S. (2019). Real-time voltage stability assessment method for the Korean power system based on estimation of thévenin equivalent impedance. *Applied Sciences* 9 (8): 1671. https://doi.org/10.3390/app9081671.

3 Zhao, J., Wang, Z., Chen, C., and Zhang, G. (2017). Robust voltage instability predictor. *IEEE Transactions on Power Systems* 32 (2): 1578–1579. https://doi.org/10.1109/TPWRS.2016.2574701.

4 Wu, Y., Musavi, M., and Lerley, P. (2016). Synchrophasor-based monitoring of critical generator buses for transient stability. *IEEE Transactions on Power Systems* 31 (1): 287–295. https://doi.org/10.1109/TPWRS.2015.2395955.

5 Zweigle, G.C. and Venkatasubramanian, V. (2016). Transient instability mitigation for complex contingencies with computationally constrained cost-based control. *IEEE Transactions on Smart Grid* 7 (4): 1961–1969. https://doi.org/10.1109/TSG.2016.2536061.

6 Wang, X. and Yaz, E.E. (2016). Smart power grid synchronization with fault tolerant nonlinear estimation. *IEEE Transactions on Power Systems* 31 (6): 4806–4816. https://doi.org/10.1109/TPWRS.2016.2517634.

7 Ayar, M., Obuz, S., Trevizan, R.D. et al. (2017). A distributed control approach for enhancing smart grid transient stability and resilience. *IEEE Transactions on Smart Grid* 8 (6): 3035–3044. https://doi.org/10.1109/TSG.2017.2714982.

8 Zhang, S. and Zhang, Y. (2017). A novel out-of-step splitting protection based on the wide area information. *IEEE Transactions on Smart Grid* 8 (1): 41–51. https://doi.org/10.1109/TSG.2016.2593908.

9 Rafferty, M., Liu, X., Laverty, D.M., and McLoone, S. (2016). Real-time multiple event detection and classification using moving window PCA. *IEEE Transactions on Smart Grid* 7 (5): 2537–2548. https://doi.org/10.1109/TSG.2016.2559444.

10 Zhou, Y., Arghandeh, R., Konstantakopoulos, I., et al. (2016). Abnormal event detection with high resolution micro-PMU data. *Power Systems Computation Conference (PSCC)*, Genoa (20–24 June 2016), 1–7. doi:https://doi.org/10.1109/PSCC.2016.7540980.

11 Zhou, Y., Arghandeh, R., and Spanos, C.J. (2017). Partial knowledge data-driven event detection for power distribution networks. *IEEE Transactions on Smart Grid* 9 (5): 5152–5162. https://doi.org/10.1109/TSG.2017.2681962.

12 Aminifar, F., Shahidehpour, M., Fotuhi-Firuzabad, M., and Kamalinia, S. (2014). Power system dynamic state estimation with synchronized phasor measurements. *IEEE Transactions on Instrumentation and Measurement* 63 (2): 352–363. https://doi.org/10.1109/TIM.2013.2278595.

13 Liang, X., Wallace, S.A., and Nguyen, D. (2017). Rule-based data-driven analytics for wide-area fault detection using synchrophasor data. *IEEE Transactions on Industry Applications* 53 (3): 1789–1798. https://doi.org/10.1109/TIA.2016.2644621.

14 Pignati, M., Zanni, L., Romano, P. et al. (2017). Fault detection and faulted line identification in active distribution networks using synchrophasors-based real-time state estimation. *IEEE Transactions on Power Delivery* 32 (1): 381–392. https://doi.org/10.1109/TPWRD.2016.2545923.

15 Ding, F., Booth, C.D., and Roscoe, A.J. (2016). Peak-ratio analysis method for enhancement of LOM protection using m-class PMUS. *IEEE Transactions on Smart Grid* 7 (1): 291–299. https://doi.org/10.1109/TSG.2015.2439512.

16 Laverty, D.M., Best, R.J., and Morrow, D.J. (2015). Loss-of-mains protection system by application of phasor measurement unit technology with experimentally assessed threshold settings. *IET Generation, Transmission & Distribution* 9 (2): 146–153. https://doi.org/10.1049/iet-gtd.2014.0106.

17 Guo, Y., Li, K., and Laverty, D. (2014). Loss-of-main monitoring and detection for distributed generations using dynamic principal component analysis. *Journal of Power and Energy Engineering* 2 (04): 423. https://doi.org/10.4236/jpee.2014.24057.

18 Ponce, C. and Bindel, D.S. (2017). Flier: practical topology update detection using sparse PMUS. *IEEE Transactions on Power Systems* 32 (6): 4222–4232. https://doi.org/10.1109/TPWRS.2017.2662002.

19 Zhou, N. (2016). A cross-coherence method for detecting oscillations. *IEEE Transactions on Power Systems* 31 (1): 623–631. https://doi.org/10.1109/TPWRS.2015.2404804.

20 Wang, X. and Turitsyn, K. (2016). Data-driven diagnostics of mechanism and source of sustained oscillations. *IEEE Transactions on Power Systems* 31 (5): 4036–4046. https://doi.org/10.1109/TPWRS.2015.2489656.

21 Guerrero, J.I., García, A., Personal, E. et al. (2017). Heterogeneous data source integration for smart grid ecosystems based on metadata mining. *Expert Systems with Applications* 79: 254–268. https://doi.org/10.1016/j.eswa.2017.03.007.

22 Moraes, R.M., Hu, Y., Stenbakken, G. et al. (2012). PMU interoperability, steady-state and dynamic performance tests. *IEEE Transactions on Smart Grid* 3 (4): 1660–1669. https://doi.org/10.1109/TSG.2012.2208482.

23 Paudel, S., Smith, P., and Zseby, T. (2016). Data integrity attacks in smart grid wide area monitoring. *ICS-CSR '16: Proceedings of the 4th International Symposium for ICS & SCADA Cyber Security Research*, Queen's University Belfast, UK (23–25 August 2016), 1–10. doi:https://doi.org/10.14236/ewic/ICS2016.9.

24 Beasley, C., Zhong, X., Deng, J., Brooks, R., and Venayagamoorthy, G.K. (2014). A survey of electric power synchrophasor network cybersecurity. *IEEE PES Innovative Smart Grid Technologies, Europe*, Istanbul (12–15 October 2014), 1–5. doi:https://doi.org/10.1109/ISGTEurope.2014.7028738.

25 Wang, X., Shi, D., Wang, J. et al. (2019). Online identification and data recovery for pmu data manipulation attack. *IEEE Transactions on Smart Grid* 10 (6): 5889–5898. https://doi.org/10.1109/TSG.2019.2892423.

26 Basumallik, S., Eftekharnejad, S., Davis, N., Nuthalapati, N., and Johnson, B.K. (2018). Cybersecurity considerations on PMU-based state estimation. *CyberSec 2018: Proceedings of the Fifth Cybersecurity Symposium*, Coeur d' Alene Idaho (April 2018), Article No.14, 1–4. doi:https://doi.org/10.1145/3212687.3212874.

27 Mao, Z., Xu, T., and Overbye, T.J. (2017). Real-time detection of malicious PMU data. *19th International Conference on Intelligent System Application to Power Systems (ISAP)*, San Antonio, TX (17–20 September 2017), 1–6. doi:https://doi.org/10.1109/ISAP.2017.8071368.

28 Farooq, S., Hussain, S., Kiran, S., and Ustun, T. (2018). Certificate based authentication mechanism for PMU communication networks based on IEC 61850-90-5. *Electronics* 7 (12): 370. https://doi.org/10.3390/electronics7120370.

29 Kim, D.-I., Chun, T.Y., Yoon, S.-H. et al. (2017). Wavelet-based event detection method using PMU data. *IEEE Transactions on Smart grid* 8 (3): 1154–1162. https://doi.org/10.1109/PESGM.2017.8274161.

30 Nadkarni, A. and Soman, S. (2018). Applications of trend-filtering to bulk pmu time-series data for wide-area operator awareness. *Power Systems Computation Conference (PSCC)*, Dublin (11–15 June 2018), 1–7. doi:https://doi.org/10.23919/PSCC.2018.8443000.

31 Ge, Y., Flueck, A.J., Kim, D.-K. et al. (2015). Power system real-time event detection and associated data archival reduction based on synchrophasors. *IEEE Transactions on Smart Grid* 6 (4): 2088–2097. https://doi.org/10.1109/ TSG.2014.2383693.

32 Gao, P., Wang, M., Ghiocel, S.G. et al. (2016). Missing data recovery by exploiting low-dimensionality in power system synchrophasor measurements. *IEEE Transactions on Power Systems* 31 (2): 1006–1013. https://doi.org/10.1109/ TPWRS.2015.2413935.

33 Nguyen, D., Barella, R., Wallace, S.A. et al. (2015). Smart grid line event classification using supervised learning over PMU data streams. *Sixth International Green and Sustainable Computing Conference (IGSC)*, Las Vegas, NV (14–16 December 2015), 1–8. doi:https://doi.org/10.1109/ IGCC.2015.7393695

34 Tashman, Z., Khalilinia, H., and Venkatasubramanian, V. (2014). Multi-dimensional fourier ringdown analysis for power systems using synchrophasors. *IEEE Transactions on Power Systems* 29 (2): 731–741. https://doi.org/10.1109/ TPWRS.2013.2285563.

35 Cui, M., Wang, J., Tan, J. et al. (2018). A novel event detection method using PMU data with high precision. *IEEE Transactions on Power Systems* 34 (1): 454–466. https://doi.org/10.1109/TPWRS.2018.2859323.

36 Pinheiro, G., Vinagre, E., Praça, I. et al. (2017). Smart grids data management: a case for cassandra. *International Symposium on Distributed Computing and Artificial Intelligence*, Porto, Porutugal (21–23 June 2017), 87–95. Springer. doi: https://doi.org/10.1007/978-3-319-62410-5_11

37 Ahmed, S., Lee, Y., Hyun, S.H., and Koo, I. (2018). Covert cyber assault detection in smart grid networks utilizing feature selection and euclidean distance-based machine learning. *Applied Sciences* 8 (5): 772. https://doi.org/10.3390/ app8050772.

38 Kaur, K.J. and Hahn, A. (2018). Exploring ensemble classifiers for detecting attacks in the smart grids. *Proceedings of the Fifth Cybersecurity Symposium*, vol 13, 1–4. doi:https://doi.org/10.1145/3212687.3212873.

39 Gomes, E.A., Vieira, J.C., Coury, D.V., and Delbem, A.C. (2018). Islanding detection of synchronous distributed generators using data mining complex correlations. *IET Generation, Transmission & Distribution* 12 (17): 3935–3942. https://doi.org/10.1049/iet-gtd.2017.1722.

40 Wang, X., McArthur, S.D., Strachan, S.M. et al. (2018). A data analytic approach to automatic fault diagnosis and prognosis for distribution automation. *IEEE Transactions on Smart Grid* 9 (6): 6265–6273.

41 Zhu, Y., Liu, C., and Sun, K. (2018). Image embedding of PMU data for deep learning towards transient disturbance classification. *IEEE International Conference on Energy Internet (ICEI)*, Beijing (21–25 May 2018), 169–174. doi:https://doi.org/10.1109/ICEI.2018.00038.

42 Ma, Y., Fan, X., Tang, R. et al. (2018). Phase identification of smart meters by spectral clustering. *2018 2nd IEEE Conference on Energy Internet and Energy System Integration*, Beijing, China (20–22 October 2018), 1–5. doi:https://doi.org/10.1109/EI2.2018.8582318.

43 Xie, L., Chen, Y., and Kumar, P.R. (2014). Dimensionality reduction of synchrophasor data for early event detection: Linearized analysis. *IEEE Transactions on Power Systems* 29 (6): 2784–2794. https://doi.org/10.1109/TPWRS.2014.2316476.

44 Gadde, P.H., Biswal, M., Brahma, S., and Cao, H. (2016). Efficient compression of pmu data in wams. *IEEE Transactions on Smart Grid* 7 (5): 2406–2413. https://doi.org/10.1109/TSG.2016.2536718.

45 Stefan, M., Lopez, J.G., Andreasen, M.H., and Olsen, R.L. (2017). Visualization techniques for electrical grid smart metering data: a survey. *IEEE Third International Conference on Big Data Computing Service and Applications (BigDataService)*, San Francisco, CA (6–9 April 2017), 165–171. doi:https://doi.org/10.1109/BigDataService.2017.26.

46 Du, J., Sheng, W., Lin, T., and Lv, G. (2018). Research on the framework and key technologies of panoramic visualization for smart distribution network. Paper presented in 6th International Conference on Computer-Aided Design, Manufacturing, Modeling and Simulation AIP Conference Proceedings, Busan, South Korea (14–15 April 2018), 20–24.

47 Liu, Y., Yao, W., Zhou, D. et al. (2016). Recent developments of FNET/GridEye – a situational awareness tool for smart grid. *CSEE Journal of Power and Energy Systems* 2: 19–27. https://doi.org/10.17775/CSEEJPES.2016.00031.

48 Liu, Y., You, S., Yao, W. et al. (2017). A distribution level wide area monitoring system for the electric power Grid-FNET/GridEye. *IEEE Access* 5: 2329–2338.

49 Anderson, D., Gkountouvas, T., Meng, M. et al. (2018). Gridcloud: Infrastructure for cloud-based wide area monitoring of bulk electric power grids. *IEEE Transactions on Smart Grid* 10 (2): 2170–2179. https://doi.org/10.1109/TSG.2018.2791021.

50 Munshi, A.A. and Mohamed, Y.A.-R.I. (2018). Data lake lambda architecture for smart grids big data analytics. *IEEE Access* 6: 40463–40471. https://doi.org/10.1109/ACCESS.2018.2858256.

51 Munshi, A.A. and Yasser, A.R.M. (2017). Big data framework for analytics in smart grids. *Electric Power Systems Research* 151: 369–380. https://doi.org/10.1016/j.epsr.2017.06.006.

52 Gegner, K.M., Overbye, T.J., Shetye, K., and Weber, J.D. (2016). Visualization of power system wide-area, time varying information. *EEE Power and Energy Conference at Illinois (PECI)*, Urbana, IL (19–20 February 2016), 1–4. doi:https://doi.org/10.1109/PECI.2016.7459263.

53 Gu, Y. (2018). Renewable energy integration in distribution system-synchrophasor sensor based big data analysis, visualization, and system operation. Preprint arXiv. arXiv:1803.06076v1.

54 Gu, Y., Jiang, H., Zhang, Y. et al. (2016). Knowledge discovery for smart grid operation, control, and situation awareness—a big data visualization platform. *North American Power Symposium (NAPS)*, Denver, CO (18–20 September 2016), 1–6. doi:https://doi.org/10.1109/NAPS.2016.7747892.

55 Ashkezari, A.D., Hosseinzadeh, N., Chebli, A., and Albadi, M. (2018). Development of an enterprise geographic information system (gis) integrated with smart grid. *Sustainable Energy, Grids and Networks* 14: 25–34. https://doi.org/10.1016/j.segan.2018.02.001.

56 Khan, M., Li, M., Ashton, P., Taylor, G., and Liu, J. (2014). Big data analytics on PMU measurements. *11th International Conference on Fuzzy Systems and Knowledge Discovery (FSKD)*, Xiamen (19–21 August 2014), 715–719. doi:https://doi.org/10.1109/FSKD.2014.6980923.

57 Khan, M., Huang, Z., Li, M. et al. (2017). Optimizing Hadoop performance for big data analytics in smart grid. *Mathematical Problems in Engineering* 2017: 1–11. https://doi.org/10.1155/2017/2198262.

58 Qu, Z., Zhu, L., and Zhang, S. (2013). Data processing of Hadoop-based wide area measurement system. *Automation of Electric Power Systems* 37 (4): 92–97. https://doi.org/10.7500/AEPS201111169.

59 Matthews, S. and Leger, A.S. (2017). Leveraging map-reduce and synchrophasors for real-time anomaly detection in the smart grid. *IEEE Transactions on Emerging Topics in Computing.* 7 (3): 392–403. https://doi.org/10.1109/TETC.2017.2694804.

60 Trachian, P. (2010). Machine learning and windowed subsecond event detection on pmu data via Hadoop and the OPENPDC. *IEEE PES General Meeting*, Minneapolis (25–29 July 2010), 1–5. doi:https://doi.org/10.1109/PES.2010.5589479.

61 Menon, V.K., Variyar, V.S., Soman, K. et al. (2018). A spark™ based client for synchrophasor data stream processing. *International Conference and Utility Exhibition on Green Energy for Sustainable Development (ICUE)*, Phuket, Thailand (24–26 October 2018), 1–9. doi:https://doi.org/10.23919/ICUE-GESD.2018.8635650.

62 Vimalkumar, K. and Radhika, N. (2017). A big data framework for intrusion detection in smart grids using apache spark. *International Conference on Advances in Computing, Communications and Informatics (ICACCI)*, Udupi (13–16 September 2017), 198–204. doi:https://doi.org/10.1109/ICACCI.2017.8125840.

63 Hazra, J., Das, K., Seetharam, D.P., and Singhee, A. (2011). Stream computing based synchrophasor application for power grids. *Proceedings of the First International Workshop on High Performance Computing, Networking and Analytics for the Power Grid*, Seattle, Washington, USA (November 2011), 43–50. doi:https://doi.org/10.1145/2096123.2096134.

64 Wang, Y., Yuan, J., Chen, X., and Bao, J. (2015). Smart grid time series big data processing system. *IEEE Advanced Information Technology, Electronic and Automation Control Conference (IAEAC)*, Chongqing (19–20 December 2015), 393–400. doi:https://doi.org/10.1109/IAEAC.2015.7428582.

65 Li, Y., Shi, F., and Zhang, H. (2018). Panoramic synchronous measurement system for wide-area power system based on the cloud computing. *13th IEEE Conference on Industrial Electronics and Applications (ICIEA)*, Wuhan (31 May to 2 June 2018), 764–768. doi:https://doi.org/10.1109/ICIEA.2018.8397816.

66 Zhang, S., Sun, J., and Wang, B. (2015). A study of real-time data encryption in the smart grid wide area measurement system based on Storm. *3rd International Conference on Machinery, Materials and Information Technology Applications*, Qingdao, China (28–29 November 2015), 532–537. Atlantis Press.

67 Zhang, S., Luo, X., Zhang, Q., Fang, X., and Litvinov, E. (2018). Big data analytics platform and its application to frequency excursion analysis. *IEEE Power & Energy Society General Meeting (PESGM)*, Portland, OR (5–9 August 2018), 1–5. doi:https://doi.org/10.1109/PESGM.2018.8586512.

68 Mo, S., Chen, H., Kothapa, U., and Zhang, L. (2018). Synchrophasor applications as a service for power system operation. *2018 North American Power Symposium (NAPS)*, Fargo, North Dakota, USA (9–11 September 2018), 1–5. doi:https://doi.org/10.1109/NAPS.2018.8600597.

69 Pegoraro, P.A., Meloni, A., Atzori, L. et al. (2017). PMU-based distribution system state estimation with adaptive accuracy exploiting local decision metrics and IoT paradigm. *IEEE Transactions on Instrumentation and Measurement* 66 (4): 704–714. https://doi.org/10.1109/TIM.2017.2657938.

70 Chakati, V., Pore, M., Banerjee, A., Pal, A., and Gupta, S.K. (2018). Impact of false data detection on cloud hosted linear state estimator performance. *2018 IEEE Power & Energy Society General Meeting (PESGM)*, Portland, OR (5–9 August 2018), 1–5. doi:https://doi.org/10.1109/PESGM.2018.8586671.

71 Shand, C., McMorran, A., Stewart, E., and Taylor, G. (2015). Exploiting massive PMU data analysis for lv distribution network model validation. *50th International Universities Power Engineering Conference (UPEC)*, Stoke on Trent, United Kingdom (1–4 September 2015), 1–4. doi:https://doi.org/10.1109/UPEC.2015.7339798.

72 Yang, Z., Chen, N., Chen, Y., and Zhou, N. (2018). A novel PMU fog based early anomaly detection for an efficient wide area PMU network. *IEEE 2nd International Conference on Fog and Edge Computing (ICFEC)*, Washington, DC, 1–10. doi:https://doi.org/10.1109/CFEC.2018.8358730.

10

Intelligent Enterprise-Level Big Data Analytics for Modeling and Management in Smart Internet of Roads

Amin Fadaeddini[1], Babak Majidi[1,2], and Mohammad Eshghi[3]

[1]Department of Computer Engineering, Faculty of Engineering, Khatam University, Tehran, Iran
[2]Emergency and Rapid Response Simulation (ADERSIM) Artificial Intelligence Group, Faculty of Liberal Arts and Professional Studies, York University, Toronto, Canada
[3]Computer Engineering Department, Shahid Beheshti University, Tehran, Iran

10.1   Introduction

The amount of machine-generated data produced by a metropolis or a regiopolis is significantly larger than a city and manual interpretation of these data is not possible. In the past decade, machine learning algorithms provided the capability of modeling the big data streams and to produce patterns which can assist the decision-makers. A metropolis is a major city combined with the smaller satellite cities around it. The increase in the population and gradual expansion of both the major central city and the smaller cities around it led to a combined urban environment which is referred to as the metropolis. As the living costs make the central urban areas unaffordable in many major cities around the world, a significant portion of the population of the metropolises prefer to live in more affordable smaller satellite cities and to commute on daily basis to the central business district of the metropolis or to the industrial precinct. In the regional areas, the same process happens to the villages and the agro-food business centers. The villages and the small regional cities are expanded and form the regiopolises. In both the metropolis and the regiopolis, there is a central business district which is the main commercial, cultural, political, and international center of the urban area.

Management of these large urban areas with populations in the scale of tens of millions is a major challenge. The energy consumption management, traffic control, planning various services, crime control, and many other essential services

Big Data Analytics for Internet of Things, First Edition. Edited by Tausifa Jan Saleem and Mohammad Ahsan Chishti.

are impossible to provide efficiently in this scale. To manage these large urban environments, the sensor and camera networks as well as many other automatic data generation tools are used to provide insight about the megacities. The analysis of the data streams created automatically by this network is a challenging problem. There has been a significant body of research on the application of big data analytics for smart environment management. Yu et al. [1] proposed a decentralized data assessment framework using Blockchain for smart city management. Zhou et al. [2] investigated big data for health services of the smart city. Nazerdeylami et al. [3] proposed a smart coastline management system using deep neural networks. Abbasi et al. [4] proposed a deep privacy-preserving framework for Internet of Robotic Things (IoRT) in the smart city. Zenkert et al. [5] proposed a new dataset for smart city mobility management. Yadav et al. [6] proposed a framework for using Internet of Things (IoT) and big data for smart city management. Xu et al. [7] proposed a framework for smart tourism using IoT and big data. Wang et al. [8] designed a framework for monitoring the conditions of the infrastructures of the smart city using big data. Shao et al. [9] proposed a framework for using the video surveillance cameras in the smart city for decision support. Muhammed et al. [10] proposed a new healthcare network for the smart city. Huang et al. [11] designed a sustainable smart city management system by adhering to the quality of service standards. Norouzi et al. [12] presented a new solution for green software-defined networking (SDN) for the smart networks.

In the next decade, gradually the intelligent autonomous vehicles are replacing the traditional manually driven cars. Therefore, the concept of smart highway and road management is of significant importance in the next years. In the smart highway system, the intelligent vehicles can recognize each other and to communicate in the vehicular Adhoc networks. However, during the transition period from traditional to intelligent highways, the manually driven cars and the intelligent vehicles drive side by side on the roads. During this transition period as the current generation vehicle does not have electronic tagging and ID systems, it is necessary for the smart vehicles to recognize the old vehicles using the traditional method of Automatic License Plate Recognition (ALPR).

ALPR has been studied extensively in the past decades. Some of the proposed methods are relying on computer vision techniques. Most of the works in license plate recognition using semantic segmentation is based on binary classification which is easier compared with multi-class classification. In the binary classification, the license plates are classified into the digits and the background. Abedin et al. [13] used contour properties and deep neural networks for license plate recognition. Abinaya et al. [14] proposed a morphological transform to rectify blurred license plate images in moving cars. Ahn et al. [15] proposed an algorithm for license plate area detection. Fomani et al. [16] presented a method for detecting license plates using adaptive morphological operations and adaptive thresholding. Ghoneim et al. [17] proposed using super-resolution enhancement for improved license plate recognition.

Computer vision and machine learning methods are powerful tools for automating complex tasks and in recent years, the accuracy of these methods are increased significantly. Recently, deep learning resulted in very good accuracies in various complex pattern recognition problems [18–21]. Image understanding in computer vision and machine learning passed an evolutionary route from finding just a single label for the whole scene and localizing objects to segmentation of every pixel in a dense image. One of the first deep learning-based models proposed for segmenting an image was by Long et al. [22] which substitute the fully connected layer in the last layer of the deep neural network with a Softmax layer to assign labels to each pixel. However, these models were not perfect for understanding global information from the images. Therefore, the models have been improved by a series of post-processing techniques like Conditional Random Field (CRF) proposed by Lafferty et al. [23], use of dilated convolutions to expand the size of filters [24], and recently introduced ReSeg network which leverages recurrent neural network (RNN) to help better fine-tune the feature maps [25].

In this chapter, the parametric model is chosen for multi-class ALPR from the low altitude angle suitable for the vision system of smart vehicles. In the proposed model, each digit is segmented using a deep neural network. The proposed algorithm is divided into two stages: (i) a model was trained for detection of the bounding box of the license plate, i.e., localization. In this stage, 500 images of cars in different environments are used for training the model. (ii) A model is trained to segment each digit in bounding boxes that computed during the first section. In this stage, a naive sample generator is used, which simultaneously generates two other outputs: ground truth for L_2 loss and ground truth for cross-entropy loss.

The rest of this chapter is organized as follows. The proposed deep ALPR framework is detailed in Section 10.2. The experimental design and the simulation scenarios are discussed in Section 10.3. Section 10.4 discusses the practical implementation of enterprise-level big data analytics for smart internet of roads. Finally, Section 10.5 concludes the chapter.

10.2 Fully Convolutional Deep Neural Network for Autonomous Vehicle Identification

The proposed framework for license plate recognition consists of several steps which will be discussed in detail in this section.

10.2.1 Detection of the Bounding Box of the License Plate

A real-life recognition application must pass two steps: detection and recognition. In this work, for detecting the license plates in the images, a convolutional neural network (CNN) is trained for semantic segmentation. The objective of this network is to find the bounding box of the license plates. The neural network is trained on roughly 500 images of cars in different environments. The output of the network is a vector containing the coordinates for the top-left location of the bounding box of the license plate and the height and width of the plate. The CNN consists of three convolution layers and three pooling layers ending with a fully connected layer. The loss is an L_2 loss that compares the output vector with the ground truth.

10.2.2 Segmentation Objective

When using a variational auto-encoder (VAE), an objective function that will be used is an L_2 loss function. Using an L_2 loss objective generally seems to be a right choice in mapping or translation in which input and output have the same statistics, but different styles. However, in semantic segmentation, there is a better choice that is called Sparse Softmax Cross-Entropy Loss which computes the cross-entropy between Logits and labels. For using this objective function, we need another ground truth that has one band, and its values are constant and in the range of category labels.

10.2.3 Spatial Invariances

Pure CNN-based models generally fail to capture all the invariances of the samples in semantic segmentation settings and because of this shortcoming, refinements are needed to solve this issue in the post-processing or convergence stages [26]. Among these methods, CRF tries to keep the connection between low-level image's information and final segmented image. This technique helps to better understand the global information in samples and therefore segmenting objects with more details.

Furthermore, semantic segmentation models are highly sensitive to spatial invariances in the input image and a small change in input distribution could reduce segmentation accuracy. Therefore, in the proposed license plate segmentation method, all the spatial invariances are removed. In the pre-processing stage, the license plates are centered in test samples. Removing spatial invariance means that the license plate centroid in images is at a constant location. The invariance strongly affects the generalization of model and forces it to perform worse on the unseen samples. Figure 10.1 shows the procedure of removing spatial invariances to improve the generalization and accuracy of semantic segmentation model.

10.2.4 Model Framework

10.2.4.1 Increasing the Layer of Transformation

Many segmentation models rely on Encoder–Decoder patterns without using any transformation stage. In VAE architectures, input samples must pass three

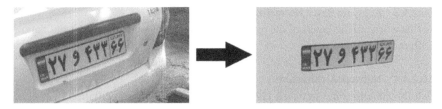

Figure 10.1 The procedure of removing spatial invariances to improve the generalization and the accuracy of semantic segmentation model.

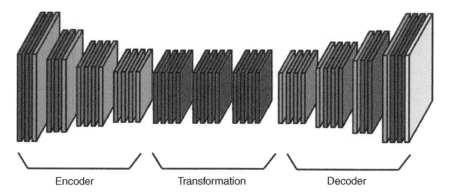

Encoder   Transformation   Decoder

Figure 10.2 The architecture according to encoding-transformation-decoding network.

stages. First, the given sample must be encoded through layers. Then, the result of encoding stage which has a lower dimensionality passes through transformation layers. This is the stage where all the transformation of input to output will happen. Finally, the decoding stage increases the dimension. In semantic segmentation, different numbers of layers in these stages affect the accuracy and results of the models. In our experiment, we saw that by increasing the number of layers in the transformation stage, considerable improvement in segmentation would occur. However, increasing the encoding and decoding layers would not necessarily lead to better-segmented objects and details. Figure 10.2 shows the proposed architecture according to encoding-transformation-decoding network.

10.2.4.2 Data Format of Sample Images

In some applications of machine learning like object classification and object detection, feeding compressed format of input samples to the model helps to reduce the overfitting problem. However, in semantic segmentation, we avoided compression in ground truth for L_2 loss to prevent losing the meaningful values.

10.2.4.3 Applying Batch Normalization

In this work, all the layers of the output are normalized using batch normalization with momentum of 0.5.

10.2.4.4 Network Architecture

A fully convolutional network [27] is used for semantic segmentation of license plates. The number of layers in encoding and decoding sections of the network is eight and three, respectively. In the transformation stage, four layers are used instead of one layer for better performance. Each layer uses leaky ReLU as the activation function and dropout with keep probability of 0.5. Sparse Softmax loss is chosen as an objective function. The optimization function is Adam optimizer which uses the learning rate of 0.0008 and beta of 0.5.

A deep neural network with convolution kernels is adopted as the core and primary architecture of this implementation. Figure 10.3 depicts the FCNN architecture. In FCNN, the convolution kernel produces either a downsized sample or performs up-sampling using the transposed convolution (de-convolution) kernels. Then, sequences of convolutions feed the learned data to a fully connected neural network for labeling.

10.2.5 Role of Data

The main purpose of using convolution on the samples is to extract features from them and based on these features find a meaningful correlation in data distributions. This is a more complicated procedure in objects like license plates where the aim is to segment each digit. This difficulty is because digits share a common texture with the background and the only correlation is the spatial location of digits. A workaround for this shortcoming is to use more data to learn the features. In our experiments, we compared the amount of data used and show that by having more data, the model has better performance in segmenting digits.

10.2.6 Synthesizing Samples

An obstacle in end-to-end learning models like deep neural networks is the preparation and pre-processing of training samples. This problem is more evident in applications like semantic segmentation. In semantic segmentation, depending on the objective function, we could have up to two or three sets of samples. For instance, by selecting sparse Softmax loss, besides the training sample which is our real RGB image input (Width, Height, 3), we need another input as ground truth with shape of (Width, Height, 1). In this input, every pixel is labeled according to the list of category labels. However, if we decide to use L_2 loss function, we need another ground truth input with shape of (Width, Height, 3) as depicted in Figure 10.4.

To resolve the difficulty in preparing samples, we decided to generate manually generated samples. In the process of sample generation, we generate a random number and a code and paste it in the license plate image. Also, two ground truth inputs are generated. Although this generated samples miss the exact details of the real counterparts, this process decreases the pre-processing time. These generated samples are presented in Figure 10.5. Although these generated samples help the model to learn more from the data, they cannot replace the real data.

10.2.7 Invariances

There are many invariances that can affect the decision of an ALPR system and make it to produce false answers when facing an outlier. By adding invariants like illumination and blurring, we can hope that our model makes a more sensible decision. However, adding many spatial invariances could make the model useless. Therefore, the only spatial invariance that added in this research is a slight rotation. Furthermore, because clarity of the background in every car's license plate is different, in data generation, we randomly changed the background opacity to include this kind of invariances.

10.2.8 Reducing Number of Features

In the training stage, our models are dealing with a broad range of features. Sometimes, many of these features not only are useless for training a model, but they make the model worse. Therefore, in license plate segmentation where our objective is to only segment digits, the surrounding area of the license plate could not be considered as a useful source of features. Thus, we replace the background with a solid color.

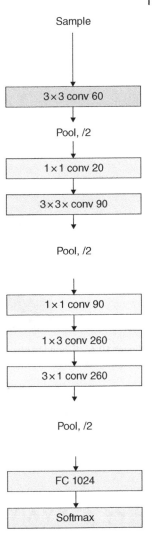

Figure 10.3 The architecture of FCNN.

(a)

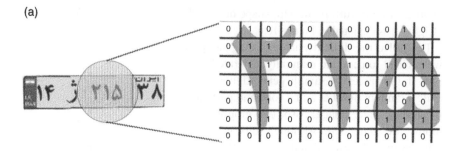

(b)

Figure 10.4 Ground truths needed for both types of objectives. (a) Sparse Softmax loss and (b) L_2 loss.

Figure 10.5 Generated samples that used to train the segmentation model.

10.2.9 Choosing Number of Classes

In multi-class ALPR methods, the license plate is segmented into N regions. Usually, N is considered as the number of digits in license. In this work, we consider N as 5. Figure 10.3b shows the results of this segmentation.

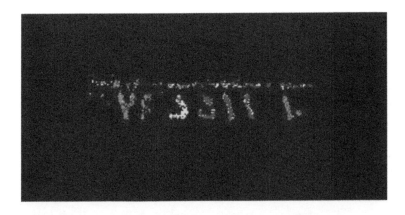

Figure 10.6 Segmented samples produced by the trained model.

10.3 Experimental Setup and Results

For a more detailed assessment of applicability of the proposed segmentation model, we first present the Softmax loss of the proposed model on unseen samples during training and then compute the mean intersection over union (M-IOU). This measure is extensively been used for semantic image segmentation assessment. Figure 10.6 presents the segmented samples produced by the proposed model after 70 iterations.

10.3.1 Sparse Softmax Loss

Table 10.1 shows the proposed model Softmax loss at final iteration.

Table 10.1 The proposed model Softmax loss at final iteration.

Dataset	Sparse Softmax
Real data sample	0.5689

Table 10.2 Representation of the proposed model M-IoU at final iteration.

Dataset	M-IoU
Unseen generated sample	0.87194
Real data sample	0.51665

10.3.2 Mean Intersection Over Union

M-IoU, which is also referred to as Jaccard index, is a better evaluation method for our application. The objective of M-IoU is to first compute the intersection of ground truth with the segmented output divided by union of each class and then averaging it over all existing classes. Table 10.2 presents the M-IoU of learned model when applied to 100 of unseen and real data samples.

$$IOU = true\ positive - \left(true\ positive + false\ positive + false\ negative\right). \quad (10.1)$$

10.4 Practical Implementation of Enterprise-Level Big Data Analytics for Smart City

In the real situation, the proposed ALPR for smart road or highway generates a large data stream of license plate numbers. The machine-generated data and log files generated by the various components of the smart city sensor network and servers play an essential role in management of the smart city. These data and log files allow the system administrators to gain knowledge of operations, executions, and security issues regarding various smart city services. Significant increase in the volume of data and log files makes manual interpretation of these files very hard and in some cases even impossible. There are various machine learning and analytics tools for autonomous interpretation of log files for optimal management of services. These tools analyze very large log files and extract the required information for service administrators from these files autonomously.

The solution provided for analyzing this vast amount of data is distributed and parallel processing. The MapReduce-based processing allows the analytics framework to split the data into smaller fractions and after processing these smaller datasets, the results are combined for the final insight into the data. The Hadoop-based data processing frameworks are one of the solutions proposed for distributed processing of the big data. A very important step in smart log analysis is searching massive machine-generated datasets. Due to the fact that searching large datasets and logs is very time consuming and usually the results of the analytics are required in real time, breaking the search into smaller searches is a vital ability for analyzing machine-generated data.

A solution for this problem is using a combination of Indexers and Forwarders. The Indexers parse the machine-generated data and the Forwarders deliver the data to the Indexers. To process large log streams in real time, the Indexing layer is distributed over several physical servers. In this distributed search environment, a Search Head has the responsibility to forward the query required by the system to parallel Indexers and each Indexer indexes a small batch of data. The Search Head also has the responsibility of aggregation of the results from the Indexers and to produce the final result. This process is demonstrated in Figure 10.7.

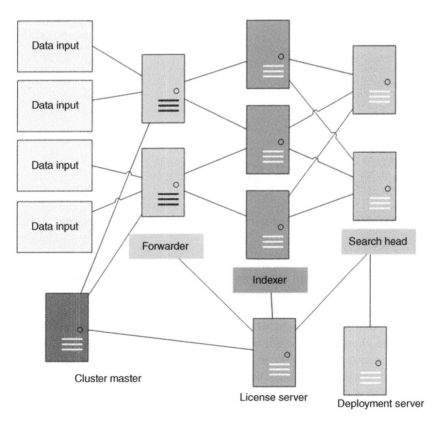

Figure 10.7 The Forwarder and Indexer architecture for parallel processing of smart city data.

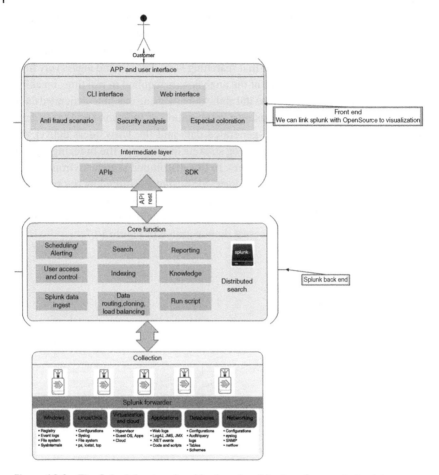

Figure 10.8   The Splunk front end and back end architecture for smart city data analytics.

A necessary requirement for processing the smart city log files is consideration of timestamps of the data. The stream of sensor logs should be considered as a time series and only in this context, many events in the city has meaning. Therefore, to extract meaningful patterns from the smart city data streams, besides the physical parallelism of the indexing process, the data can be also parallel-processed in time. The parallelism of indexing on data size refers to as Spatial MapReduce and the parallelism in time is called Temporal MapReduce. This solution is proposed by the Splunk for enterprise-level data analytics. The back end of the Splunk uses massively parallel searching and indexing for finding patterns in the big data. The front end can be linked to various machine learning, visualization, and data analytics platforms for decision support for the

Figure 10.9 The clustering results for the traffic in Dublin.

Figure 10.10 The classification results for the air pollution hotspots in Dublin.

smart city. The architecture of the Splunk front and back ends is demonstrated in Figure 10.8.

As a case study, a dataset from the city of Dublin, Ireland, is used. Two main scenarios are considered for evaluation of the big data analytics for smart city management. These scenarios are:

- Service planning based on the commuter population.
- Air pollution management.

The scenarios are investigated using the traffic in the city of Dublin. Clustering the traffic shows the sections which have highest traffic and therefore require optimized management. The output of this scenario is presented in Figure 10.9.

The next scenario is the classification results for the air pollution hotspots in Dublin. The output of this scenario is presented in Figure 10.10.

One of the most important issues in providing the results of the data analytics to the smart city management is visualization and dashboards. The dashboards should simulate the city map and offer insights to the management using the sensor network data. The large volume of data gathered by the sensor network of the smart metropolis should be presented in a comprehensible format for the human managers. A map-based dashboard is necessary to address this issue.

10.5 Conclusion

In this chapter, an enterprise-level big data analytics framework for autonomous management of road traffic in the smart cities is presented. The proposed framework is capable of providing decision support using large data streams. A deep neural network is used for segmentation of vehicle license plates in a complex and multi-language environment. The performance of the proposed algorithm is evaluated using a dataset of real and manually generated data. The experimental results show that the proposed framework can detect and segment the license plates in complex scenarios and the results can be used in smart road applications.

References

1 Yu, H., Yang, Z., and Sinnott, R.O. (2018). Decentralized big data auditing for smart city environments leveraging blockchain technology. *IEEE Access* 7: 6288–6296.

2 Zhou, M. and Chen, X. (2018). Application of dendrobium compound preparation in the health environment of big data smart city. Paper presented at the 2018 International Conference on Intelligent Transportation, Big Data & Smart City (ICITBS), Xiamen, China.

3 Nazerdeylami, A., Majidi, B., and Movaghar, A. (2019). Smart coastline environment management using deep detection of manmade pollution and hazards. Paper presented at the 2019 5th Conference on Knowledge Based Engineering and Innovation (KBEI), Tehran, Iran.

4 Abbasi, M.H., Majidi, B., Eshghi, M., and Abbasi, E.H. (2019). Deep visual privacy preserving for internet of robotic things. Paper presented at the 2019 5th Conference on Knowledge Based Engineering and Innovation (KBEI), Tehran, Iran.

5 Zenkert, J., Dornhofer, M., Weber, C., Ngoukam, C., and Fathi, M. (2018). Big data analytics in smart mobility: modeling and analysis of the Aarhus smart city dataset. Paper presented at the 2018 IEEE Industrial Cyber-Physical Systems (ICPS), St. Petersburg, Russia.

6 Yadav, P., and Vishwakarma, S. (2018). Application of Internet of Things and big data towards a smart city. Paper presented at the 2018 3rd International Conference On Internet of Things: Smart Innovation and Usages (IoT-SIU), Bhimtal, India.

7 Xu, C., Huang, X., Zhu, J., and Zhang, K. (2018). Research on the construction of sanya smart tourism city based on internet and big data. Paper presented at the 2018 International Conference on Intelligent Transportation, Big Data & Smart City (ICITBS), Xiamen, China.

8 Wang, T., Bhuiyan, M.Z.A., Wang, G. et al. (2018). Big data reduction for a smart city's critical infrastructural health monitoring. *IEEE Communications Magazine* 56 (3): 128–133.

9 Shao, Z., Cai, J., and Wang, Z. (2017). Smart monitoring cameras driven intelligent processing to big surveillance video data. *IEEE Transactions on Big Data* 4 (1): 105–116.

10 Muhammed, T., Mehmood, R., Albeshri, A., and Katib, I. (2018). UbeHealth: a personalized ubiquitous cloud and edge-enabled networked healthcare system for smart cities. *IEEE Access* 6: 32258–32285.

11 Huang, J., Xing, C.-C., Shin, S.Y. et al. (2017). Optimizing M2M communications and quality of services in the IoT for sustainable smart cities. *IEEE Transactions on Sustainable Computing* 3 (1): 4–15.

12 Norouzi, A., Majidi, B., and Movaghar, A. (2018). Reliable and energy-efficient routing for green software defined networking. Paper presented at the 2018 9th International Symposium on Telecommunications (IST), Tehran, Iran.

13 Abedin, M.Z., Nath, A.C., Dhar, P., Deb, K., and Hossain, M.S. (2017). License plate recognition system based on contour properties and deep learning model. Paper presented at the 2017 IEEE Region 10 Humanitarian Technology Conference (R10-HTC), Dhaka, Bangladesh.

14 Abinaya, G., Banumathi, R., Seshasri, V., and Kumar, A.N. (2017). Rectify the blurred license plate image from fast moving vehicles using morphological process. Paper presented at the 2017 IEEE International Conference on Electrical, Instrumentation and Communication Engineering (ICEICE), Karur, India.

15 Ahn, C.S., Lee, B.-G., Yang, S.-S., and Park, S.-C. (2017). Design of car license plate area detection algorithm for enhanced recognition plate. Paper presented at the 2017 4th International Conference on Computer Applications and Information Processing Technology (CAIPT), Kuta Bali, Indonesia.

16 Fomani, B.A. and Shahbahrami, A. (2017). License plate detection using adaptive morphological closing and local adaptive thresholding. Paper presented at the 2017 3rd International Conference on Pattern Recognition and Image Analysis (IPRIA), Shahrekord, Iran.

17 Ghoneim, M., Rehan, M., and Othman, H. (2017). Using super resolution to enhance license plates recognition accuracy. Paper presented at the 2017 12th International Conference on Computer Engineering and Systems (ICCES), Cairo, Egypt.

18 Fadaeddini, A., Majidi, B., and Eshghi, M. (2019). Privacy preserved decentralized deep learning: a blockchain based solution for secure ai-driven enterprise. Paper presented at the International Congress on High-Performance Computing and Big Data Analysis, Tehran, Iran.

19 Anbari, S., Majidi, B., and Movaghar, A. (2019). 3D modeling of urban environment for efficient renewable energy production in the smart city. Paper presented at the 2019 7th Iranian Joint Congress on Fuzzy and Intelligent Systems (CFIS), Bojnord, Iran.

20 Majidi, B., Patra, J.C., and Zheng, J. (2014). Modular interpretation of low altitude aerial images of non-urban environment. *Digital Signal Processing* 26: 127–141.

21 Sanaei, S., Majidi, B., and Akhtarkavan, E. (2018). Deep multisensor dashboard for composition layer of web of things in the smart city. Paper presented at the 2018 9th International Symposium on Telecommunications (IST), Tehran, Iran.

22 Long, J., Shelhamer, E., and Darrell, T. (2015). Fully convolutional networks for semantic segmentation. Paper presented at the Proceedings of the IEEE Conference on Computer Vision and Pattern Recognition, Boston, USA.

23 Lafferty, J., McCallum, A., and Pereira, F.C. (2001). Conditional random fields: probabilistic models for segmenting and labeling sequence data. Paper presented at the Proceedings of the Eighteenth International Conference on Machine Learning, Williamstown, USA.

24 Chen, L.-C., Papandreou, G., Kokkinos, I. et al. (2017). Deeplab: semantic image segmentation with deep convolutional nets, atrous convolution, and fully connected crfs. *IEEE Transactions on Pattern Analysis and Machine Intelligence* 40 (4): 834–848.

25 Visin, F., Ciccone, M., Romero, A., et al. (2016). Reseg: a recurrent neural network-based model for semantic segmentation. Paper presented at the Proceedings of the IEEE Conference on Computer Vision and Pattern Recognition Workshops, Las Vegas, USA.

26 Garcia-Garcia, A., Orts-Escolano, S., Oprea, S. et al. (2018). A survey on deep learning techniques for image and video semantic segmentation. *Applied Soft Computing* 70: 41–65.

27 Shelhamer, E., Long, J., and Darrell, T. (2017). Fully convolutional networks for semantic segmentation. *IEEE Transactions on Pattern Analysis and Machine Intelligence* 39 (4): 640–651.

11

Predictive Analysis of Intelligent Sensing and Cloud-Based Integrated Water Management System

Tanuja Patgar and Ripal Patel

Dr. Ambedkar Institution of Technology, Bengaluru, Karnataka, India

11.1 Introduction

For human survival, water is considered one of the vital resources. Based on population increase and water consumption rate, the water sector companies are rounding the clock around the smart water management system. The World Health Organization stated that water consumption by each person in one day is 150 liters, and one family with four members exceeds 600 liters water consumption in a single day. An average Indian family of four, living in an apartment, utilizes more than 1000 liters of water every day. Close to 25–30% of the world's water supply is wasted due to leaking. A trickle flow from the leakage tank has the potential to leak upward of 600 liters of water every day. Such leaks can be detected by installing a meter at every point to track its consumption. Almost 4% of fresh water is available, as compared to other countries, with India covering only 2.6% of the world's surface area. India is the highest groundwater extracting country in the world, followed by China and the United States. But water usage of the United States and China put together, India's water consumption is still higher; by global standards, it is a water-scarce country.

Looking at annual per capita freshwater availability, the indicator used to gauge water scarcity across countries, India's per capita annual consumption during independence in 1947 was above 5000 cu.m. But in the race for pulling groundwater for cash crops, electrification, a shift to cultivating water-intensive crops, and creating boreholes slowly saw the water availability getting reduced with each decade.

In recent years, urban water supply has been monitored, controlled, and distribution crisis make big news worldwide. Urban water supply networks are in

Big Data Analytics for Internet of Things, First Edition. Edited by Tausifa Jan Saleem and Mohammad Ahsan Chishti.

demand because of their bonding between drinking water supply and water utilized by consumers. The demand and supply scale distribution networks are required for the existence of urban people, in maintaining healthy steps of economic development, and for the unconditional operation of industries.

In the Indian scenario, the smart city project took the front row for five years. It includes urban water distribution infrastructure and must assure water sources, water quality checks, quality control, monitoring, water supply mode, and technological process parameters. The restrictions are imposed by environmental conditions, hydrological conditions, water availability, tank storage capacity, water towers, water supply infrastructure, and increasing water consumption. Nowadays in cities, water is supplied to the desired station with the help of the main water pipeline along with some manpower. The modern wireless communication technologies such as GSM and ICT-based water managing systems are more reliable for transmitting and receiving information efficiently from traditional-based water managing systems.

By providing sensors with cloud-based analytical platform are used as water leakage detection mostly in cooling towers, commercial and industrial buildings. By deploying IoT-enabled sensors to capture flow data from water infrastructure and communicating to a cloud-hosted analytical engine, the analytical platform uses proprietary analytical codes to monitor flow pattern and identifies equipment malfunction; when abnormal flow behavior is identified, the operating staff is alerted instantly with step-by-step instructions about the location of the problem and corrective measures to rectify the error. The user can monitor water flow by strapping a sensor device around the existing water meter. With mobile application through, a user can view real-time water usage data, detect abnormal water usage, live leaks, and send a notification to the plumbing services provider also.

11.2 Literature Survey

Accurate updates of user water meter are essential for the successful revenue water management system. The traditional data collection, manpower involvement, inefficiency, and inaccurate water bill took the last bench in many developed countries. The research [1] proposes a sensor and ICT-based solution by deriving the automatic water meter reading with the GSM module and solenoid control valve. The volumetric flow is considered when water flows through the flow sensor and generates pulses. The water flow measurement is calculated and sends SMS for generation of the water bill.

Recent research has been carried out to find a sensor-based solution to the water management system. Some of them have proposed an automatic water meter reading system with a cost-effective and more accurate meter reading. The water distribution and consumption system play major roles in the water-scarce country like Kuwait. The research proposes [2] how to find the flow rate of the water in the pipe with the newly designed algorithm "Water-WOLF." The technique converts continuous water flow meter into wireless flow rate. The pipe distribution network for detecting a magnetic field generated by flow meters is generated using tunneling magneto-resistive (TMR) materials. In this way, flow sensing can increase leak detection accuracy.

The better relationship is balanced between water supplier and customers only when water distribution controls water consumption [3]. The consumer awareness platform is required for their water consumption measurement. In addition to that, a metering system is also adopted to identify water leakage. The designed parameters such as water leakage detection, water consumption monitoring, generating the smart bill are the building block of automating water systems. The platform is to discuss the pros and cons of the traditional meter system and maintaining the smart billing process faster with a reduction in water wastage. The chapter proposes GSM and ZigBee-based technology for consumption data collection, for diagnoses, and transferring that data to a central database for billing. The remote monitoring system saves the providers time and energy without visiting every location for a meter reading. Similarly, customers are satisfied with billing based on current consumption water compared to predicted consumption.

Several systems address a variety of issues in the water management sector. It includes a water flow rate, water wastage modes, and its conservation [4]. The author had implemented the Internet of Things platform for control and monitoring the water management system by deploying pH and conductivity sensors in every home to measure the quality of water distributed. The supply and demand for water are increasing day by day from global population growth. The water pipe infrastructure has been leaking due to aging as well as environmental conditions [5]. Under such situations, a new technology is required to maintain low-cost water infrastructure for more secure distribution; the NEC Group is partnering with Imperial College London to implement Smart Water Management (SWM) system based on ICT.

The SWM system can be improved by hybrid ICT products and be more accurately used for continuous monitoring, detecting anomalies, and water distribution network optimization. The paper in [6] proposes an IoT-based SWM. The fundamental three-tier architecture for IoT is discussed. The first (bottom) layer consists of sensing input devices. The second layer is the bridge between the reliable transfer of information generated. The data processing, analysis, and displaying are at the application layer (Top layer). It includes a sensor network for data acquisition, Wi-Fi or Internet for data distribution, and finally for processing, storing, visual display using cloud and web-based tools.

Wireless technologies such as sensors, RFID, Bluetooth, Wi-Fi, ZigBee, and its addressing schemes used in designing IoT have been discussed [7]. ZigBee has more desirable characteristics when compared to other wireless technologies in terms of operating schedule, distance coverage, and power requirement. The time required for sleeping mode to wake up is 15 ms for ZigBee, but Bluetooth 3–4 second. ZigBee's three-tier security protocols such as ACL or AES are used in the IoT platform.

The major challenge in data transmission is wireless devices' power requirement. The comparison chart for wireless devices' power requirement specification and also some sort of comparison between Raspberry Pi and Arduino related to cost, processing, and computer functionalities are listed in [8]. It also discussed the ESP8266 Wi-Fi module that can build any real-time application. The microcontrollers will communicate to the Wi-Fi module through UART but via GPIO, the sensors will communicate. The cloud platform such as IBM Bluemix, Carriots, Nimbits, Thingspeak, and Ubidots supporting IoT platforms is discussed [9]. The application platform such as platform as a service (PaaS) helps the developers to deploy the least flexible application when the whole structure is invisible.

11.3 Proposed Six-Tier Data Framework

The proposed data analytical cycle includes different spectrums such as data acquisition, data integration, data distribution, data processing, data storing, and display. The technology tools used are sensor networks, Wi-Fi, cloud platform, modeling, and analysis using the processor and web-based tools. The fundamental six-tier data flow architecture for the IoT platform is depicted in Figure 11.1. The bottom layer 1 is data source which can connect the main resource (water tank). The data selector is layer 2 which can connect IoT devices such as sensing input devices. The data pre-processor is layer 3 and data modeling designed layer 4, which connect with the processor for the algorithm. The middle layer 5 is the bridge between the reliable transfer of information generated and is stored using cloud technology. The top layer is the application layer where data are analyzed and displayed.

The research proposes a six-tier data analytical cycle concept to the smart water meter management system. Some of the issues are excess water distribution and consumption by particular users by theft using extra pipes and motor-pump sets. By monitoring the water distribution channel mostly in metropolitan cities, it is mandatory to implement embedded-based automatic water control and monitoring and theft prevention system. With this type of advanced technology, each consumer water report is generated using the IoT platform. Especially, embedded-based water flow monitoring system consists of a controller, flow sensor for water flow rate, and exchange the information remote monitoring station using a wireless transmitter.

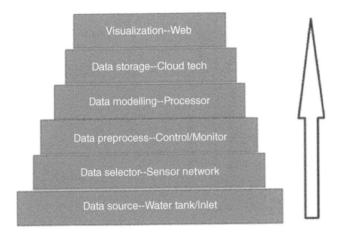

Figure 11.1   Six-tier data analytical cycle.

With wireless advancement technology and ICT, booming makes public comfort for online water-related information such as water consumption, water bill, payment gateway, and water supply pattern. Figure 11.2 represents the proposed framework for smart water meter.

The system comprises the following design parameters.

11.3.1   Primary Components

The system consists of an electrically operated solenoid valve as the controlling element in supplying water to customers. For a particular time, the control valve is turned open and closed by the server to distribute the water. Suppose the water flow rate exceeds, then automatically the solenoid valve is instructed to stop the water supply. The water supplied rate and distribution mode-related information are transferred to recommended users and officers for immediate action using wireless communication (GSM MODEM).

The sensor (YF-S210) is used as the primary IoT device which can measure the water flow rate. It is mainly deployed on the water pipeline. The integrated magnetic Hall–Effect sensor will generate output in the form of electrical pulses. The revolution is proportional to the water instantaneous flow rate. The generated output is connected to the controller for monitoring water consumption and balancing water in the tank. When the water is passed through the pipeline, the flow rate is calculated by sensing devices. Any variation in water flow due to the intervention of external pumping of water will be detected as the sensor operates under certain predefined value.

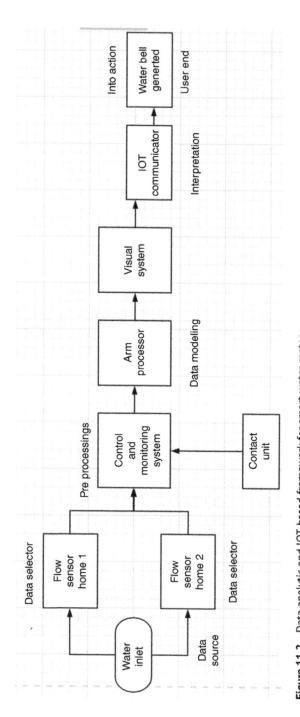

Figure 11.2 Data analytic and IOT-based framework for smart water meter.

11.3.2 Contact Unit (FC-37)

The contact water sensor module is used to measure water intensity as well as water drop falls. The module is also named as the detection-control-monitor module. The chapter proposes this module for water leakage detection, controlling the solenoid valve, and monitoring the water bill.

11.3.3 Internet of Things Communicator (ESP8266)

All wireless networking solutions are designed using the IoT platform. The communication is designed in two types of application mode such as host application and specific receiving application mode. Using GPIO pins, it can connect to sensors and other devices. It gives a platform to specific features includes integration, processing, storing, and displaying.

11.3.4 GSM-Based ARM and Control System

Automatic meter reading system is a smart system where data are collected remotely. It includes various technologies such as wireless sensor networks, GSM networks, ZigBee, and RFID. The GSM-based AMR system conveys the water meter information to consumers and water distributors automatically. The information includes water flow rate, home water meter ID, distribution time, and consumption amount, date, and time. The data are stored in the cloud server for processing and monitoring. At the end of the month, the server will calculate the bill amount automatically depending on information and send the bill through SMS gateway. On request from the customer, the bill can be generated on an hourly, daily, and monthly basis.

The method replaces manpower, data acquire period, water leakage, and payment delay. It improves data security, customer satisfaction, and more profit. The system will broadly highlight the information such as the control valve is in the on–off state and any leakage in the water pipe.

11.3.5 Methodology

In the traditional water distribution system, the control valves of each society or whole area are turned on/off depending on water requirements to each endpoint, even though some endpoints do not receive the water or receive water at low water pressure. To control and monitor such a problem, the embedded device having a control valve can be turned on/off. The flow sensor sends data to the cloud using the controller via the Internet. The generated data are stored using cloud technology and control valves are automatically turned on/off for water distribution. The flow rate is measured using a water flow sensor.

Let us consider some of the designed parameters related to the flowmeter.

V = Water velocity, Q = Water flow rate volume

M = Mass flow rate, A = Water pipe cross-sectional area

ρ = Water density

The chapter proposes water flow through two pipes which are of constant cross-sectional area A, constant pressure, constant density, respectively. Let the pressure, velocity, cross-sectional area, volume and mass water flow rate, and water density are expressed as $p1$, $v1$, $A1$, $Q1$, $m1$, and $\rho1$ for water pipe 1(Home 1) and the corresponding values for water pipe 2 (Home 2) are $p2$, $v2$, $A2$, $Q2$, $m2$, and $\rho2$, respectively. We also assume that the water flowing is incompressible.

For constant parameters such as,

$$P1=P2=P, A1=A2=A, \rho1=\rho2=\rho.$$

When water flow is in continuous mode,

$$Q = A2V2 = AV2 = A1V1 = AV1. \tag{11.1}$$

Assume there is no water leakage in both the water pipes and consider the full volumetric distribution of water. Also, we can design using the pressure of water flow consideration,

$$Q \propto P^2$$

$$P = \left(\rho \times Q^2\right) / \left(A1A2\right), \tag{11.2}$$

where P is the pressure at the output side, Q is the volume flow rate, water density is 1 (Constant), and A is an area of pipe (Constant). This equation is for an ideal volume flow rate.

With this technique, the water valve can control and monitor each endpoint of the water supply. Similarly, the necessary step can be immediately taken for such issues. The control over each Home model is that, if any Home model does not receive water with more pressure, then set different timings for water distribution to that particular Home. When a specific Home model is considered, and then automatically turns off the control valve of other areas or society. Generally, flow rate becomes maximum when the full amount of water is passing with no internal leakages.

Figure 11.3 shows the architecture of the automatic meter reading and control system. The water meter reading is stored and monitored continuously. The digital water meter is integrated with the controller unit for data collection, and interfaced with GSM for communication of messages.

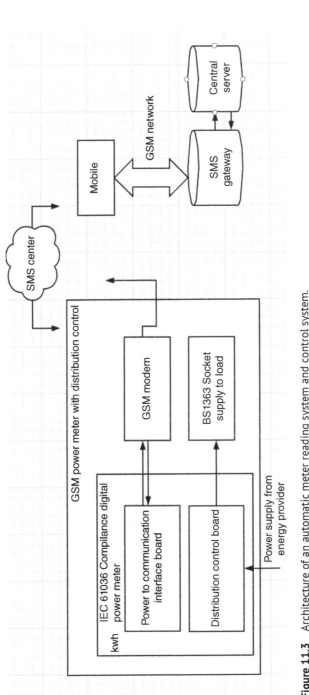

Figure 11.3 Architecture of an automatic meter reading system and control system.

The GSM modem will transfer all acquired data to AMR on request. When the water distributor puts a request for a water consumption bill, then the SMS gateway will respond automatically to all the information. The information includes water flow rate, Home water meter ID, distribution time, and consumption amount, date, and time. The data are stored in the cloud server for processing and monitoring. The month-end server will calculate the bill amount automatically depending on information and sends the bill through SMS gateway. In such a way, user-friendly technology with transparency between customer and water provider is maintained. Figure 11.4 shows a water meter block schematic for measurement of meter reading from a remote location.

The server will send a meter reading request message to the GSM modem. The Master microcontroller acknowledges the message from the SIM card and communicates with the Slave microcontroller. It calculates continuously from the meter through a serial peripheral interface. The Slave communicates with the Master and sends a current reading. The message is replied on request of the server and is received by the server's modem. The modem will forward the message to the database for the data collection system. Figure 11.5 shows an experimental setup for cloud-based water meter. The controller unit disintegrates with the water meter system and the GSM module.

11.3.6   Proposed Algorithm

- Initialize the Controller and GSM module, Set the GSM to transmit volume after every one minute.
- Set the Home 1 control valve, the Home 2 control valve ON (Open).
- Read control valve value, check rotor of Flow sensor 1 and 2 roll.

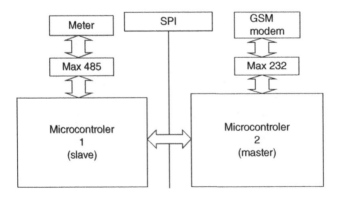

Figure 11.4   Block schematic for GSM-based water meter reading and control system.

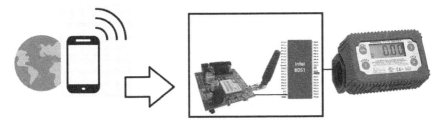

Figure 11.5   Design of GSM-based ARM and control system water flow sensor.

- Display water flow value and turn OFF control valve.
- Generate warming message and Upload flow sensor value to database host.
- Generate the water bill and control valve automatically closed.

11.4   Implementation and Result Analysis

To analyze water flow measurement, a control valve test is proposed for different rates. At various frequencies and times, the supply of water for Home 1 and Home 2 is listed. By setting the initial assumption that when the water valve is in closed state, no water flows through the pipe and zero liters of water consumption. Similarly, when the water valve is opened, the corresponding water flow rate is calculated and displayed. When the control valve is just opened, the least amount of water is flowing with frequency measurement of 8 Hz. The highest frequency measurement of 64 Hz is considered when the control valve is completely opened. The comparison chart between frequency in Hz and time (in a sec) taken to distribute water to Home 1 and Home 2 is shown in Table 11.1.

Figure 11.6 shows the graphical representation of water flow rate for two Homes 1 and 2. The water distribution for the two homes is different for the same time period.

The research proposes the GSM Module SIM900 for communication. The operating frequency band is 850, 900, 1800, and 1900 MHz, respectively. Through UART, the module communicates with specific devices. Speculative analysis for water flow rate and its corresponding bill generation using the GSM module is marked in Table 11.2. By setting the initial assumption that when the water valve is in the OFF state, no water flows through the pipe and zero liters of water consumption are displayed in the bill. Similarly, when the water valve is in the ON state, the corresponding water flow rate is calculated and displayed. Table 11.2

Table 11.1 Time taken to supply at different frequencies and the resulting flow rate.

Time (S)	Frequency (Hz)	Flow rate (l/min) Home 1	Flow rate (l/min) Home 2
80	8	5.36	9.24
70	16	7.68	12.56
60	24	9.23	14.26
50	32	11.78	16.81
40	40	13.67	18.20
30	48	14.89	21.33
20	56	16.45	25.03
10	64	17.68	26.667

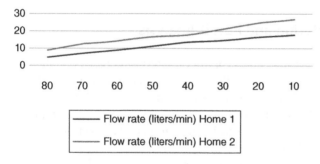

Figure 11.6 Time vs. water flow rate of Home 1 and Home 2.

Table 11.2 Flow rate vs. bill generated.

Flow rate (l/min) Home 1	Flow rate (l/min) Home 2	Bill generated for Home 1 (in rupees)	Bill generated for Home 2 (in rupees)
5.36	9.24	3.10	8.45
7.68	12.56	5.27	10.05
9.23	14.26	7.13	11.18
11.78	16.81	8.41	12.67
13.67	18.20	9.10	14.13
14.89	21.33	10.15	16.08
16.45	25.03	11,24	19.41
17.68	26.667	13.33	22.133

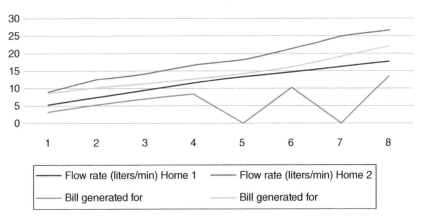

Figure 11.7   Water flow rate with corresponding bill generation.

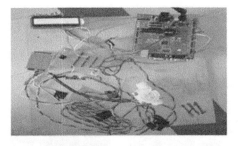

Figure 11.8   Experimental setup module of Home 1.

shows for every 10 second, how the flow rate is measured, and its corresponding bill generation for Home 1 and Home 2 module (Figure 11.7).

A solenoid valve has been used as a control valve in this work. The valve opens or closes to either allow water to pass through or prevent it from flowing. The opening and closing of the valve are under the control of the water supply utility.

Figure 11.9 Flow rate and corresponding bill of Home 1.

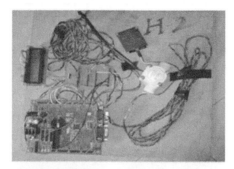

Figure 11.10 Experimental setup module of Home 2.

It gives access to water for authorized customers who are in good standing in terms of service payments. Depending on the control valve On–Off conditions, a message is sent to the user whether water consumption is calculated during the opening and closing of the control valve. Several tests have been carried to test the water consumption rate with the message sent.

Figure 11.8 shows the hardware module of Home 1. Initially, when the power supply is switched on, then the Home 1 module is activated and LCDs are

Figure 11.11 Flow rate and its corresponding bill.

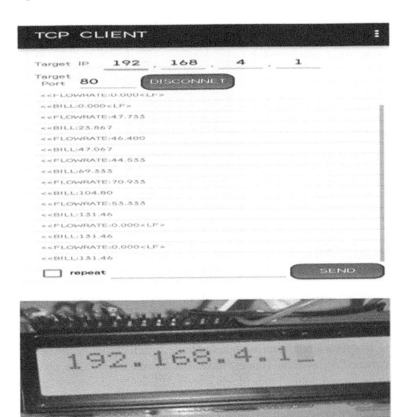

Figure 11.12 Data distribution using Wi-Fi IOT communicator app.

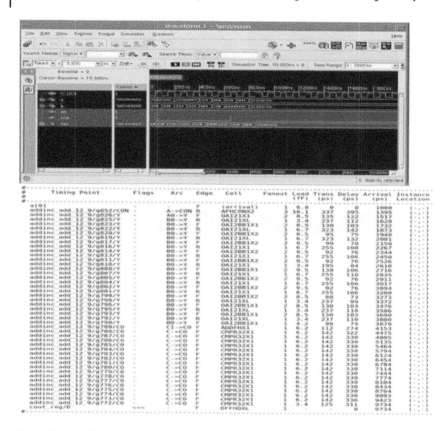

Figure 11.13 Water report of both the houses.

connected with its corresponding home number. When the control valve is open, LCDs' amount of water flows through the flow sensor (Home 1). The bill is generated based on water supply and consumption (Figure 11.9).

Figure 11.10 shows a hardware module of Home 2. Initially, when the power supply is switched on, then the Home 2 module is activated, and LCDs are connected with its corresponding home number.

When the control valve is open, then LCDs' water flow and calculates that the quantity of water flows through the Home 1 water flow sensor. The bill is generated based on the quantity of water supply flowed in and consumption.

This process keeps on repeating for modules of Home 1 and Home 2 and the bill is generated simultaneously based on flow rate of different values as shown in Figure 11.11. The data are displayed in the LCD display and sent to the mobile app through Wi-Fi IOT communicator as shown in Figure 11.12.

11.4.1   Water Report for Home 1 and Home 2 Modules

The result analysis includes the water consumption of both modules. Figure 11.13 suggests the design and implementation of water flow rate, pressure report, area of pipe, control valve report, and timing report. The water report shows the water bill has been reduced as consumption or flow rate reduced as this one is a comparative study of bill reduction and the area report shows the net area as well as total area.

11.5   Conclusion

A cost-effective and more accurate cloud-based smart water meter management system has been implemented. The proposed data analytical cycle includes different spectrums such as data acquisition, data integration, data distribution, data processing, data storing, and display. The technology tools used are sensor networks, Wi-Fi, cloud platform, modeling, and analysis using the processor and web-based tools. The water consumption is displayed and transmitted to a mobile phone through GSM Network to iOS and Android-compatible applications. Two sets of modules Home 1 and Home 2 metering systems are designed and tested. The test results show, for flow rate of 17.688 and 26.667, the bill generated is 13.33 and 22.133, respectively.

References

1 Mwangi, P. (2016). A low cost water meter system based on the global system for mobile communications. *International Journal of Computer Applications* (0975 – 8887) 142 (12): 7–12.

2 Schantz, C., Donnal, J., Leeb, S. et al. (2014). WaterWOLF: water watch on load flow. *WIT Transactions on The Built Environment* 139: 12.

3 Raykar, M.M., Vinod, P., and Vinod, P. (2015). Preethi K. M, L. Jain,"automated water billing with detection and control of water leakage using flow conservation,". *International Journal of Engineering Development and Research* 3 (2): 285–287.

4 Amatulla, P., Navnath, B.P., Yogesh, B.P., and Ashwini, Z.S. (2017). IOT based water management system for smart city. *International Journal of Advanced Research Ideas and Innovation in Technology* 3 (2): 379–383.

5 Kumura, T., Suzuki, N., Takahashi, M. et al. (2015). Smart water management technology with intelligent sensing and ICT for the integrated water systems. *Special Issue on Solutions for Society-Creating a Safer and More Secure Society* 9: 1.

6 Shahanas, K.M. and Sivakumar, P.B. (2016). Framework for a smart water management system in the context of smart city initiatives in India. *Procedia Computer Science* 92: 142–147.

7 Kumar, A., Kamal, K., Arshad, M. O., Mathavan, S., & Vadamala, T. (2014). Smart irrigation using low-cost moisture sensors and XBee-based communication. *IEEE Global Humanitarian Technology Conference (GHTC 2014)*, pp. 333–337. IEEE.

8 Mutchek, M. and Williams, E. (2014). Moving towards sustainable and resilient smart water grids. *Challenges* 5 (1): 123–137.

9 Khalifa, T., Naik, K., and Nayak, A. (2010). A survey of communication protocols for automatic meter reading applications. *IEEE Communication Surveys and Tutorials* 13 (2): 168–182.

12

Data Security in the Internet of Things

Challenges and Opportunities

Shashwati Banerjea, Shashank Srivastava, and Sachin Kumar

Department of Computer Science and Engineering, Motilal Nehru National Institute of Technology Alslahabad, Prayagraj, Uttar Pradesh, India

12.1 Introduction

The Internet of Things (IoT) [1] covers an environment where every object, be it a smart device (for example, a smart phone) or a dumb noncommunicating object (such as a street light), is able to connect to the Internet and communicate information. These devices may not only communicate the received data, but in some cases also have the capability to collect the data from nearby devices, store the data, and further carry out some initial processing on it. The producers and consumers in this ecosystem include human beings and number of different machines. The number of devices connected to the Internet has exceeded the human population in 2010 [2] and it is expected that by the end of 2020, it would cross 50 billion [3].

However, with the unprecedented growth in the number of connected devices, providing security, privacy, authorization, authentication, trust establishment, and access control becomes a major challenge [4, 5]. For example, a smart healthcare system allows a doctor to monitor the health status of a patient sitting remotely. However, in the absence of proper security mechanisms, the health record can be leaked and further modified, which can lead to dangerous outcomes. The adoption of smart IoT-based systems greatly depends on the security assurance provided by the application developer [6].

The IoT ecosystem generates a huge amount of data, often referred to as big data. The analysis of this collected data helps in predicting business trends, predict deadly diseases at an early stage, predict natural calamities, etc. Furthermore,

Big Data Analytics for Internet of Things, First Edition. Edited by Tausifa Jan Saleem and Mohammad Ahsan Chishti.
© 2021 John Wiley & Sons, Inc. Published 2021 by John Wiley & Sons, Inc.

this technological trend brings in a new generation of data-intensive application and automation system ranging from driverless cars, remote monitoring of patients, monitoring the crop production, etc. This automation technology improves the lives of human beings; however, as the reliance on technology increases, the security and privacy of the data becomes more critical. Any damage or misuse of the data not only affects an individual, but may have everlasting impact on a large population.

The chapter presents a survey on the IoT security threats and vulnerabilities. The chapter begins with a brief introduction of IoT (Section 12.2) and the challenges relevant in securing the IoT (Section 12.2.1). We discuss the common three-layered architecture of IoT and vulnerabilities associated in each layer (Section 12.2.2). Then, we present a taxonomy of the IoT security on the basis of authentication, authorization, and communication (Section 12.3). We move on to discuss the data security in IoT context, the requirements and relevant work done in the domain (Section 12.4). Finally, we have concluded our work in Section 12.5.

12.2 IoT: Brief Introduction

With the proliferation of handheld devices and high-speed networks, the IoT has evolved as a domain of huge potential, impact, and growth. It is an extension of the Internet obtained by the fusion of mobile networks, wireless sensor networks, adhoc networks, and intelligent devices to provide better services to the applications or users. Figure 12.1 presents an overview of the IoT.

This huge diversity in IoT makes it vulnerable to numerous varieties of attacks such as availability, service integrity, secure communication, data confidentiality, privacy protection, etc. The IoT architecture can be broadly classified into three layers. The lowest layer in the hierarchy is the sensing layer, which comprises the sensing devices which collect data from environment. Above the sensing layer lies the network layer which acts as communication backbone. The topmost layer is the interface layer, which acts as user interface and deals with a variety of applications ranging from smart home, to smart transportation, to RFID tag tracking, etc. The sensing devices have limited resources such as battery, computing capability, and storage and communication capability. Thus, complex security-providing algorithms cannot be executed on this layer. In the network layer, the devices use different communication strategies such as Bluetooth, Zigbee, Wi-Fi, etc. Each of the protocols has a different security requirement which facilitates eavesdropping, man-in-the-middle (MitM) attack, Denial-of-Service (DoS), etc. At the upper layer, providing lightweight authentication, data aggregation, designing encryption technique becomes challenging. The layer-wise security requirements have been discussed in Section 12.2.2.

Figure 12.1   Overview of IoT.

12.2.1   Challenges in a Secure IoT

The major challenges in a secure IoT are as follows:

a) Different IoT devices use different protocols and technologies for communication. Moreover, each IoT device has a different processing, transmission and computing capability that create complex configurations.

b) The technologies used IoT environments are still evolving and they lack maturity.

c) IoT encompasses a big space. A single solution requires security at machine, communication, and cloud levels to understand actual vulnerabilities.

d) The lower layer devices in IoT solutions are generally produced in mass with the same configurations. Absence of proper installation makes them vulnerable to attacks.

e) The same IoT device may be deployed in different environments having different security profiles. Consider the example of a temperature sensor. The same sensor might be deployed to monitor room temperature or may have been embedded in a nuclear reactor. Both the applications have different data protection and security requirements.

f) Gateway devices are usually the first point of contact for sensors and actuators. However, these legacy devices were generally not connected; thus, these devices do not possess the basic minimum protection mechanisms.

g) With the growing number of IoT-based applications, the number of IoT devices deployed is increasing at an unprecedented rate. Therefore, it is not possible to reboot the deployed IoT devices. This makes them prone to several types of threats.

12.2.2 Security Requirements in IoT Architecture

A typical requirement of IoT system is to collect, transmit, and process the information received from end devices through the network to provide services to users. Furthermore, the applications should have strong security protection. As discussed in Section 12.2, there are three layers in IoT: sensing, network, and interface layer [7]. Figure 12.2 presents the layered architecture. Each layer has its own security requirement. The vulnerabilities at each layer and the possible security requirement are discussed in the following sections.

12.2.2.1 Sensing Layer

The devices in the sensing layer sense the environment and transmit data to the designated gateway. The different hardware devices used in any IoT application include RFID tags, Bluetooth, and Zigbee-enabled devices. The RFID tags establish a wireless communication with the reader to exchange information [1]. The possible threats on RFID are tracking, DoS, repudiation, spoofing, eavesdropping, and counterfeiting [5].

Zigbee consists of radio and microcontroller. The Zigbee-enabled devices are small-sized, have limited power consumption, and are cheap. However, these

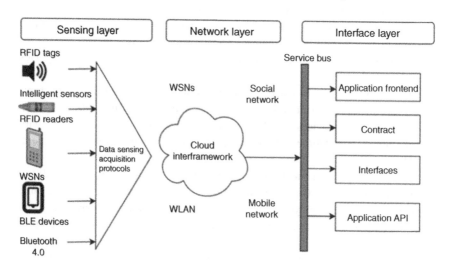

Figure 12.2 Layered architecture of IoT.

devices are vulnerable to threats and attacks, such as packet manipulation, hacking, key exchange, KillerBee, and Scapy [8]. Bluetooth too is also prone to eavesdropping, DoS, Bluesnarfing, Bluejacking, car whisperer, and Bluebugging attacks [9].

The major issues are as follows:

1) Resource constraints: Many devices at the sensing layer are battery operated, which limits the power availability. Moreover, the resources have limited computational, transmission, and computing capability.
2) Cost and size: The devices should have low cost and require minimum space.
3) Deployment: The deployment is application specific. For example, the IoT devices deployed in atomic power plants have to be sustainable to high temperature and pressure. Similarly, the IoT devices deployed for undersea application, should have high transmission capability, consume low power, etc.

The potential vulnerabilities are presented in Table 12.1.

Following measures can be taken to ensure security at this layer:

1) Implementation of security standards at the IoT devices.
2) Trustworthy data sensing system.
3) Secure software or firmware at the IoT node.

12.2.2.2 Network Layer

The conventional wireless networks are different from IoT networks. The IoT devices are set up on Low Power and Lossy Networks (LLN). LLNs have limited

Table 12.1 Vulnerabilities in sensing layer.

Name of the threat	Description
Physical damage	The devices may be damaged by attackers
Unauthorized access	The device may captured to reveal sensitive information
Availability	The device may stop working since it has been physically or logically been acquired by the attacker
Spoofing attack	The attacker masquerades as IoT end device, gateway, or end node by falsifying data
Selfish threat	The IoT device might stop working to save resource such as bandwidth, power, or battery
Botnet-based attack	Insert malicious codes in the IoT device which are later controlled by some centralized server. These devices can be used to carry out distributed denial of service (DDoS) attack

memory and processing power and often have dynamic topology. These factors are not considered for standard internet. Figure 12.3 presents the difference between the Internet and LLN. Furthermore, the networks suffer from data loss due to node impersonation. For example, during transmission of data, if an attacker is able to establish a valid connection, it can pretend as an authentic node. As an example, consider a smart meter application. The attacker may modify the readings and send erroneous messages [10].

The major concerns in network layer are as follows:

- MitM attack: The attacker earwigs the message meant for some other receiver, and relay it to the receiver. The receiver unknown of the attack, believes that the message is coming from the authentic sender. The attacker may simply read the message and redirect it, or may modify it.
- Fake network message: Attacker could send wrong/false information to transform the working of the device.
- Communication security: Integrity and confidentiality of signaling in communication.

The threats in the network layer are discussed in Table 12.2.

In order to incorporate security in the network layer, following actions should be taken:

a) Secure transport encryption.
b) Authentication/authorization.

Figure 12.3 Internet and LLN.

Table 12.2   Threats in network layer.

Name of the threat	Description
Denial-of-service (DoS) attack	An attempt to make the service unavailable to users
Data breach	Secure information being captured to untrusted authorities
Transmission threats	Threats during transmission of data such as blocking, data manipulation, forgery, etc.
Routing attack	Attack on routing path

12.2.2.3   Interface Layer

Currently, IoT is being deployed in so many smart applications such as smart city, transportation, health care system, etc. As discussed in the previous sections, each IoT application requires different hardware devices which, in turn, are based on different communication protocols. Ensuring security in such situation becomes challenging. The smart gateways such as smart meters deployed in smart city applications are prone to different kinds of malicious activities [11].

Another application which involves sensitive information is smart healthcare system. The health records of patients are vulnerable to thefts, misuse, hacking, and cyberattacks. Furthermore, eavesdropping and MitM attacks may lead to serious repercussions.

Intelligent transportation is another application of IoT. Smart traffic controlling, parking easy to access public transportation, etc., definitely make the human lives easy. However, smart transportation system is also vulnerable to different cyberattacks, DoS, and other such type of malicious attacks.

Security at the application layer has the following concerns:

1) Integrity and confidentiality for transmission between layers.
2) Cross layer authentication and authorization.
3) Security during software downloading and updating, software patches, etc.

Table 12.3 presents the vulnerabilities at this layer.

12.2.3   Common Attacks in IoT

- **DoS attack:** A DoS is a kind of attack where the attacker tries to disrupt the services being provided by a machine by flooding it with unnecessary requests with the intention to overload the system and prevent the legitimate requests from getting fulfilled. The attack is carried out by forming a network of bots.

Table 12.3 Vulnerabilities in interface layer.

Name of threat	Description
Mis-configuration	Mis-configuration at remote IoT end node, end device, or gateway
Remote configuration	Fails to configure at interfaces
Management system	Failure of management system

- **Selective forwarding attack:** In this attack, the malicious node drops certain packets randomly and allows the rest to flow as it is in the network. The dropped packets may contain necessary data for processing.
- **Witch attack:** In case a legitimate node fails, a compromised node takes benefit of this situation. A route through the compromised node is used for subsequent communication that leads to loss of data.
- **HELLO flood attacks:** This attack is mini version of DoS attack. A malicious node broadcasts a hello packet to all its neighbor legitimate node regularly. When legitimate nodes get this hello packet, then not able to process actual packet and busy in this processing, this useless hello packet then affects availability of nodes.
- **Sinkhole attack:** The compromised node extracts data from all the surrounding nodes. This generally takes place if the sensors are left unattended and carelessly for long time.

12.3 IoT Security Classification

Primarily, the security domain of IoT can be classified into three categories: application, architecture, and communication [12]. Different IoT-based applications are developed that have different security requirements. Figure 12.4 presents the IoT security taxonomy.

12.3.1 Application Domain

The most common security features in application domain are authentication, authorization, trust establishment, exhaustion of resources, etc.

12.3.1.1 Authentication
Authentication validates that the data coming from the source is the connected peer device. Before exchanging actual information, the devices in communication guarantee that the data are coming from valid source. Several research works have been proposed in this regard.

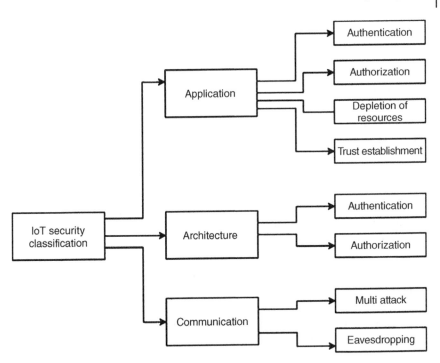

Figure 12.4 IoT security classification.

The authors in [13] proposed a hashing-based authentication scheme. The element extraction process has some lightweight irreversible features that ensure security of connected devices. The element being extracted is shared along with a hash function to prevent any kind of attack. Theoretical analysis has been done by the authors; however, no supporting practical implementation has been given.

The authors in [14] proposed measures to ensure both authentication and access control in IoT. The authors proposed elliptic curve cryptography (ECC)-based key establishment methodology for device authentication. Access control policy was implemented using Role-Based Access Control. The implementation results prove that the sensor devices had a high communication overhead.

On similar lines, the authors in [15] also proposed a scheme that ensures authentication in IoT. The authors have used ECC to generate a session key to be exchanged between the communicating parties. The authors have focused on providing authentication; however, the proposed work does not shed any light on the access control policy to be followed by the devices.

The authors in [16] proposed a lightweight security approach suitable for IoT devices with limited capability. The authentication model is based on identity-based approach. In order to ensure protection from MitM attacks, a timestamp is also exchanged among the communicating devices as a part of authentication message. There are three steps in the authentication phase: key generation, establishment, and implementation. In the key generation stage, ECC-Diffie–Hellman is used to generate a session key. In the next phase, identity of the device is established by one-way or two-way authentication protocol. The last phase grants access control to devices so that communication takes place.

Security of RFID has been proposed in [17]. Establishing secure communication between RFID reader and tags is even more challenging because of the limited computational capability of these devices. The authors proposed a lightweight authentication protocol which can be applied on constrained devices.

IoT devices are generally resource-constrained. These devices require lightweight authentication protocols. On these lines, [18] proposed an ECC-based lightweight no-pairing attribute-based encryption (ABE) scheme to address the security and privacy aspects of data. Undoubtedly, this mechanism decreases the computation and communication overhead in the IoT. However, ABE has limited scalability which makes it difficult to deploy multi-authority applications.

The authors in [19] and Raza et al. [20, 21] proposed a mechanism for channel establishment to ensure confidentiality and preserve the data content during data transmission. The scheme is able to analyze the confidential value of the dataset. However, the scheme does not consider multidimensional sensor data. It cannot measure the privacy degree on the basis of multivariate data.

12.3.1.2 Authorization

Authorization ensures that the data are available only to users with appropriate privileges. Authorization is another important aspect of security, since the data must be accessible to authentic users. Furthermore, in IoT, the data consumers may be human users, servers, machines, etc. Thus, it is very important to ensure that the data are made available to the right user.

The authors in [22] proposed identity authentication at IoT devices. The nodes in communication share matrix among them. Each message is encrypted using a key, node ID, and timestamp. Timestamp attached with the message is used for authentication. Changing the password constantly improves the security of the specific IoT framework.

12.3.1.3 Depletion of Resources

IoT devices have limited computing, storage, and transmission capability. Also, the devices have limited battery power. The resource depletion/exhaustion attacks aim to drain the capability of the targeted devices. Resource-exhaustion is a type of attack that causes unnecessary resource allocation, or consumption of resources when it is not needed leading to wastage of the resource. The attack is typically performed by introducing loops in the path, thus extending the path during packet transmission. Another common strategy adopted by attackers is to transmit packets that require intervention by the target nodes. The attacker may send large number of packets from one or more attack nodes. The goal of these attackers is to drain the batteries of the target sensor nodes. The target node may be either required to retransmit the packet, or send back some response within a limited time frame. These attacks have more devastating effects than DoS attacks, since a large number of nodes may be unavailable at the same time and resource depletion can make these nodes worthless.

12.3.1.4 Establishment of Trust

Trust management involves data protection of IoT users. Establishment of trust is necessary because it is used for authentication, both on the IoT devices as well as data and during communication of data among different IoT applications. Trust management becomes especially difficult in IoT application because it involves heterogeneous nodes each of which has a different parameter for measuring trust.

The authors in [23, 24] proposed a secured IoT architecture. The authors studied the complexity of establishing trust among heterogeneous entities. The proposed work aims to analyze the security challenges and threats imposed by IoT.

The authors in [25, 26] proposed a model for trust establishment on a heterogeneous network. The authors studied the technologies for controlling heterogeneous devices in IoT.

The work presented in [27] proposed a model for an entity on how to access the trust data from a secure system service. However, the authors have not proposed any particular network architecture to evaluate the authentication of parameters and determine how it can be used in IoT.

The authors in [28] proposed a model that establishes trust between two devices that enables communication among them. However, the proposed work did not focus on the development of trust negotiation tool that handles security and privacy requirements in heterogeneous IoT environments.

12.3.2 Architectural Domain

An important feature of IoT is the diversity of connected objects. Each connected device has a different computational capability and uses a different protocol for data transmission. Currently, there is no universally adopted protocol stack that is considered as standard for communication. The proposed research works target authentication and authorization in IoT architecture.

12.3.2.1 Authentication in IoT Architecture

SDN-based architecture has been proposed by [29]. SDN allows the controller to have a global view of the network. This, in turn, gives the administrators the flexibility to control the network according to the application requirements. It allows programmability of the network functions. However, the SDN controller is prone to central point of failure. Lack of proper security mechanism at the controller may crash the working of the entire network.

The authors in [9] proposed an architecture for IoT-based smart healthcare systems. The architecture uses public key-based authentication and ECC-based primitives for data authentication and integrity. The Elliptic Curve Digital Signature Algorithm (ECDSA) is used for key exchange while Elliptic Curve Diffie–Hellman (ECDH) is used for secure data exchange among the communicating parties. Both the ECC-based primitives are well suited in a constrained environment. DoS attack is one of the major drawback of the proposed architecture. Healthcare systems store sensitive health-related information of users. Any breach can result in the leakage of sensitive information. Furthermore, the architecture does not specify any algorithm that ensures privacy of the stored information.

12.3.2.2 Authorization in IoT Architecture

The authors in [30] demonstrated several IoT security issues in a three-layered IoT system architecture. The three-layered architecture comprises of perception, network, and application layer. The proposed work identifies security aspects at each of the three layers. The common devices in the perception layer include RFID, Zigbee, sensor devices, etc. The network layer too is prone to many types of attacks such as MitM, eavesdropping, etc. For every application, the security requirements are different which makes security more cumbersome. A secured-unified architecture is still required. Proper authentication mechanism, public key infrastructure (PKI), security routing, and intrusion detection must be set up.

The authors in [31] proposed a software-defined IoT model (SDIoT) for managing the security aspects of IoT. The proposed work integrates the SDN, SDNStore, and SDNSec in a single software-defined control model. This model handles the problems faced in traditional architecture. However, the problems such as compatibility, security, and interoperability still exit. The authors have not developed an implementation of SDIoT to test the different forms of IoT topologies.

12.3.3 Communication Channel

Different IoT devices share and exchange information among them. There is no standard communication protocol being followed by all the devices. Every device might be using a different communication technology. Thus, the data being transferred becomes prone to several types of attacks such as MitM, eavesdropping, etc. In MitM attack, the attacker captures the packet targeted for some other recipient, makes modifications in the information, and sends it to the receiver. The recipient unknown of the attack has the perception that the data has been sent by an authentic sender. MitM is a serious security threat because it gives the attacker the authority to control the communication channel. The attacker forms a link between the two communicating nodes, captures the packet intended for a different recipient, and possibly modifies the data and redirects the changed data to receiver. MitM is especially dangerous for healthcare, financial, military, and other such type of sensitive applications. Several research efforts have been proposed to provide security against MitM attacks. The attacker usually takes benefit of the flaws in the authentication protocol to carry out the MitM attacks [32].

Eavesdropping is the mechanism to passively listen to the communication between two entities. The data packets are sniffed and subsequently analyzed to generate information [33]. Privacy is used as the mechanism to provide security [34].

12.4   Security in IoT Data

Data security is not a new problem. Addressing this problem dates back to early 1970s [35]. The work [36] provides a background of the research efforts on access control techniques, which form the foundation stone of data security. However, back in those days, the access control and privacy techniques were designed for data stored in corporate database systems.

In the present day, many applications may need to access the data at the record level and thus, providing access at the abstraction level may not be a viable solution. Thus, we need to incorporate both the old and new methodologies to provide complete data protection.

IoT data security necessitates developing new solutions as well as extending existing security solutions to fit in the IoT requirements. The solutions must ensure protection while data are being transmitted and processed at the devices. Also, data availability is important, thus minimizing data loss must also be incorporated.

12.4.1   IoT Data Security: Requirements

The IoT devices generate a vast data. This data is often termed as big data. Big data is characterized by three features: volume, variety, and velocity. Volume means the size of data ranges from terabytes to zetabytes. Variety means the data does not have a fixed structure. Moreover, the type of data can be image file, audio clip, video recording, etc., which are more difficult to process. Velocity means that the data are arriving continuously from different applications at very high frequencies. It is important to understand here that, the time to analyze this streaming data is very small, since a large volume of data is continuously being streamed.

The biggest challenge arises when there is huge volume of structured, unstructured, semi-structured data arriving continuously at a very high speed from different sources. Addressing such challenge requires a breakthrough privacy-enhancing and security techniques.

Universally, there are three basic security requirements: confidentiality, integrity, and availability.

Confidentiality refers to protection of resources from unauthorized viewing or access. Integrity is protecting the data from modifications by unauthorized sources and availability guarantees data availability to authorized users. The three requirements are more prevalent in today's world since the scope of attack has expanded due to heterogeneous sources of data sharing.

Additionally, data privacy has also evolved as a critical requirement. The terms data privacy and confidentiality are often considered as a single entity. Data confidentiality is to be enforced in order to ensure data privacy. However, privacy requirements may vary from person to person. For example, a person may be comfortable to share his medical details with a researcher working in the particular domain. However, it may not be necessary that all the patients have the same perspective. Thus, it is important to consider privacy preferences of persons as well. The privacy preferences may also change over time. Therefore, in addition to access control policies, the system should be able to support user preferences and legal regulations.

12.4.1.1 Data: Confidentiality, Integrity, and Authentication

Data confidentiality is an important requirement for data privacy. The most common technique for ensuring data confidentiality is access control and encryption. This section presents a survey of research efforts done in this domain.

Encryption mechanism forms an important research area. However, just developing a new encryption technique is not sufficient to ensure data security [37]. Strong security can only be achieved if cryptographic protocols are deployed and implemented in an appropriate manner. However, the constrained computing resources and their heterogeneous nature make implementation again a critical issue. Furthermore, in case of large IoT systems, encryption key also plays a very significant role. This section presents the research efforts done in this area.

The authors in [38] proposed a certificate-less, sign encryption protocol where the protocol supports both message encryption and authentication without requiring key certificates. The proposed protocol has been compared with several other protocols and on resource-constrained devices such as Raspberry-Pi and Android. Since, the protocol is free from expensive pairing operations, it is better in comparison to other such type of protocols.

The authors in [39] proposed protocols for authentication in networked vehicles. The main challenge in this case is execution of multiple concurrent authentications. The response time is very important in this scenario, because, if a vehicle has to stop suddenly, this message is propagated to all nearby devices so that all the other devices have enough time to stop. Thus, it is necessary that the authentications procedure at both the sender and receiver has minimal overhead. Under such circumstances, the authentication operations are executed on the system-on chips GPU embedded in vehicles.

The authors in [40] have focused on encryption protocols for networks of sensors and drones. The sensors accumulate data from the deployment site, while drones hover over these sensors and collect data from them. There are two challenges in this case: energy saving and less time for key generation. The authors have used low-power listening techniques at the sensors and dual radio channels at drone. This helps in timely generation of keys, while drones approach sensors.

The authors in [41] have used a two-way authentication scheme for IoT. Specifically, Datagram Transport Layer Security (DTLS) protocol is used which lies between the transport and application layer. This architecture is designed for 6LoWPAN. The evaluation proves that the architecture provides message integrity, confidentiality, and authenticity with low memory overhead.

In order to ensure confidentiality and integrity, the authors in [42] analyzed the existing key management systems and their applicability in the IoT context. The Key Management System (KMS) protocols can be broadly classified into four categories: key pool framework, mathematical framework, negotiation framework, and public key framework. It has been presented that most of the KMS are not suitable for IoT.

12.4.1.2 Data Privacy

Data privacy is an important pillar of IoT security. IoT is being deployed in many applications such as smart healthcare, smart transportation, smart city, forensics, etc. For every application, users expect that their lifestyle-related habit, human interactions, etc., remain private. The section presents the efforts done in this area.

The authors in [43] proposed a privacy-preserving access control protocol that is user-governed. The protocol is based on context-aware k-anonymity policy. The user has full authority to control, who is accessing the data, what data is being accessed, at what time it is being accessed, etc.

In [44], the authors have divided the privacy mechanism into two classes: discretionary and limited access. The former works to check replication of sensitive data, while the latter works to limit unauthorized attacks.

The authors in [45] presented a cluster-based privacy-preserving technique for streaming data, which were once applicable on static data. The technique is named as Continuously Anonymizing STreaming data via adaptive cLustEring (CASTLE). The model defines k-anonymized cluster, by using the quasi-identifier attributes of tuples to preserve data privacy.

12.4.2 IoT Data Security: Research Directions

Among the different security requirements, data trustworthiness is an important concern in IoT security. The main reason for this concern is the data are being acquired from different devices, which may have poor calibration, errors, susceptible to physical attacks, etc. The data fusion techniques need to be extended to deal with dynamic, large-scale heterogeneous data sources. Deployment and configuration of security tools for IoT is again a challenging task, as there is a tradeoff between the associated risk and cost and energy consumption.

Another significant research direction is data protection from insider threat. Protection against insider threat requires a combination of different techniques such as context-based access control, detection of anomaly in accessing data and usage, and most importantly, user behavior monitoring. Monitoring the behavior of user also includes user privacy; therefore, understanding the tradeoff between user privacy and security risk is necessary.

The software engineering principles need to be extended to incorporate privacy assurance. The software being developed must be able to identify the portions of code dealing with sensitive data. The applications must be able to work on anonymized data, handle lack of permissions, etc.

Another research area which is in trend these days is data privacy in social networks. In social networks, collaborative approaches are required for access control. Consider an example of an image being shared on social media. The image may be related to multiple social groups, and therefore all such related users must be able to express their privacy concerns when sharing the image.

12.5 Conclusion

With the recent advancements in sensing and communication technology, the applications of IoT are growing at a rapid rate. The IoT technology provides integration of various sensors and devices to communicate with each other without requiring intervention from users. However, large-scale deployment increases the security risks and concerns. The present work highlights the IoT security threats and vulnerabilities. We have categorized the IoT security in the context of application, architecture, and communication. Furthermore, this work discusses the research directions in confidentiality, privacy, and IoT data security.

References

1 Atzori, L., Iera, A., and Morabito, G. (2010). The internet of things: a survey. *Computer Networks* 54 (15): 2787–2805.

2 Al-Fuqaha, A., Guizani, A., Mohammadi, M. et al. (2015). Internet of Things: a survey on enabling technologies, protocols, and applications. *IEEE Communication Surveys and Tutorials* 17 (4): 2347–2376.

3 Wyman, O. (2015). The Internet of Things: disrupting traditional business models. *Oliver Wyman*. http://www.oliverwyman.com/content/dam/oliver-wyman/global/en/2015/jun/Internet-ofThings_Report.pdf.

4 Gu, X.Q.J. and Wang, J. (2012). Research on trust model of sensor nodes. *WSNs Procedia Engineering* 29: 909–913.

5 Jing, Q., Vasilakos, A.V., Wan, J. et al. (2014). Security of the Internet of Things: perspectives and challenges. *Wireless Networks* 20 (8): 2481–2501.

6 Sicari, S., Rizzardi, A., Grieco, L.A., and Coen-Porisini, A. (2015). Security, privacy and trust in Internet of Things: the road ahead. *Computer Networks* 76: 146–164.

7 Li, S. (2017). Chapter 2 - Security architecture in the internet of things. In: *Securing the Internet of Things* (eds. S. Li and L.D. Xu), 27–48. Syngress. ISBN: 9780128044582.

8 Lu, C. (2014). Overview of security and privacy issues in the Internet of Things. *Internet of Things (IoT): A vision, Architectural Elements, and Future Directions*, pp. 1–11.

9 Moosavi, S.R., Gia, T.N., Rahmani, A.M. et al. (2015). SEA: a secure and efficient authentication and authorization architecture for IoT-based healthcare using smart gateways. *Procedia Computer Science* 52 (1): 452–459.

10 Suo, H., Wan, J., Zou, C., Liu, J. (2012). Security in the Internet of Things: a review. *Proceedings of 2012 International Conference on Computer Science and Electronics Engineering, ICCSEE 2012s*, Vol. 3, pp. 648–651. http://www.cse.wustl.edu/~jain/cse574-14/ftp/security.pdf.

11 Barreto, L., Celesti, A., Villari, M. et al. (2015). An authentication model for IoT Clouds. *Proceedings of the IEEE/ACM International Conference on Advances in Social Networks Analysis and Mining 2015 (ASONAM)*, pp. 1032–1035.

12 Alaba, F.A., Othman, M., Hashem, I.A.T., and Alotaibi, F. (2017). Internet of Things security: a survey. *Journal of Network and Computer Applications* 88: 10–28.

13 Gubbi, J., Buyya, R., Marusic, S., and Palaniswami, M. (2013). Internet of Things (IoT): a vision, architectural elements, and future directions. *Future Generation Computer Systems* 29 (7): 1645–1660.

14 Ndibanje, B., Lee, H.J., and Lee, S.G. (2014). Security analysis and improvements of authentication and access control in the internet of things. *Sensors* 14 (8): 14786–14805.

15 Ye, N., Zhu, Y., Wang, R.C. et al. (2014). An efficient authentication and access control scheme for perception layer of internet of things. *Applied Mathematics & Information Sciences* 8: 4.

16 Neisse, R., Steri, G., Fovino, I.N., and Baldini, G. (2015). SecKit: a model-based security toolkit for the internet of things. *Computers & Security* 54: 60–76.

17 Al-Turjman, F. and Gunay, M. (2016). CAR Approach for the Internet of Things. *Canadian Journal of Electrical and Computer Engineering* 39 (1): 11–18.

18 Yao, X., Chen, Z., and Tian, Y. (2015). A lightweight attribute-based encryption scheme for the Internet of Things. *Future Generation Computer Systems* 49: 104–112.

19 Bose, T., Bandyopadhyay, S., Ukil, A., et al. (2015). Why not keep your personal data secure yet private in IoT?: Our lightweight approach. *2015 IEEE Tenth International Conference on Intelligent Sensors, Sensor Networks and Information Processing (ISSNIP)*, pp. 1–6. IEEE.

20 Raza, S., Shafagh, H., Hewage, K. et al. (2013). Lithe: lightweight secure CoAP for the internet of things. *IEEE Sensors Journal* 13 (10): 3711–3720.

21 Raza, S., Wallgren, L., and Voigt, T. (2013). SVELTE: real-time intrusion detection in the Internet of Things. *Ad Hoc Networks* 11 (8): 2661–2674.

22 Gaur, A., Scotney, B., Parr, G. et al. (2015). Smart city architecture and its applications based on IoT. *Procedia computer science* 52: 1089–1094.

23 Chen, D., Chang, G., Jin, L., et al. (2011). A novel secure architecture for the internet of things. *2011 Fifth International Conference on Genetic and Evolutionary Computing*, pp. 311–314. IEEE.

24 Chen, M., Lai, C.F., and Wang, H. (2011). Mobile multimedia sensor networks: architecture and routing. *EURASIP Journal on Wireless Communications and Networking* 2011 (1): 159.

25 Ma, H.D. (2011). Internet of things: objectives and scientific challenges. *Journal of Computer Science and Technology* 26 (6): 919–924.

26 Yan, Z., Zhang, P., and Vasilakos, A.V. (2014). A survey on trust management for Internet of Things. *Journal of Network and Computer Applications* 42: 120–134.

27 Bahtiyar, Ş. and Çağlayan, M.U. (2012). Extracting trust information from security system of a service. *Journal of Network and Computer Applications* 35 (1): 480–490.

28 Sicari, S., Rizzardi, A., Miorandi, D. et al. (2016). A secure and quality-aware prototypical architecture for the Internet of Things. *Information Systems* 58: 43–55.

29 Valdivieso Caraguay, Á.L., Benito Peral, A., Barona Lopez, L.I. et al. (2014). SDN: evolution and opportunities in the development IoT applications. *International Journal of Distributed Sensor Networks* 10 (5): 735142.

30 Zhao, K., & Ge, L. (2013). A survey on the internet of things security. *2013 Ninth International Conference on Computational Intelligence and Security*, pp. 663–667. IEEE.

31 Jararweh, Y., Al-Ayyoub, M., Benkhelifa, E. et al. (2015). SDIoT: a software defined based internet of things framework. *Journal of Ambient Intelligence and Humanized Computing* 6 (4): 453–461.

32 Mahmood, K., Chaudhry, S.A., Naqvi, H. et al. (2016). A lightweight message authentication scheme for Smart Grid communications in power sector. *Computers and Electrical Engineering* 52: 114–124.

33 Pongle, P., & Chavan, G. (2015). A survey: attacks on RPL and 6LoWPAN in IoT. *2015 International conference on pervasive computing* (ICPC), pp. 1–6. IEEE.

34 Vučinić, M., Tourancheau, B., Rousseau, F. et al. (2015). OSCAR: object security architecture for the Internet of Things. *Ad Hoc Networks* 32: 3–16.

35 Denning, D.E. and Denning, P.J. (1979). Data security. *ACM Computing Surveys (CSUR)* 11 (3): 227–249.

36 Bertino, E. and Sandhu, R. (2005). Database security-concepts, approaches, and challenges. *IEEE Transactions on Dependable and Secure Computing* 2 (1): 2–19.

37 Schneier, B. (2016). Cryptography is harder than it looks. *IEEE Security and Privacy* 14 (1): 87–88.

38 Seo, S.H., Won, J., and Bertino, E. (2016). pCLSC-TKEM: a pairing-free certificateless signcryption-tag key encapsulation mechanism for a privacy-preserving IoT. *Transactions on Data Privacy* 9 (2): 101–130.

39 Singla, A., Mudgerikar, A., Papapanagiotou, I., et al. (2015). Haa: Hardware-accelerated authentication for internet of things in mission critical vehicular networks. *MILCOM 2015-2015 IEEE Military Communications Conference*, pp. 1298–1304. IEEE.

40 Won, J., Seo, S. H., & Bertino, E. (2015). A secure communication perotocol for drones and smart objects. In: *Proceedings of the 10th ACM Symposium on Information, Computer and Communications Security*, 249–260.

41 Kothmayr, T., Schmitt, C., Hu, W. et al. (2013). DTLS based security and two-way authentication for the Internet of Things. *Ad Hoc Networks* 11 (8): 2710–2723.

42 Roman, R., Alcaraz, C., Lopez, J., and Sklavos, N. (2011). Key management systems for sensor networks in the context of the Internet of Things. *Computers and Electrical Engineering* 37 (2): 147–159.

43 Huang, X., Fu, R., Chen, B. et al. (2012). User interactive internet of things privacy preserved access control. *2012 International Conference for Internet Technology and Secured Transactions*, pp. 597–602. IEEE.

44 Yang, J.C. and Fang, B.X. (2011). Security model and key technologies for the Internet of things. *The Journal of China Universities of Posts and Telecommunications* 18: 109–112.

45 Cao, J., Carminati, B., Ferrari, E., and Tan, K.L. (2010). Castle: continuously anonymizing data streams. *IEEE Transactions on Dependable and Secure Computing* 8 (3): 337–352.

13

DDoS Attacks

Tools, Mitigation Approaches, and Probable Impact on Private Cloud Environment

R. K. Deka[1], D. K. Bhattacharyya[2], and J. K. Kalita[3]

[1] *Department of Computer Science and Engineering, Assam Don Bosco University, Guwahati, Assam, India*
[2] *Department of Computer Science and Engineering, School of Engineering, Tezpur University, Tezpur, Assam, India*
[3] *Department of Computer Science, College of Engineering and Applied Science, University of Colorado, Boulder, CO, USA*

13.1   Introduction

The cloud computing infrastructure allows a service provider on the Internet to provide the use of computing resources to fulfill the necessary demands of users. Due to virtualization, it is possible to provide services using optimal resources. Khorshed et al. [1] define cloud computing as "a system of shared resources of a data centre using virtualization technology. Such systems provide elasticity based on demand and ask for charges based on customer usage."

Scanning, DoS, and penetration [2] can occur in a live network of computers. The Arbor Networks[1] reported the largest (at that time) DDoS attack of 400 Gbps in 2014. In Figure 13.1, we can see DDoS attack trends in 2020[2]. In particular, large-scale DDoS attack frequency has continued to trend upward, as shown in Figure 13.2.

The Mirai botnet attack is launched using IoT devices such as DVR players and digital cameras. The victims were the servers of Dyn, a company that controls much of the Internet's Domain Name System (DNS) infrastructure. It was hit on 21 October

---

1 http://www.arbornetworks.com accessed June 2020.
2 https://blog.cloudflare.com/network-layer-ddos-attack-trends-for-q1-2020/ accessed June 2020.

Big Data Analytics for Internet of Things, First Edition. Edited by Tausifa Jan Saleem and Mohammad Ahsan Chishti.

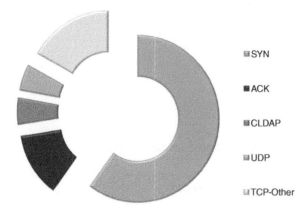

Figure 13.1 Types of intrusion scenario in 2020.

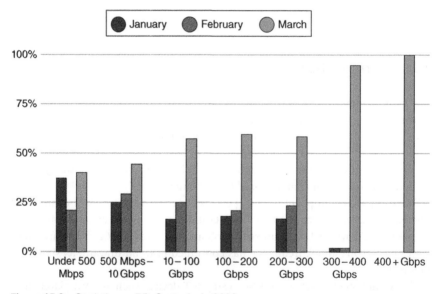

Figure 13.2 Statistics on DDoS attacks in 2020.

2016 with an extraordinary attack strength of around 1.2 Tbps and remained under sustained assault for most of the day, bringing down many sites including Twitter, The Guardian, Netflix, Reddit, CNN, and many others in Europe and United States[3].

---

3 https://www.theguardian.com/technology/2016/oct/26/ddos-attack-dyn-mirai-botnet accessed February 2020.

13.1.1   State of the Art

Research on DDoS attacks and defense in the cloud environment is still at an evolving stage. These days, researchers are very much concerned about services in the cloud and cloud security. Sabahi [3], Pitropakis et al. [4], and Grover and Sharma [5] discuss efforts to secure user data in the cloud. Rather than storing the information locally at the client's infrastructure, data are stored in the cloud provider's location. It is evident that in such a situation, people are worried about the security of their data. Thus, cloud organizations should provide adequate protection for the customer and also for the safety of their own.

In the context of the cloud, requests for resources like virtual machines (VM) can be made by any user through the Internet. As a result, a network of zombies can quickly launch DDoS attacks by sending fake requests for resources. Modi et al. [6] provide a survey of different types of intrusions which can take place in the cloud environment. Khorshed et al. [1], and Subashini and Kavitha [7] focus on flaws, challenges, and security concerns in different service layers.

In this article, we discuss the seriousness of the threats posed by DDoS attacks in the context of the cloud, particularly in the personal private cloud. We present a discussion of different approaches which used to defend or mitigate DDoS attacks in general network architecture, and also some methods that consider cloud computing technology in particular. Unlike [6], we highlight challenges and issues faced particularly by the private cloud environment when facing DDoS attacks in a general way. We have discussed about a generic framework to defend against DDoS attacks in an individual private cloud environment looking after different challenges and issues.

The first report on DDoS attack was in 1999 against servers of the University of Minnesota. In the early 2000s, many famous and major Websites like Yahoo!, eBay, CNN, and Amazon.com were assaulted by DDoS attacks [8]. Their systems were down for hours, and users were denied access to services [9]. These attacks were able to create a disaster because of the use of botnets. Stone-Gross et al. [10] and Hoque et al. [11] provide a detailed investigation of botnets [12], a network of compromised machines under the control of a master. Khorshed et al. [1] provide a survey of challenges related to the cloud and present a proactive approach toward detection of attacks in the cloud.

In [13], the methods or approaches are based on supervised learning, unsupervised learning, probabilistic learning, and soft computing. Yu et al. [14] and Xiang et al. [15] present detection methods depending on rates of traffic.

There has been some work on mitigating or tolerating DDoS attacks in the cloud environment. With the increased sophistication of attackers, protection of open systems is increasingly challenging. Nguyen and Sood [16] opine that intrusion tolerance should be a part of overall in-depth security. They compare three types of intrusion-tolerant system architectures. Lua and Yow [17] propose a method in which an intelligent large swarm network is used against the attack to mitigate it. The swarm network constantly reconfigures itself through the use of a parallel optimization algorithm i.e., the intelligent water drop mechanism [18]. Amazon has created a technique called cloudWatch[4] to monitor resources and to mitigate the situation according to the attack. Yu et al. [19] attempt to provide the theory of optimal resource allocation in a cloud platform when defending a DDoS attack. Wang et al. [20] have also developed a method on optimal resource allocation, which is adaptable to the cloud scenario.

In Table 13.1, a comparison is provided among a few existing survey papers with our work. For comparison, we choose four parameters, inclusion of attacks, description of defense solutions, issues and challenges, and addition of recommendations in these papers.

Security and complications with data privacy and data protection continue to restrict the growth of the cloud market, and these survey papers are more specific to the security issues that have been raised due to the nature of the service delivery system of a cloud environment. Sabahi [3] also pose the same concern about the cloud environment. Comparison between the benefits and risks of cloud computing is necessary for a full evaluation of the viability of cloud computing. Some critical issues that clients need to consider as they contemplate moving to cloud computing. Sabahi summarizes reliability, availability, and security issues faced by cloud computing, and proposed feasible and available solutions for some of them [21].

Zhang et al. [22], Wong and Tan [23], Kumar and Gohil [24], Chiba et al. [25], and Mishra et al. [26] present different survey-work focusing on various IDSs developed in the last few years concerning the cloud environment. Basu et al. [27] mentioned that there are differences between mappings of different challenges/ issues regarding cloud security with their own solutions. Few researchers present the virtualization challenges/issues and resolution mechanisms while others focus on techniques of the control procedure. Dong et al. [28] showed details about the DDoS attack in SDN and cloud environments. Their works also pointed out the open research problems in the identification and mitigation of DDoS attacks.

13.1.2 Contribution

This chapter presents an organized survey concerning security in the network infrastructure of cloud computing; specifically the impact of DoS and DDoS attacks on the networking services of a cloud environment. It begins with a description of types of cloud environments and then different types of DDoS attacks. It also highlights the seriousness of DDoS attacks in private clouds. We

---

4 https://aws.amazon.com/cloudwatch accessed August 2019.

Table 13.1 Comparison with existing survey articles.

Authors	Year	Attacks included	Defense solutions	Issues and challenges	Recommendations
Subashini and Kavitha [7]	2010	√	×	√	×
Sabahi [3]	2011	√	×	√	×
Khorshed et al. [1]	2012	√	×	√	×
Modi et al. [6]	2013	√	×	×	×
Wong and Tan [23]	2014	√	√	√	×
Kumar and Gohil [24]	2015	√	√	×	×
Chiba et al. [25]	2016	√	√	√	×
Mishra et al. [26]	2017	√	√	√	×
Basu et al. [27]	2018	√	√	√	×
Dong et al. [28]	2019	√	√	√	×
Our survey	2020	√	√	√	√

present an in-depth discussion of the challenges and issues in defending such attacks. The significant contributions of this survey are the following:

- Our presentation is specific to the security of cloud computing.
- There are just a handful of surveys on cloud security, and published reviews do not emphasize the impact of DDoS attacks on individual private clouds. We present challenges and issues to help the researcher in creating a defense theory and in building a defense system against DDoS attacks.
- Pros and cons analysis of a large number of detection and mitigation methods is included.
- We also discuss trending concepts such as the role of big data and software-defined networking in cloud security.
- A generic framework for device defense mechanism in a cloud-based environment is also presented.

13.1.3 Organization

The rest of the chapter is organized as follows. Different deployment models of clouds, DDoS attacks, and types of DDoS attacks along with the probable impact on private clouds are discussed in Section 13.2. Different existing approaches and potential solutions are briefed, and some recommendations for developing a defense model are presented in Section 13.3. In Section 13.4, challenges and issues related to a private cloud in defending against DDoS attacks are mentioned. A generic framework to defend against DDoS attacks is discussed in Section 13.5. Finally, we conclude in Section 13.6. In Figure 13.3, the taxonomy of terms and concepts used in the entire article is provided for better understanding for the reader.

13.2 Cloud and DDoS Attack

13.2.1 Cloud Deployment Models

A cloud node can provide three basic services to customers: IaaS, PaaS, and SaaS (Figure 13.4). The deployment differences can be seen in Figure 13.5, and an explanation of different deployment models are given below.

a) **Public cloud**: The cloud is created for the general public where free or rental services are provided. This can be accessed by any authorized user. Examples of public clouds include Amazon Elastic Compute Cloud (EC2), Google AppEngine, and Windows Azure Services Platform. A public cloud provides abstractions for resources using virtualization techniques on a large scale. It

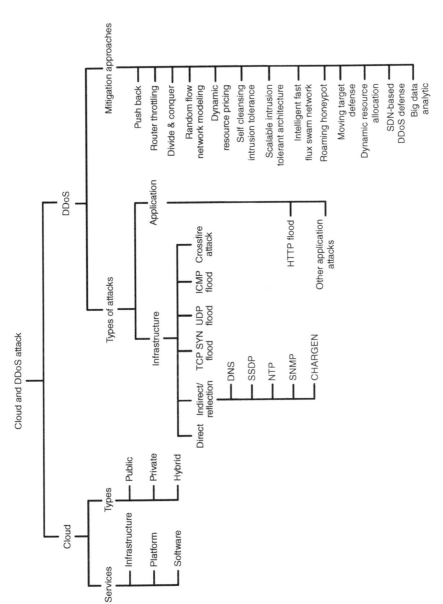

Figure 13.3 Taxonomy.

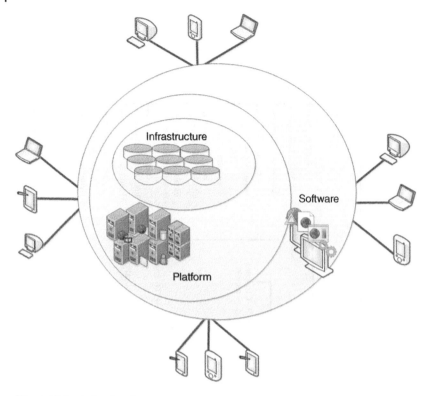

Figure 13.4 A cloud node.

benefits the user by delivering for backup and access to secure resources by synchronizing, replicating, and allocating the resources throughout the network.

b) **Private cloud**: A private cloud is designed to offer the same features and benefits of public cloud systems, usually with limited resources for maintaining the cloud environment. Unlike a public cloud, a private cloud remains within the corporate firewall, which means the private cloud is privately managed by a company for the private use of its individual users and not for the public on pay per use basis. Also, a private cloud can be used by a company. In this scenario, in that cloud, sensitive data can be stored internally and also provide the advantages of cloud computing infrastructure to their business. For example, as per the demand, Apache CloudStack, OpenStack, VMware vCloud Suite, etc., allocate the resources to the clients. Individual private cloud customers as well as the provider.

c) **Community cloud**: Few groups of users or organizations may have shared concerns (e.g. mission, security requirements, policy, and compliance

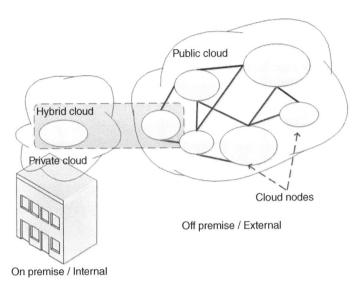

Figure 13.5 Deployment models.

considerations). Community cloud can be controlled by one or more organizations of that community, or a third party, or mixture of both. Some examples are Dimension Data, Layer- Stack, and Zoolz.

d) **Hybrid cloud**: It is a mixture of two or more linked cloud deployment models with a provision to transfer data between them. The combination may include both private and public clouds. For example, a company can maximize its efficiency by deploying public cloud services for all non-sensitive operations, but only deploy private cloud when it needs to store sensitive operations as it is surrounded by a firewall, and ensure that all of their platforms are seamlessly integrated. Some examples are Amazon Web Services, Rackspace Hybrid Cloud, EMC Hybrid Cloud, and HP Hybrid Cloud.

13.2.1.1 Differences Between Private Cloud and Public Cloud

In Table 13.2, differences between private and public clouds are enumerated. A cloud has shared general features, whether private or public. As clouds have evolved on and from the Internet, we can build defense models based on research that has been conducted on general defense solutions against DDoS attacks and features of clouds. We can then proceed to discuss individual private cloud defense. Private clouds require more attention because they have limited resources, and the cost is high during an attack compared to a public cloud. We know that a private cloud is accessed by authorized users or private organizations paying money as per need. Both ends (customer and service provider) heavily rely on

Table 13.2 Differences between private and public cloud.

Key points	Private	Public
Use of Technology	Old	New
Capital expenses	Not shifted	Shifted to operational expenses
Utilization rate	Low	High
Infrastructure cost	High	Low
Elasticity	Less	More
Economies of sale	Less	High
Business attraction	Low	High
Security	Less	High
Perimeter complacency	Suffer	Not suffer
Skill level	Unknown	Usually high
Penetration testing	Insufficient	Sufficient
Business focus	Deeply in data center	Out of data center

security. A DDoS attack can cripple the whole private cloud and jeopardize entire businesses. So, DDoS attack is more threatening to individual private cloud customers than a public cloud's customers.

13.2.2 DDoS Attacks

13.2.2.1 Attacks on Infrastructure Level

In a Dos attack, legitimate users are denied access to the resources over the network. A botnet or a network of attackers inflicts severe damage on the victim. This distributed and coordinated attack can be called as DDoS attack. Nowadays, a lot of resources are in cloud in concentrated way and also a large number of users shared the same infrastructure. In this scenario, a DDoS attack will create huge loss [29].

Resources to compute, resource to transmit, and resources to route can be considered in the category of infrastructure. During infrastructure-level attacks, attackers can overwhelm the capacity of a limited infrastructure of individual or private cloud. Attackers send a large numbers of fake requests to access the server so that the performance of the servers can be degraded.

a) **Direct**: An example of direct infrastructure-level attack can be visualized in Figure 13.6 [30]. A DDoS attack includes an overwhelming quantity of packets sent from multiple attack sites to a victim site. These packets arrive in such a

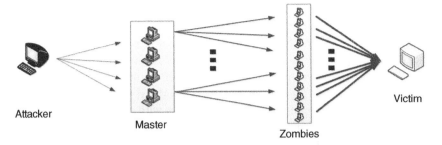

Figure 13.6 Direct DDoS attack.

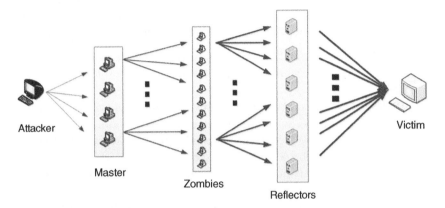

Figure 13.7 Indirect DDoS attack.

high quantity that some key resource at the victim is quickly exhausted. The victim either crashes or spends so much time handling the attack traffic that it cannot attend to its real work.

b) **Indirect**: This DoS attack compromises victim machines so that they unwillingly participate in a DDoS attack. Flashes of requests to the victim host machines are redirected or reflected from the victim hosts to the target. Some reflection or indirect-based attacks are mentioned below. The general approach is as shown in Figure 13.7.

- Domain Name Service (DNS) reflection or amplification attacks use botnets that send a large number of DNS queries to open DNS resolver using spoofed IP addresses of victims. Thus, such an attack can do a lot of damage as it is difficult to stop this type of attack at an early stage.
- Simple Service Discovery Protocol (SSDP) reflection attacks are created using the Simple Object Access Protocol (SOAP) to deliver control messages to universal plug and play (UPnP) devices and to communicate information. These

requests are created to elicit responses, which reflect and amplify a packet and redirect responses toward a target.

- Network Time Protocol (NTP) reflection attacks are created by the attacker to send a crafted packet in which requests for a large amount of data are sent to the host. NTP is used to synchronize the time between client and server.
- In an Simple Network Management Protocol (SNMP) reflection attack, the culprits send out a huge number of SNMP queries with forged IP addresses to numerous victim machines. SNMP is a network management protocol for configuring and collecting information from servers.
- Character Generator Protocol (CHARGEN) is often misused when attackers use the testing features of the protocol to create malicious payloads and reflect them by spoofing the address of the source to direct them to the target. CHARGEN is a debugging and measurement tool and also a character generator service.

c) **TCP SYN flood**: Manipulating the three-way handshake in a TCP connection, lots of SYN fragments are flooded by attackers so that legitimate users are denied.

d) **UDP flood**: Massive numbers of datagrams are transmitted to random opened ports of the victim side. Sometimes, ports remain open without knowledge of administrators, causing the server to respond. A response to each UDP packet with an IMCP unreachable reply to the spoofed source IP address makes the situation worse by overwhelming the network environment of the victimized IP addresses.

e) **ICMP flood**: ICMP flood is a ping-based DoS attack that sends large numbers of ICMP packets to a server and attempts to crash the TCP/IP stack on the server and cause it to stop responding to incoming TCP/IP requests.

f) **Crossfire attack**: A botnet controller can compute a large set of IP addresses whose advertised routes cross the same link, and then direct its bots to send low-intensity traffic toward these addresses. This type of attack is called the Crossfire attack [31].

13.2.2.2 Attacks on Application Level

a) **Common application-layer DDoS attack types**: When a heavy amount of legitimate application-layer requests or normal requests that consume heavy resources.

b) **HTTP flood attacks**: Application layer attacks come in the form of GET floods. HTTP request attacks are those attacks where attackers send HTTP GETs and POSTs to Web servers in an attempt to flood them by consuming a large amount of resources. The HTTP POST method enables attackers to POST large amounts of data to the application layer at the victim side, and it happens to be the second most popular approach among the application layer attacks.

13.2.3 DoS/DDoS Attack on Cloud: Probable Impact

The public cloud infrastructure stands a better chance against DDoS attack because a public cloud usually has a lot of resources that make it easy to counter the attack dynamically. It is almost impossible to shut down such clouds by attacking them. But, if an intense DDoS attack occurs on customers of an individual private cloud like a data center with limited resources, it cannot escape from such attack, and it becomes a battle of survival using all the resources there are to confront [32, 33]. If we allocate necessary and sufficient resources on mitigation process efficiently, then we can defeat DDoS attack on cloud platform without much caring about efficient detection and prevention mechanism [34].

Cloud Service Provider (CSP) provides two plans for the customers, i.e. for short duration and for long duration or both [35]. Economic Denial of Sustainability (EDoS) can exploit this business model of resource allocation [36–38]. Initially, the allocated resources for any application in these models are limited. Thus, it will lead to a severe DDoS attack [39, 40], whether it is spot instance allocation or any reservation of resources for maximum use.

Some possible examples of DDoS attacks in cloud environments are Smurf attack, IP spoofing attack, Tear drop attack, SYN flood attack, ping of death attack, Buffer overflow attack, LAND attack, etc., as shown in Figure 13.8 [41, 42]. From news report we can state that large-scale IoT-enabled DDOS attacks will continue to dominate enterprise security. Darwish et al. [43] discuss DDoS attacks as attacks that target the resources of these services, lowering their ability to provide optimum usage of the network infrastructure, due to the nature of cloud computing, the methodologies for preventing or stopping.

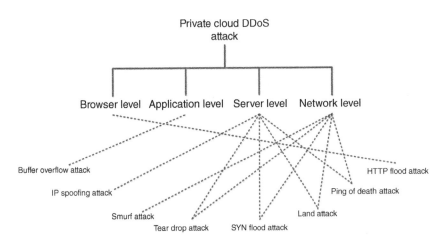

Figure 13.8 DDoS attack types in cloud.

If we compare DDoS attacks in cloud with traditional network infrastructure, we will find quite differences. But, the new approaches to tackle DDoS attack in cloud which are published through various research platforms are actually the updated or adapted versions of old or traditional ones.

We summarize below the security concerns in the private cloud against DDoS attacks in the following:

- The customers or clients associated with cloud infrastructure are in large number. Thus, it has high potential to launch a severe and huge DDoS attack.
- Characteristics of DoS/DDoS attacks are unusual slow network performance, nonavailability of resources or inability to access the servers of websites and increase in spam attack dramatically.
- The patterns of DDoS attack are always changing. Attack growth, intensity, and penetration time change fast along with the Internet world.
- In the resource-constrained environment of a private cloud network, it is essential to handle a DDoS attack as quickly as possible.
- It is usually a battle for survival with all the resources the private cloud can muster.
- In an individual private cloud, deft resource management is necessary and most definitive way to defend against a DDoS attack. Putting the best detection or filtering algorithm may not always work. But tolerating the attack by optimal resource utilization may resist the attack and may help counter the DDoS attack.
- Virtualization of resources gives some edge in a cloud environment to defeat DDoS attack.

13.3 Mitigation Approaches

DDoS attack mitigation is a classic problem. However, in the cloud environment, it becomes a more significant challenge [44]. We also cannot separate a cloud environment from the traditional network infrastructure. All approaches presented in this section have some advantages, which can be adapted for private cloud-like environment. Some promising new approaches have also been developed in the context of the cloud. These include like SDN-based ideas and ideas from the big data analytic point of view [45]. A defense approach can be deployed in the network itself or in the host (victim) environment. We analyze different existing approaches, and based on features of the approaches such as the level of operation, time to respond, and time to cooperate with other devices, we divide active response into two main categories, as shown in Figure 13.9.

In a **proactive approach**, a step taken to control potential incident activity before it happens rather than waiting for it to happen.

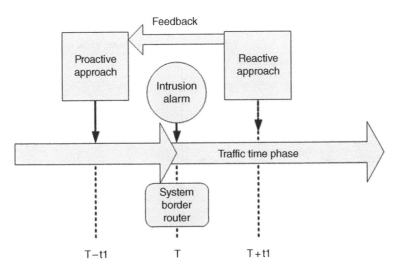

Figure 13.9 Attack response scenario.

A **reactive approach** detects the abnormality and informs the security administrator or automatically takes a responsive counter-action immediately, i.e. in real time. A reactive response reacts only after the intrusion is detected.

In the rest of this section, a few prominent approaches are discussed. The models developed by different authors based on these approaches are analyzed. Each of the methods can be included either in the proactive or the reactive category. It very much remains open to debate which type of category will work best in the individual private cloud environment.

Push-back: To mitigate the DDoS attack, the congestion in the network traffic needs to be controlled. Because essentially DDoS attack mitigation is a congestion control problem. The congestion occurs due to not obeying the traditional end-to-end congestion policies by the malicious host. Most researchers think that the problem needs to be handled by routers. To detect and to drop malicious packets as per the preference, the functionality can be added in the router. Those dropped packets might belong to an attack. A push-back mechanism based on managing congestion at the routers has been implemented by Ioannidis and Bellovin [46].

Router Throttling: Participating routers can regulate the packet rate destined for a server. Yau et al. [47] propose and simulate a router throttling model to establish the efficacy of the concept, as shown in Figure 13.10. This idea can also increase the service reliability for legal users. Using the improvised K-level max-min fairness theory [48], Yau et al. find that the throttling mechanism is highly effective in countering an aggressive attacker. They efficiently regulate the server load to a level below its design limit amid a DDoS attack.

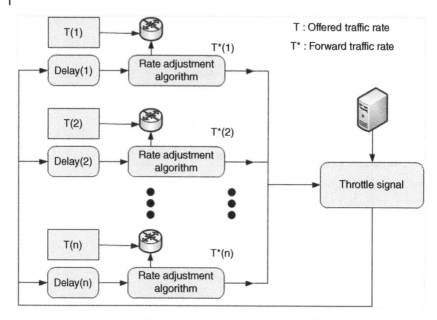

Figure 13.10 Router throttling model proposed by Yau et al. [47]. *Source:* Based on Yau, D. K., Lui, J. C., Liang, F., and Yam, Y. (2005). Defending against distributed denial-of-service attacks with max-min fair server-centric router throttles. *IEEE/ACM Transactions on Networking*, 13(1), 29–42.

Divide and Conquer: Chen et al. [49] use the divide and conquer strategy to actively throttle the attacking traffic. They present a diagnosis and attack mitigation scheme that combines the concepts of push-back and packet marking. Detection of attacks is executed near the source-end. Initially, IDS detects the attack on the victim side. The traceback scheme is carried out till the source end is reached. We believe that this idea can be adapted to the cloud environment.

Random Flow Network Modeling: This approach adapts the theoretical concept represented by the max-flow min-cut theorem of [50] concerning flow in a network. Kong et al. [51] rely on this theory in designing a random flow network model to mitigate DDoS attacks. They show that this mitigation problem can be reduced to an instance of the maximum flow problem. We know that a DDoS attacker heavily pumps the flow of traffic toward the sink. The strategy depends on the fact that the maximum achievable flow value from the source to the sink is equal to the capacity of a certain cut in the flow network. This method is suitable for any kind of computing environment because it does not depend on the end infrastructure; rather it is concerned with the intermediate network infrastructure.

Self-Cleansing Intrusion Tolerance (SCIT): SCIT [52], a method based on virtualization technology, tries to achieve mitigation by constantly cleansing the

Figure 13.11 A high-level view of SCIT model.

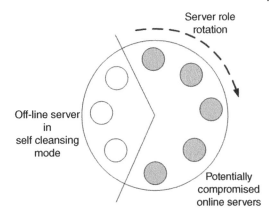

servers and rotating the roles of individual servers, as shown in Figure 13.11. We know that virtualization is a key technique in a cloud-based environment. If a server is initiated, SCIT places a pristine, malware-free copy of the server's operating system into a VM. To coordinate among server modes, rotations can be performed with the help of a central controller or a distributed control mechanism using the Cluster Communication Protocol (CCP) [53]. In the rotation process, online servers are set offline. Afterward, the system is rebooted to initiate cleansing procedures.

Dynamic Resource Pricing: Mankins et al. [54] discuss the applicability of dynamic resource pricing to discriminate well from bad traffic. They implement a dynamic pricing strategy that favors good user behavior and punishes aggressive adversarial behavior. They propose a distributed gateway framework and a payment protocol. The idea is to impose dynamically changing prices on both network servers and information resources so that the approach can push the cost of initiating service requests, in terms of monetary payments and/or computational burdens, to requesting clients. Thus, the architecture can provide for service quality discrimination to separate good client behavior from adversarial behavior in a private cloud environment serving a large set of heterogeneous consumers.

Intelligent Fast-Flux Swarm Network: Lua et al. [17] describe an autonomous intelligent coordinated network of swarm nodes to mitigate DDoS attacks. This swarm network ensures autonomous co-ordination among nodes and allocation of swarm nodes (deploying nodes densely) to maintain connection. A load-balancing process checks the health of nodes and removes those that are unresponsive. However, when a DDoS attack is in progress, it may not be robust. For better optimization, they use IWD [18]. It is a nature-inspired algorithm. The algorithm mimics how water drops behave in the flow of a river, i.e. the dynamic behavior of a river.

Roaming Honeypot: Khattab et al. [55] and Sardana and Joshi [56] propose the concept of roaming honeypots, changing the locations of the honeypots continuously and disguising them within a server pool. The roaming honeypot mitigates attacks from behind the firewall by dropping all connections when a server switches from acting as honeypot to become an active server. So, if we can adapt this approach to the individual private cloud environment, a roaming honeypot may be a very good defender for that environment with limited resources for legitimate users.

Target Defense Moving: Researchers have proposed an innovative way to defend DDoS attack. Aspects of the systems to present the attackers are changed and created a varied surface for the attacker. Thus, it becomes more difficult to exploit the vulnerability. In general, the attacker looks for exploiting the drawbacks or loopholes that exist in a system. But, while analyzing and learning the vulnerabilities by an attacker, the system will change its aspects so that the required time to launch an attack and to disrupt the functionality of the system is reduced. In that time, the system has changed to more or less a new system [57, 58]. This approach may provide an effective defense solution in context of private cloud environment as well.

Dynamic Resource Allocation: In addition the traditional defense approaches, we need to explore resource allocation and utilization strategies for defending DDoS attacks in the cloud. Yau et al. [47] contend that DDoS defense is a resource management problem. Every day the attack patterns keep changing. It will be a fruitless waste of time and resources to try to defend against DDoS attacks by just looking at patterns learned earlier. In addition, it is important to not only defend against an attack but also make services available during an attack. To beat DDoS attacks in the cloud, Yu et al. [19] propose a dynamic resource allocation procedure within an individual cloud, as shown in Figure 13.12. It is a simple methodology of cloning Intrusion Prevention Servers from idle resources to filter out attack packets quickly and provide general services simultaneously. Some other specific resource allocation approaches have been proposed as well.

Virtualization is a key concept in resource provisioning and management in the cloud. Virtualization provides a view of resources used to instantiate VMs. Isolating and migrating the state of a machine help improve optimization of resource allocation. Live VM migration transfers the "state" of a VM from one physical machine to another, and can mitigate overload conditions and enable uninterrupted maintenance activities. Mishra et al. [59] incorporate dynamic resource management in a virtual environment. Their approach answers basic questions such as when to migrate, how to migrate, types of migration, and where to migrate. It also treats the migration of resources differently in different network architectures, e.g. local area networks (LAN) and wide area networks (WAN).

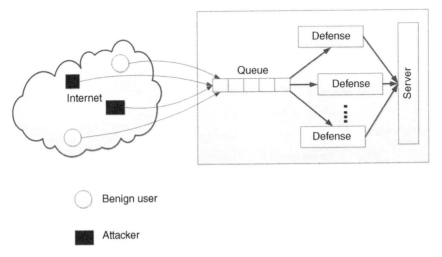

Figure 13.12 Dynamic resource allocation strategy by Yu et al. [19]. *Source:* Based on Yu, S., Tian, Y., Guo, S., and Wu, D. O. (2013). Can we beat DDoS attacks in clouds? *IEEE Transactions on Parallel and Distributed Systems*, 25(9), 2245–2254.

The cloud environment can be described as probabilistic in nature. So there is a need to assess the performance of a cloud center for resource provisioning. The probabilistic nature of the cloud can be represented in terms of stochastic processes [60–64]. Shawky [65] introduces an approach to model and analyze the performance of the resource allocation process using stochastic process algebra.

SDN-based DDoS Defense: An attacker can infect a sufficient number of machines in a short time frame in traditional networks. Attackers are also known to use cloud as **Malware as a Service** by renting different VMs and using them as bots [66]. Separation of the control plane from the data plane enables one to establish easily large-scale attack and defense experiments. A logical centralized controller of an SDN permits a system defender to build consistent security policies and to monitor or analyze traffic patterns for potential security threats. A programmable intermediate network architecture can be setup easily in an SDN.

The cloud networks face challenges such as guaranteed performance of applications when applications are moved from on-premise to the cloud facility, flexible deployment of appliances (e.g. intrusion detection systems or firewalls), and security and privacy protection. An environment, providing good programmable, flexible, and secure infrastructure is needed. SDNs are evolving as the key technology that can improve cloud manageability, scalability, controllability, and dynamism [67]. In the past few years, several innovative SDN-based defense solutions have been introduced. These solutions belong to the three basic types of

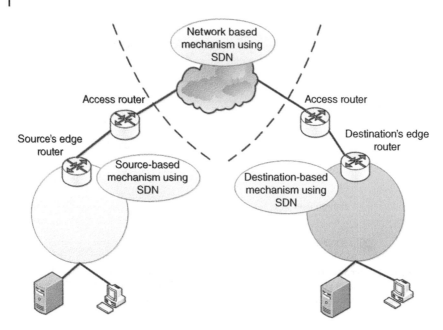

Figure 13.13 SDN to defend DDoS attack.

SDN-based DDoS defense mechanisms as shown in Figure 13.13. In [68], include a detailed discussion of SDNs, SDN-based clouds, and autonomous defense in clouds.

Lin et al. [69] refer to SDNs as an emerging wave to transform network industries. They discuss SDNs and standardization in terms of components such as controllers, applications, service chains, network function virtualization, and interfaces. Braga et al. [70] presents a lightweight method for DDoS attack detection based on traffic flow features, in which the extraction of such information is made with a very low overhead compared to traditional approaches. This is possible due to the use of the NOX platform [71], which provides a programmatic interface to facilitate the handling of switch information. Shin and Gu [72] show a new attack to fingerprint SDN networks and further launch efficient resource consumption attacks. This attack demonstrates that SDNs also introduce new security issues that may not be ignored. Flow Table Overloading in Software-Defined Networks is a vulnerability to be handled carefully. Yuan et al. [73] point out this issue and provide a security service in an SDN using QoS-aware mitigation strategy, namely, peer support strategy, integrating the available idle flow table resource of the whole SDN system to mitigate such an attack on a single switch of the system.

Nguyen et al. [74] propose a SDN-based approach, Whack-a-Mole. It is a cloud resource management procedure using network obfuscation to help CSPs. This approach protects critical services proactively against a DDoS attack and putting very less service interruption. It deploys VM spawning model to assign random address space by creating multiple replicated VM instances for the services which are critical. They have shown its effectiveness using such optimized VM spawning based on real Service-level Agreements and implemented the whole approach using SDN/OpenFlow controllers over Open vSwitches on a GENI testbed. Xu et al. [75] devised a defensive approach by classifying the traffic using SDNFV for flexibility. Thus, it reduces load on SDN.

SDNs have been accepted as a new paradigm to provide an entire set of virtualization and control mechanisms to meet defense challenges in cloud networking [28, 76–78]. Thus, exploring the use of SDNs in providing better DDoS defense solutions in the cloud computing environment is likely to be beneficial.

Big Data Analytics: For detecting DDoS attacks, Jiao et al. [79] identifies FSIA and RSIA for extraction of TCP traffic features and better classification through Big Data analytics using two decision tree classifiers [80].

Vieira et al. [81] propose the Intrusion Responsive Autonomic System (IRAS) to analyze real-time traffic to detect intrusion and mitigate attacks in the cloud platform, as shown in Figure 13.14. IRAS is an autonomous intrusion response technique endowed with self-awareness, self-optimization, and self-healing properties.

Figure 13.14 Intrusion responsive autonomic system (IRAS).

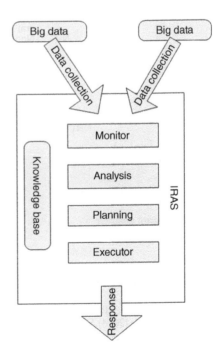

Internet evolves and the computing infrastructure changes rapidly. All these are happening because more processing power produces more data at every opportunity. Researchers have coined the concept of "Big Data" to refer to data handled by large enterprises like Google, Facebook, IBM, and so on [82]. Lee et al. [83] propose a method to analyze Internet traffic using the MapReduce [84] framework within the cloud computing platform. They compare their results with Hadoop [85] and other tools concluding 72% improvement in computational efficiency. Tripathi et al. [86] also study characteristics of DDoS attacks in the cloud and developed a scheme to detect such attacks in a Hadoop-based environment. Lee et al. [87] also provide two algorithms to detect DDoS attacks using packet tracing method in a MapReduce environment.

Govinda and Sathiyamoorthy [88] introduce a process of clustering the traffic into different groups. These groups are flash traffic, interactive traffic, latency sensitive traffic, non-real time traffic, and unknown traffic, as shown in Figure 13.15. They use Hadoop technology to analyze big data traffic. If any of these packets is categorized as unknown traffic, it is identified as a part of DDoS attack and eliminated by the packet analyzer.

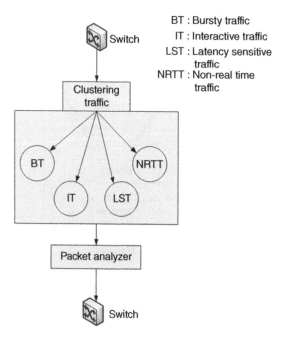

Figure 13.15 Traffic cluster analysis by [88]. *Source:* Based on Govinda, K., and Sathiyamoorthy, E. (2014). Secure traffic management in cluster environment to handle DDoS attack. *World Applied Sciences Journal*, 32(9), 1828–1834.

Table 13.3 Selected approaches handling DDoS attacks.

Authors	Key points	Real-time	High rate/ Low rate
Lua and Yow [17]	• Create a network of intelligent nodes like fast-flux swarm • Use of IWD algorithm [18] • Balance the lodes	Yes	Does not depend on rate
Wang et al. [20]	• Service minimization during attack • Use dynamic fault tolerance architecture	Yes	Not mentioned
Ioannidis and Bellovin [46]	• Attack signature detection • Detect and drop malicious packet at router • Rate limitation	Yes	High rate
Yau et al. [47]	• Traffic throttling at upstream routers • To forestall an impending attack • Uses k-level max-min fairness theory [48]	Yes	High rate
Chen et al. [49]	• Marking malicious packet at upstream routers • Trace backing • At the source side routers, packets are dropped	Yes	Does not depend on rate
Kong et al. [51]	• Attack mitigation • Generalization of DDoS attack as max- flow min cut theorem [50]	Yes	Not mentioned
Bangalore and Sood [52]	• Virtualization • Servers are cleaned and changed the roles • Tolerant or mitigate the attack • Exposed time of the servers gets reduced	Yes	Does not depend on rate
Mankins et al. [54]	• Dynamic pricing strategy in terms of payments and/or computational load of each user	Yes	Not mentioned

(Continued)

Table 13.3 (Continued)

Authors	Key points	Real-time	High rate/ Low rate
Khattab et al. [55]	• Roaming honeypot technique • Changing locations • Provide general service using a subset of the servers • Rest of the idle servers act as honeypots • Detects the attack and tries mitigation	Yes	High rate
Yu et al. [19]	• Resources are allocated dynamically • Reallocate or de-allocate of resources • Intrusions are prevented at the servers	Yes	Does not depend on rate
Nguyen et al. [74]	• VM spawning model • Random address space • Replicate VM instances of critical services • Cloud environment	Yes	Not mentioned
Xu et al. [75]	• Traffic classification • Using SDNFV	Yes	Not mentioned
Jiao et al. [79]	• FSIA and RSIA detection • TCP traffic feature extraction • Big data analytics • Decision tree classifiers	Yes	Not mentioned
Vieira et al. [81]	• Big Data analytics • Attack mitigation • Cloud environment	Yes	Not mentioned
Tripathi et al. [86]	• Packet/traffic analysis • Cloud/Hadoop environment • DDoS attack detection	Yes	Not mentioned
Lee et al. [87]	• Packet tracing • MapReduce environment	Yes	Flow/Rate analysis
Govinda and Sathiyamoorthy [88]	• Big Data • Clustering technique • Traffic classification • Unknown traffic packet	Yes	Traffic analysis

13.3.1 Discussion

The approaches discussed in this section are presented compactly in Table 13.3. We can summarize our discussions in the following observations:

- It is necessary to build a real-time defense system, whether it is network based or host based.
- Incorporating dynamic behavior in the solution can provide adaptability to the defense.
- The discussed methods employ the tolerance approach. Thus, allocating and utilizing resources effectively can provide a good defense.
- As cloud computing systems incorporate traditional network topology and also new resource sharing methods, defense solutions against DDoS in the individual private cloud environment need to evolve to adapt to both.
- Resource utilization in a virtualized cloud computing environment is important. So, resource sharing and utilization need to be smooth enough to provide services along with security.
- In a large infrastructure network, the converging network traffic will be always high enough for analysis. New data analysis techniques need to be adapted for better defense.

13.4 Challenges and Issues with Recommendations

A service provider usually has adequate amount of resources for specific service seekers. Challenges and issues regarding DDoS defense in limited resource environment of cloud are listed below.

- For a cyber-defense tool, effectiveness should be measured in terms of time taken and accuracy of detection obtained in real time. Lack of efficient performance can be a roadblock to large-scale adoption of any real-time defense mechanism.
- A mitigation technique for flooding attacks must take into account in system and protocol design to ensure an effective and successful implementation.
- The service provider must ensure that its DDoS attack defense operations neither affect nor are affected by other cloud activities.
- If the cloud provider has only the resources required to provide services to its customers but not much more to defend, this may encourage undesirable DDoS attacks if attackers can guess the situation.
- In a private cloud environment, we need to build the defense strategy using virtualization technology [89].

- Resource allocation and VM migration processes are fast-paced. So, for any approach to defend DDoS attack in such scenario needs to adapt the dynamisms of a network and adapt the topological changes. Along with that, it has to maintain high detection rate showing smooth reaction capability. In other words, a successful defense mechanism must be dynamic and adaptive.
- Patterns for different attacks are different. It is obvious that one cannot build defensive approaches for each type of attack in a private cloud with a particular amount of resource dedicated to each attack.
- No security precautions can guarantee that a system will never be intruded and so at the critical moment when the system is designed, applications still need to provide minimal services to the legal customers.

13.5   A Generic Framework

Based on the recommendations presented earlier, we believe that an automatic host-based approach emphasizing tolerance can provide better utilization of resources in the cloud environment to respond to DDoS attacks in an individual private cloud. With limited resources, it is necessary to develop a procedure to defend against DDoS attacks and to provide general service. A generic conceptual framework is shown in Figure 13.16. It is a combination of different phases and components. The whole defense module is just a conceptual depiction of cloud-based defense solution against DDoS attacks adapting concepts borrowed from existing techniques, adapted to a new environment. Detection and

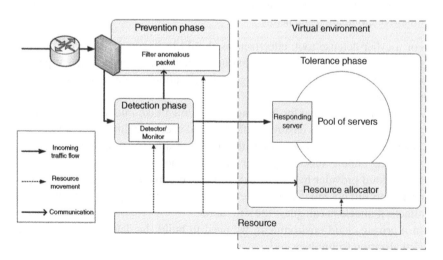

Figure 13.16   A generic cloud-based defense framework.

prevention phases of this framework incorporate some ideas of the traditional Internet and also tolerance techniques to the cloud environment. The framework, which we discuss below, abides by the recommendations discussed previously.

i) **Detection Phase**: In this phase, the monitor component analyzes the behavior of the traffic coming to a responding server which handles incoming requests. If the incoming traffic shows any abnormality, the monitor catches it automatically and sends an alert message to the resource allocator. The alert message contains the threat level, how to act to tackle the abnormality, and when to initiate the migration stage. The monitor will also communicate with the prevention component with alert messages about the incoming traffic. This component needs to detect abnormal changes in network traffic in real time.

ii) **Tolerance Phase**: In this phase, we can utilize the resources effectively using a virtualization technique available in the cloud environment. For example, the data center or the private cloud provider has the ability to provide the resources to users using virtualization. So, the utilization of the resources should be appropriate for the security needed in a crisis situation since resources are always limited in private cloud environment.

- The resource allocator maintains a queue of fresh server copies to provide services that must be rendered by the responding server. Depending on the level of the alert message, it tries to maintain a dynamic queue of spare resources to fight back if a rapid change in service demands occurs because of any high traffic attack. It can also push unnecessary resources back to the resource pool when the state becomes normal.

- A responding server may be detached from service depending on two things, time and computation load. A server needs to be exposed only for a limited amount of time and if the computation load exceeds a threshold level due to malicious activity, it can be switched to inactive status. Before switching, using live migration we can copy the necessary states of the server to an incoming fresh server so that the usual services can be resumed with minimum delay.

iii) **Prevention Phase**: An adaptive and dynamic mapping intrusion response system for effective prevention of DDoS attacks in real time is essential. In the prevention phase, alert messages coming from the detection component need to be analyzed and correlated to discover patterns or strategy in attacks. Using these, we can filter out matching incoming packets later with a low false-positive rate. If the traffic flow is high, the amount of alert messages may be high enough to analyze. In such a situation, we need to use newly developed data analysis techniques, such as big data analytics to analyze the patterns.

13.6 Conclusion and Future Work

We can definitely say that in the near future, most computing activities and resources will migrate to the cloud and security will be a prime concern. DDoS attacks may be resisted with generic solutions to survive and to provide best services under the circumstances. However, to be successful, more than the usual is necessary in the cloud context. In this chapter, we have discussed issues in handling DDoS attacks, specifically in a private cloud environment. We have highlighted issues and challenges faced in the private cloud environment when providing defense solutions against DDoS attacks. Some useful approaches developed by researchers to address these issues have been presented and analyzed in this chapter. The importance of mitigating the attack by tolerating it and by optimized use of resources in the private cloud scenario has been emphasized. Finally, the role of big data analytics in defending DDoS attacks in the cloud has been introduced.

In the near future, we plan to deploy the conceptual cloud framework in a testbed to demonstrate and analyze the effectiveness of our proposed framework. It is important to know how far this framework can resolve different issues and challenges when defending against DDoS attacks in an individual private cloud environment.

References

1 Khorshed, M.T., Ali, A.S., and Wasimi, S.A. (2012). A survey on gaps, threat remediation challenges and some thoughts for proactive attack detection in cloud computing. *Future Generation Computer Systems* 28 (6): 833–851.

2 Deka, R.K., Kalita, K.P., Bhattacharya, D.K., and Kalita, J.K. (2015). Network defence: approaches, methods and techniques. *Journal of Network and Computer Applications* 57: 71–84.

3 Sabahi, F. (2011). Cloud computing security threats and responses. *Proceedings of 2011 IEEE 3rd International Conference on Communication Software and Networks*, Xi'an, China (27–29 May 2011), pp. 245–249. IEEE.

4 Pitropakis, N., Darra, E., Vrakas, N., and Lambrinoudakis, C. (2013). It's all in the cloud: reviewing cloud security. *Proceedings of the 2013 IEEE 10th International Conference on Ubiquitous Intelligence and Computing and 2013 IEEE 10th International Conference on Autonomic and Trusted Computing*, Italy (18–21 December 2013), 355–362. IEEE.

5 Grover, J., and Sharma, M. (2014). Cloud computing and its security issues—a review. *Fifth International Conference on Computing, Communications and Networking Technologies (ICCCNT)*, Hefei, China (11–13 July 2014), 1–5. IEEE.

6 Modi, C., Patel, D., Borisaniya, B. et al. (2013). A survey of intrusion detection techniques in cloud. *Journal of Network and Computer Applications* 36 (1): 42–57.

7 Subashini, S. and Kavitha, V. (2011). A survey on security issues in service delivery models of cloud computing. *Journal of Network and Computer Applications* 34 (1): 1–11.

8 Lau, F., Rubin, S. H., Smith, M. H., and Trajkovic, L. (2000). Distributed denial of service attacks. *SMC 2000 Conference Proceedings 2000 IEEE International Conference on Systems, Man and Cybernetic:. Cybernetics Evolving to Systems, Humans, Organizations, and Their Complex Interactions* Sheraton Music City Hotel, Nashville, Tennessee, USA (8–11 October 2000), 2275–2280. IEEE.

9 Peng, T., Leckie, C., and Ramamohanarao, K. (2007). Survey of network-based defence mechanisms countering the DoS and DDoS problems. *ACM Computing Surveys (CSUR)* 39 (1): 3-es.

10 Stone-Gross, B., Cova, M., Cavallaro, L., Gilbert, B., Szydlowski, M., Kemmerer, R., and Vigna, G. (2009). Your botnet is my botnet: analysis of a botnet takeover. *Proceedings of the 16th ACM conference on Computer and Communications Security*, CCS 2009, Chicago, Illinois, USA (9–13 November 2009), 635–647. ACM.

11 Hoque, N., Bhuyan, M.H., Baishya, R.C. et al. (2014). Network attacks: taxonomy, tools and systems. *Journal of Network and Computer Applications* 40: 307–324.

12 Fabian, M. A. R. J. Z., and Terzis, M. A. (2007). My botnet is bigger than yours (maybe, better than yours): why size estimates remain challenging. *Proceedings of the 1st USENIX Workshop on Hot Topics in Understanding Botnets*, HotBots'07, Cambridge, MA, USA (10 April 2007), 18. USENIX Association.

13 Bhattacharyya, D.K. and Kalita, J.K. (2013). *Network Anomaly Detection: A Machine Learning Perspective*. CRC Press.

14 Yu, S., Zhou, W., Jia, W. et al. (2011). Discriminating DDoS attacks from flash crowds using flow correlation coefficient. *IEEE Transactions on Parallel and Distributed Systems* 23 (6): 1073–1080.

15 Xiang, Y., Li, K., and Zhou, W. (2011). Low-rate DDoS attacks detection and traceback by using new information metrics. *IEEE Transactions on Information Forensics and Security* 6 (2): 426–437.

16 Nguyen, Q. and Sood, A. (2010). A comparison of intrusion tolerant system architectures. *IEEE Security and Privacy* 9 (4): 24–31.

17 Lua, R. and Yow, K.C. (2011). Mitigating DDoS attacks with transparent and intelligent fast-flux swarm network. *IEEE Network* 25 (4): 28–33.

18 Shah-Hosseini, H. (2009). The intelligent water drops algorithm: a nature-inspired swarm-based optimization algorithm. *International Journal of Bio-inspired computation* 1 (1-2): 71–79.

19 Yu, S., Tian, Y., Guo, S., and Wu, D.O. (2013). Can we beat DDoS attacks in clouds. *IEEE Transactions on Parallel and Distributed Systems* 25 (9): 2245–2254.

20 Wang, H., Wang, F., Liu, J., and Groen, J. (2012). Measurement and utilization of customer-provided resources for cloud computing. *2012 Proceedings IEEE INFOCOM 2012*, Orlando, FL, USA (25–30 March 2012), 442–450. IEEE.

21 Bhadauria, R., Chaki, R., Chaki, N., and Sanyal, S. (2011). A survey on security issues in cloud computing. *ArXiv* 1109 (5388): 1–15.

22 Zhang, Z., Wu, C., and Cheung, D.W. (2013). A survey on cloud interoperability: taxonomies, standards, and practice. *ACM SIGMETRICS Performance Evaluation Review* 40 (4): 13–22.

23 Wong, F. and Tan, C.X. (2014). A survey of trends in massive DDoS attacks and cloud-based mitigations. *International Journal of Network Security & Its Applications* 6 (3): 57.

24 Kumar, U. and Gohil, B.N. (2015). A survey on intrusion detection systems for cloud computing environment. *International Journal of Computer Applications* 109 (1).

25 Chiba, Z., Abghour, N., Moussaid, K., El Omri, A., and Rida, M. (2016). A survey of intrusion detection systems for cloud computing environment. *2016 International Conference on Engineering & MIS (ICEMIS)*, Ibn Zohr University, Agadir, Morocco (22–24 September 2016), 1–13. IEEE.

26 Mishra, P., Pilli, E.S., Varadharajan, V., and Tupakula, U. (2017). Intrusion detection techniques in cloud environment: a survey. *Journal of Network and Computer Applications* 77: 18–47.

27 Basu, S., Bardhan, A., Gupta, K., Saha, P., Pal, M., Bose, M., and Sarkar, P. (2018). Cloud computing security challenges & solutions-A survey. *2018 IEEE 8th Annual Computing and Communication Workshop and Conference (CCWC)*, Las Vegas, NV, USA (8–10 January 2018), 347–356. IEEE.

28 Dong, S., Abbas, K., and Jain, R. (2019). A survey on distributed denial of service (DDoS) attacks in SDN and cloud computing environments. *IEEE Access* 7: 80813–80828.

29 Devarapalli, S.J. and Joshi, P.S. (2013). Domain name system security extensions (DNSSEC) for global server load balancing. Google Patents, US Patent 8,549,147, 1 October 2013.

30 Mirkovic, J. and Reiher, P. (2004). A Taxonomy of DDoS attack and DDoS defence mechanisms. *ACM SIGCOMM Computer Communication Review* 34 (2): 39–53.

31 Kang, M. S., Lee, S. B., and Gligor, V. D. (2013). The crossfire attack. *2013 IEEE Symposium on Security and Privacy, SP 2013*, Berkeley, CA, USA (19–22 May 2013), 127–141. IEEE Computer Society.

32 Miao, R., Potharaju, R., Yu, M., and Jain, N. (2015). The dark menace: Characterizing network-based attacks in the cloud. In: *Proceedings of the 2015 Internet Measurement Conference, IMC 2015*, Tokyo, Japan (28–30 October 2015), 169–182. ACM.

33 Peng, C., Kim, M., Zhang, Z., and Lei, H. (2012). VDN: Virtual machine image distribution network for cloud data centres. *2012 Proceedings IEEE INFOCOM 2012*, Orlando, FL, USA (25–30 March 2012), 181–189. IEEE.

34 Vissers, T., Van Goethem, T., Joosen, W., and Nikiforakis, N. (2015). Maneuvering around clouds: Bypassing cloud-based security providers. *Proceedings of the 22nd ACM SIGSAC Conference on Computer and Communications Security*, Denver, CO, USA (12–16 October 2015), 1530–1541. ACM.

35 Chaisiri, S., Lee, B.S., and Niyato, D. (2011). Optimization of resource provisioning cost in cloud computing. *IEEE Transactions on Services Computing* 5 (2): 164–177.

36 Idziorek, J., Tannian, M.F., and Jacobson, D. (2012). The insecurity of cloud utility models. *IT Professional* 15 (2): 22–27.

37 Somani, G., Gaur, M. S., and Sanghi, D. (2015). DDoS/EDoS attack in cloud: affecting everyone out there! *Proceedings of the 8th International Conference on Security of Information and Networks, SIN 2015*, Sochi, Russian Federation (8–10 September 2015), 169–176. ACM.

38 Sqalli, M. H., Al-Haidari, F., and Salah, K. (2011). EDoS-shield-a two-steps mitigation technique against EDoS attacks in cloud computing. *2011 Fourth IEEE International Conference on Utility and Cloud Computing, UCC 2011*, Melbourne, Australia (5–8 December 2011), 49–56. IEEE Computer Society.

39 Wang, Q., Ren, K., and Meng, X. (2012). When cloud meets Ebay: towards effective pricing for cloud computing. *2012 Proceedings IEEE INFOCOM*, Orlando, FL, USA (25–30 March 2012), 936–944. IEEE.

40 Yi, S., Andrzejak, A., and Kondo, D. (2011). Monetary cost-aware check pointing and migration on amazon cloud spot instances. *IEEE Transactions on Services Computing* 5 (4): 512–524.

41 Yan, Q. and Yu, F.R. (2015). Distributed denial of service attacks in software-defined networking with cloud computing. *IEEE Communications Magazine* 53 (4): 52–59.

42 Deshmukh, R.V. and Devadkar, K.K. (2015). Understanding DDoS attack and its effect in cloud environment. *Procedia Computer Science* 49: 202–210.

43 Darwish, M., Ouda, A., and Capretz, L. F. (2013). Cloud-based DDoS attacks and defences. In: *International Conference on Information Society (i-Society 2013)* (pp. 67–71). IEEE.

44 Anwar, Z. and Malik, A.W. (2014). Can a DDoS attack meltdown my data centre? A simulation study and defence strategies. *IEEE Communications Letters* 18 (7): 1175–1178.

45 Fayaz, S. K., Tobioka, Y., Sekar, V., and Bailey, M. (2015). Bohatei: Flexible and elastic DDoS defence. In: *24th {USENIX} Security Symposium ({USENIX} Security 15)*, Washington, DC, USA (12–14 August 2015), 817–832. USENIX Association.

46 Ioannidis, J. and Bellovin, S.M. (2002). Implementing pushback: router-based defence against DDoS attacks. *Proceedings of the Network and Distributed System*

Security Symposium, NDSS 2002, San Diego, California, USA. doi: 10.7916/ D8R78MXV. The Internet Society 2002.

47 Yau, D.K., Lui, J.C., Liang, F., and Yam, Y. (2005). Defending against distributed denial-of-service attacks with max-min fair server-centric router throttles. *IEEE/ ACM Transactions on Networking* 13 (1): 29–42.

48 Nace, D. and Pióro, M. (2008). Max-min fairness and its applications to routing and load-balancing in communication networks: a tutorial. *IEEE Communication Surveys and Tutorials* 10 (4): 5–17.

49 Chen, R., Park, J.M., and Marchany, R. (2007). A divide-and-conquer strategy for thwarting distributed denial-of-service attacks. *IEEE Transactions on Parallel and Distributed Systems* 18 (5): 577–588.

50 Dantzig, G. and Fulkerson, D.R. (2003). On the max flow min cut theorem of networks. *Linear Inequalities and Related Systems* 38: 225–231.

51 Kong, J., Mirza, M., Shu, J. et al. (2003). Random flow network modeling and simulations for DDoS attack mitigation. *IEEE International Conference on Communications, 2003, ICC'03*, Anchorage, Alaska, USA (11–15 May 2003), 487–491. IEEE.

52 Bangalore, A. K., & Sood, A. K. (2009). Securing web servers using self-cleansing intrusion tolerance (scit). *2009 Second International Conference on Dependability*, Athens/Glyfada, Greece (18–23 June 2009), 60–65. IEEE.

53 Huang, Y., Arsenault, D., and Sood, A. (2010). U.S. Patent No. 7,680,955. Washington, DC: U.S. Patent and Trademark Office.

54 Mankins, D., Krishnan, R., Boyd, C., Zao, J., and Frentz, M. (2001). Mitigating distributed denial of service attacks with dynamic resource pricing. *Seventeenth Annual Computer Security Applications Conference (ACSAC 2001)*, New Orleans, Louisiana, USA (11–14 December 2001), 411–421. IEEE Computer Society.

55 Khattab, S. M., Sangpachatanaruk, C., Mossé, D., Melhem, R., and Znati, T. (2004). Roaming honeypots for mitigating service-level denial-of-service attacks. *Proceedings of the 24th International Conference on Distributed Computing Systems (ICDCS 2004)*, Hachioji, Tokyo, Japan (24–26 March 2004), 328–337. IEEE Computer Society.

56 Sardana, A. and Joshi, R. (2009). An auto-responsive honeypot architecture for dynamic resource allocation and QoS adaptation in DDoS attacked networks. *Computer Communications* 32 (12): 1384–1399.

57 Touch, J. D., Finn, G. G., Wang, Y. S., and Eggert, L. (2003). DynaBone: dynamic defence using multi-layer Internet overlays. In: *Proceedings DARPA Information Survivability Conference and Exposition (DISCEX-III 2003)*, Washington, DC, USA (22–24 April 2003), 271–276. IEEE Computer Society.

58 Venkatesan, S., Albanese, M., Amin, K., Jajodia, S., and Wright, M. (2016). A moving target defence approach to mitigate DDoS attacks against proxy-based architectures. *2016 IEEE Conference on Communications and Network Security (CNS)*, Philadelphia, PA, USA (17–19 October 2016), 198–206. IEEE.

59 Mishra, M., Das, A., Kulkarni, P., and Sahoo, A. (2012). Dynamic resource management using virtual machine migrations. *IEEE Communications Magazine* 50 (9): 34–40.

60 Doob, J.L. (1953). *Stochastic Processes*, vol. 101. New York;: Wiley.

61 Coşgun, Ö. and Büyüktahtakın, İ.E. (2018). Stochastic dynamic resource allocation for HIV prevention and treatment: an approximate dynamic programming approach. *Computers & Industrial Engineering* 118: 423–439.

62 Cui, J., Liu, Y., and Nallanathan, A. (2019). Multi-agent reinforcement learning-based resource allocation for UAV networks. *IEEE Transactions on Wireless Communications* 19 (2): 729–743.

63 Mireslami, S., Rakai, L., Wang, M., and Far, B.H. (2019). *Dynamic Cloud Resource Allocation Considering Demand Uncertainty*. IEEE Transactions on Cloud Computing.

64 Jyoti, A. and Shrimali, M. (2020). Dynamic provisioning of resources based on load balancing and service broker policy in cloud computing. *Cluster Computing* 23 (1): 377–395.

65 Shawky, D. M. (2013). Performance evaluation of dynamic resource allocation in cloud computing platforms using Stochastic Process Algebra. *2013 8th International Conference on Computer Engineering & Systems (ICCES)*, Cairo, Egypt (26–28 November 2013), 39–44. IEEE.

66 Banikazemi, M., Olshefski, D., Shaikh, A. et al. (2013). Meridian: an SDN platform for cloud network services. *IEEE Communications Magazine* 51 (2): 120–127.

67 Azodolmolky, S., Wieder, P., and Yahyapour, R. (2013). SDN-based cloud computing networking. *2013 15th International Conference on Transparent Optical Networks (ICTON)*, Cartagena, Spain (23–27 June 2013), 1–4. IEEE.

68 Yan, Q., Yu, F.R., Gong, Q., and Li, J. (2015). Software-defined networking (SDN) and distributed denial of service (DDoS) attacks in cloud computing environments: a survey, some research issues, and challenges. *IEEE Communication Surveys and Tutorials* 18 (1): 602–622.

69 Lin, Y.D., Pitt, D., Hausheer, D. et al. (2014). Software-defined networking: standardization for cloud computing second wave. *Computer* 47 (11): 19–21.

70 Braga, R., Mota, E., and Passito, A. (2010). Lightweight DDoS flooding attack detection using NOX/OpenFlow. *IEEE Local Computer Network Conference*, Denver, Colorado, USA (10–14 October 2010), 408–415. IEEE Computer Society.

71 Gude, N., Koponen, T., Pettit, J. et al. (2008). NOX: towards an operating system for networks. *ACM SIGCOMM Computer Communication Review* 38 (3): 105–110.

72 Shin, S., and Gu, G. (2013). Attacking software-defined networks: a first feasibility study. In: *Proceedings of the Second ACM SIGCOMM Workshop on Hot Topics in Software Defined Networking, HotSDN 2013*, The Chinese University of Hong Kong, Hong Kong, China (16 August 2013), 165–166. ACM.

73 Yuan, B., Zou, D., Yu, S. et al. (2016). Defending against flow table overloading attack in software-defined networks. *IEEE Transactions on Services Computing* 12 (2): 231–246.

74 Nguyen, M., Pal, A., and Debroy, S. (2018). Whack-a-mole: software-defined networking driven multi-level DDoS defence for cloud environments. *2018 IEEE 43rd Conference on Local Computer Networks LCN 2018*, Chicago, IL, USA (1–4 October 2018), 493–501. IEEE.

75 Xu, C., Lin, H., Wu, Y. et al. (2019). An SDNFV-based DDoS defence technology for smart cities. *IEEE Access* 7: 137856–137874.

76 Sahoo, K.S., Puthal, D., Tiwary, M. et al. (2018). An early detection of low rate DDoS attack to SDN based data centre networks using information distance metrics. *Future Generation Computer Systems* 89: 685–697.

77 Bhushan, K. and Gupta, B.B. (2019). Distributed denial of service (DDoS) attack mitigation in software defined network (SDN)-based cloud computing environment. *Journal of Ambient Intelligence and Humanized Computing* 10 (5): 1985–1997.

78 Chen, Z., Jiang, F., Cheng, Y., Gu, X., Liu, W., and Peng, J. (2018). XGBoost classifier for DDoS attack detection and analysis in SDN-based cloud. In: *2018 IEEE International Conference on Big Data and Smart Computing, BigComp 2018*, Shanghai, China (15–17 January 2018), 251–256. IEEE Computer Society.

79 Jiao, J., Ye, B., Zhao, Y., Stones, R. J., Wang, G., Liu, X., Wang, S. & Xie, G. (2017). Detecting TCP-based DDoS attacks in Baidu cloud computing data centres. In *2017 IEEE 36th Symposium on Reliable Distributed Systems SRDS 2017*, Hong Kong, Hong Kong (26–29 September 2017), 256–258. IEEE Computer Society.

80 Fontugne, R., Mazel, J., and Fukuda, K. (2014). Hashdoop: a MapReduce framework for network anomaly detection. In: *2014 IEEE Conference on Computer Communications Workshops (INFOCOM WKSHPS)*, Toronto, ON, Canada (27 April to 2 May 2014), 494–499. IEEE.

81 Vieira, K. M., Schubert, F., Geronimo, G. A., de Souza Mendes, R., and Westphall, C. B. (2014). Autonomic intrusion detection system in cloud computing with big data. *Proceedings of the International Conference on Security and Management (SAM)*, Las Vegas, USA (July 2014), 173–178. WorldComp.

82 Lohr, S. (2012). The age of big data. *New York; Times* 11 (2012).

83 Lee, Y., Kang, W., and Son, H. (2010). An internet traffic analysis method with MapReduce. *2010 IEEE/IFIP Network Operations and Management Symposium Workshops, NOMS 2010*, Osaka, Japan (19–23 April 2010), 357–361. IEEE.

84 Dean, J. and Ghemawat, S. (2008). MapReduce: simplified data processing on large clusters. *Communications of the ACM* 51 (1): 107–113.

85 White, T. (2012). *Hadoop: The Definitive Guide*. O'Reilly Media, Inc.

86 Tripathi, S., Gupta, B., Almomani, A. et al. (2013). Hadoop based defence solution to handle distributed denial of service (DDoS) attacks. *Journal of Information Security* 04 (03): 150–164.

87 Lee, Y., Kang, W., and Lee, Y. (2011). A Hadoop-based packet trace processing tool. In: *International Workshop on Traffic Monitoring and Analysis*, 51–63. Berlin, Heidelberg: Springer.

88 Govinda, K. and Sathiyamoorthy, E. (2014). Secure traffic management in cluster environment to handle DDoS attack. *World Applied Sciences Journal* 32 (9): 1828–1834.

89 Clark, C., Fraser, K., Hand, S., Hansen, J. G., Jul, E., Limpach, C., Pratt, I., & Warfield, A. (2005). Live migration of virtual machines. In: *Proceedings of the 2nd conference on Symposium on Networked Systems Design & Implementation*. Vol. 2 (pp. 273–286).

14

Securing the Defense Data for Making Better Decisions Using Data Fusion

Syed Rameem Zahra

Department of Computer Science and Engineering, National Institute of Technology Srinagar, Jammu and Kashmir, India

14.1 Introduction

IoT and big data make the two sides of the same coin as they only feed into the success of each other. Their development is exponentially increasing and has an effect on all categories of business and technologies. As they provide huge benefits on both organizational and individual levels, the two are being studied immensely. The organizations take advantage of huge quantities of data that are generated by IoT devices by analyzing them and offering personalized solutions to various problems [1]. Therefore, IoT big data analytics tries to help business organizations to understand the data in a better manner so that they can make better decisions. Big data analytics help in:

- Analyzing large quantities of unstructured data that cannot be harnessed by using old existing tools [2].
- Using data mining techniques to dig out the useful information from the available data [3] so that companies can benefit from it.

Huge numbers of sensors are deployed that vary in nature as well as type and hence IoT data is very different from the normal big data obtained through various systems [4]. This IoT big data is heterogeneous, has variety, and is increasing at an alarming rate. IoT is striving to make our environment an intelligent one that has smart cities, smart energy, smart industries, smart healthcare, smart homes, smart offices, etc., by deploying a plethora of sensors everywhere [5]. These devices are able to communicate with each other and also to the central authoritative bodies by using technologies such as ZigBee, Wi-Fi, Bluetooth, GSM, etc. [6].

Big Data Analytics for Internet of Things, First Edition. Edited by Tausifa Jan Saleem and Mohammad Ahsan Chishti.

14.2   Analysis of Big Data

Analysis of big data is a three-step process:

- Search the database,
- Mine the data obtained, and
- Analyze the data to obtain the required targets.

Since the amounts of data produced are huge, big data analytics requires new tools and technologies since such quantities cannot be handled using old technologies. The algorithms underlying the new technologies meant for analysis of big data must be able to find out the hidden patterns, trends, and relations over different time horizons in the data [7]. Visualization of achieved results follows data analysis. Also, analysis of big data may prove to be a complex process in application where the data is highly unstructured. The underlying algorithms performing these operations must also be scalable. Moreover, different sources of data have different formats which make integration of sources for analytical solutions critical. Therefore, performance of the existing big data analytics algorithms must be analyzed. Figure 14.1 presents the various aspects of IoT big data analytics.

14.2.1   Existing IoT Big Data Analytics Systems

Based on the requirements of specific IoT applications, a number of different analytical systems are used [8]. They can be classified into five categories:

- **Real-time Analytical Systems:** This type of analysis is performed when there is a continuous change in data and in the situation where fast data analytics is required to provide results quickly. It is specially used for data obtained through sensors. Two architectures exist for the real-time analytics of data-memory-based computing platform and parallel processing clusters employing old relational databases. Famous examples of real-time architectures include GreenPlum and Hana.
- **Off-Line Analytics Systems:** This type of analysis is used when a quick response is not needed [8]. The advantage of using off-line analytics is that it can lower the prices of format conversions. Examples of such architectures include Kafka, Scribe, Time Tunnel, and Chukwa. All these make data acquisition quite efficient.
- **Memory-Level Analytics Systems:** Mongo-DB architecture is typically used when the cluster's memory exceeds that of the size of data (currently the cluster memory is at TB level). It can also be used for carrying out the real-time analysis.
- **Business Intelligence (BI) analytics:** It is used when the data size is greater than the memory level, i.e. opposite of the memory-level analytic

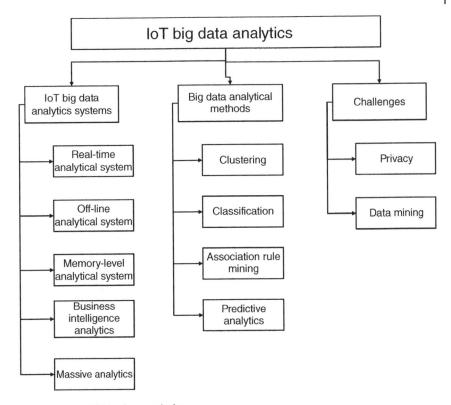

Figure 14.1 IoT big data analytics.

systems. Here, import of data to BI environment is possible. This type of analytics also supports TB level of data. Advantages of conducting BI analytics include:

a) Discover various potential business opportunities by analyzing the pool of data.
b) Simple and efficient interpretation of volumes of data.

- **Massive Analytics:** It is adopted when the data size is larger than the full capacity of BI analysis product and old databases. Hadoop distributed file system is employed in massive analytics for storage of data while as MapReduce is used for analysis of that data. Advantages of massive analytics are:

a) Effective servicing.
b) Obtaining of useful values from data that can be used to create business foundations and have scalability in the market.

14.2.2 Big Data Analytical Methods

The methods for big data analytics can be classified into four categories:

- **Clustering:** It is a data mining technique that forms groups of objects on the basis of different characteristics. The objects that display same features are put in one cluster. Clustering employs an unsupervised learning method for creating the clusters [9]. The most famous clustering techniques are the hierarchical clustering and partitioning.
- **Classification:** It also classifies the objects into groups but the method used for that is supervised learning [9]. The classification methods are provided by Bayesian networks, K-nearest neighbor method, and support vector machine (SVM). Bayesian networks are typically directed acyclic graphs in which random variables denote the nodes and conditional dependencies reveal edges. Employing statistical learning approach, SVM can be an efficient method for classifying data into groups. K-nearest neighbor method is specially designed to find the hidden patterns from large sets of data, ensuring that the objects that are retrieved fall in the same category as was predefined [9]. Classification is one of the most popular and highly employed data mining techniques for analyzing big IoT data.
- **Association Rule Mining:** Used in the applications of decision-making and market analysis, it includes recognizing the relationships among various events, entities, or objects to help improve decision-making capabilities of business and offer better analysis of market [9].
- **Predictive Analytics:** This method uses training data which is actually previous historical data so as to reveal the trends in data [9].

14.2.3 Challenges in IoT Big Data Analytics

This section describes the major challenges that need addressing at first hand in the analytics of IoT big data. These include:

- **Privacy:** It is one of the core issues in IoT and big data analytics. With increased penetration of IoT devices, the privacy of individuals is put to risk and hence people still feel reluctant to let IoT creep into their lives [4, 10]. Though many solutions to ensure data privacy are available like anonymity, temporary identification, etc., decisions need to be taken keeping in thought the ethics such as why use big IoT data, what and how to use it, etc. Heterogeneity that marks the essence of IoT defines a huge security risk [11, 12]. As the devices and the data generated by them are highly heterogeneous, it is a challenge for security professionals since they are new to heterogeneity and security. Most of the solutions relating to security and privacy in IoT and big data analysis rely on trusted third parties. Therefore, when the amount of data grows, difficulties arise in securing every portion of vital and sensitive data [4]. These algorithms also are incapable of dealing with the dynamism of data and hence cannot be applied effectively in IoT and big data analysis. Specific security and privacy problems that can emerge with respect to data stamped out by IoT devices include:

a) Attack management: it refers to the recognition of doubtful traffic patterns in the legitimate networks and also the probable failure to catch unidentifiable attacks [5].
b) It is difficult to keep the system updated [5].
c) Interoperability: since there are many proprietary procedures present, they can create problems in identifying the hidden or zero-day attacks [4].
d) Convergence of IPv4 and IPv6: IPv6 is still not completely deployed and hence the introduction of security rules over IPv4 may not be applicable to protecting IPv6.

Data fusion that is illustrated in the next section can be used as one of the solutions to handle security issues of IoT.

- **Data mining challenges:** Numerous challenges exist in the way big data in IoT is mined. These challenges are volume of data, accessibility of data, correctness of data, data heterogeneity, etc.
- **Visualization issues:** The visualization issues include:
  Visual noise: Usually the objects in the datasets are closely linked to each other, and hence users may see different results of same type [7].
  Information loss: Information can be lost by applying reduction methods on visible sets of data.
  Observation of huge images: The inherent problems present in visualization tools for aspect ratio, physical perception limits, and device resolution.
  Frequently changing image: Rapid changes in output may not be perceptible to the users.
- **Heterogeneity issues:** The data generated from IoT devices are heterogeneous in nature for it may be structured, semi-structured, or unstructured. As such, the issue of integration creeps into the analytics of big data. For this purpose, we have identified data fusion as one of the probable solutions to deal with this problem.

14.3 Data Fusion

It refers to the method employed for making the optimum use of huge volumes of data coming from different sensors [13], e.g. multi-sensor fusion aims to join the information coming from different sensors to obtain inferences which are otherwise not possible to obtain from a single sensor [14]. Hence, reliable and exact information is obtained which is crucial for decision-making purposes.

14.3.1 Opportunities Provided by Data Fusion

- It makes the incoming information smarter by combining data from multiple sensors. Per sensor information might not make much sense.
- It is a fact that high-accuracy sensors consume a lot of power but an open issue in IoT is to device sensors that do not require a lot of energy. So, data fusion paves the way for that by allowing the use of low-power sensors whose information would then be used to produce highly accurate information.
- It also helps to conceal the semantic details responsible for fused results. It can be employed in critical areas such as military, medical areas, etc.

14.3.2 Data Fusion Challenges

- Imperfection of data: Data fusion algorithms must have the ability to deal with imperfect data [14].
- Ambiguities and inconsistencies: Outliers must be detected. Also replacement and data imputation can play a vital role in IoT atmosphere.
- Conflicting data must be treated well.
- Alignment problem: It arises when transformation of sensor data from local frame to common frame occurs prior to the fusion. It must be avoided.
- Both trivial and important data come from the sensors deployed in various environments. The accuracy of data fusion can be affected if it focuses on the processing of the trivial information. Hence, only the most relevant features should be selected before striving for data fusion [14].

14.3.3 Stages at Which Data Fusion Can Happen

- Decision-level data fusion
- Feature-level data fusion
- Pixel-level data fusion, and
- Signal-level data fusion.

Also, there can be single-hub data fusion as or multi-hub data fusion. This is depicted in Figure 14.2.

14.3.4 Mathematical Methods for Data Fusion

- Probability-based methods, e.g. Bayesian analysis, statistics, and recursive operations [15].

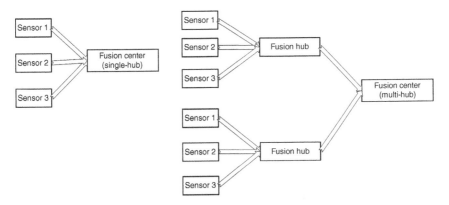

Figure 14.2 Levels of data fusion.

- Artificial intelligence-based methods, e.g. fuzzy logic, artificial neural networks (ANN), machine learning, and genetic evaluation.
- Data fusion method based on theory of evidence.
- Probability-based methods: Even though they are less efficient than integral techniques, they are the most widely adopted ones. Most commonly used ones are the Monte-Carlo method, Morkov chain, and Bayesian theory.
- AI-based data fusion algorithms: These include [15]:

a) Supervised machine learning.
b) Artificial neural networks: they have the ability to derive meaning from imprecise and complex data. They have an extraordinary ability to extract hidden patterns from complex sets of data. Their fault tolerance is very high. They are used in wired speed forecasting because the nature of wind speed is highly complex, ANN can do it. The wind power generation is directly proportional to the cube of wind speed. Therefore, accurate estimation of wind power output is crucial. Here ANNs are trained and tested by several wind datasets.
c) Fuzzy logic: It is used to handle the idea of half-truth whose range lies between complete truth and complete lie. Deep learning can be used for data fusion [16].

14.4 Data Fusion for IoT Security

Although, the applications of IoT range from baby monitors to high-end military devices [17], the European Research Cluster on the Internet of Things (IERC) identifies the following five as its key application domains viz. smart cities, smart home, smart industry, smart transport, and smart health. This section identifies

Table 14.1 Data and sensor characteristics of various IoT application domains.

Characteristic	Smart city	Smart home	Smart Industry	Smart transport	Smart health
Security requirements [4]	Availability Confidentiality Authentication Integrity	• Confidentiality • Context awareness	Availability Confidentiality Authentication Integrity	Timeliness Availability Non-repudiation	Availability Confidentiality Authentication Integrity
Sensitivity to privacy [4]	Medium	High	Medium	Medium	High
Distribution of sensors [18]	Wide area	Small area	Small area	Medium area	Small area
Multi-modality of data [18]	Huge heterogeneity	Less	Less	Medium	Medium
Imprecise data [18, 19]	Inconsistent, imperfect, and ambiguous data available	Ambiguous data available	Inconsistent and ambiguous data available	Inconsistent, imperfect, and ambiguous data available	Inconsistent, imperfect, and ambiguous data available

the sensor characteristics and data features of these application domains. For example, the home-related data are particularly vulnerable, privacy-sensitive, and critical when taking a decision or inferring the context. The carbon-dioxide contents measured by the sensor inside the house can also let the hacker know the number of individuals present inside a building. The light's data can tell him if the owner is home or not. As such, heavy fusion accuracy and confidentiality of data are required in the smart home IoT applications. Table 14.1 provides a gist of the data and sensor characteristics for these application scenarios.

14.4.1 Defense Use Case

There are four types of defense operations viz. Battlefield, Urban Warfare, Other than War, and Force Protection. These scenarios define the size and type of the environment which, in turn, define the size and the requirement for the Internet of Things sensors. A battlefield, for example, refers to a large-scale, non-manually deployed Internet of Things network. An urban warfare and other than warfare operations mean a medium scale (up to several hundred nodes) manually deployed network while as an other than war is an "any-scale" both manually and non-manually deployed network.

The types of sensors used for carrying out these operations are [20]:

- Presence and Intrusion sensors based on the combination of infra-red, photo-electric, laser, acoustic, vibration, etc.
- Chemical, biological, radiological, nuclear, explosive, and toxic industrial material detectors.
- Ranging sensors (RADAR, LIDAR, etc.).
- Imaging sensors (Infra-red and LADAR imaging).
- Noise sensors.
- Soldier-worn sensors that detect serious distress or risks of fatality (force protection scenario).

Because of the various types of constraints on IoT devices and the data generated by them, relying on the data coming from a single sensor in the area of defense is nonsensical. Therefore, the idea is to fuse the results obtained from all these type of sensors and allow the forces to take a much better decision.

14.5 Conclusion

In recent times, big data analytics has obtained significant consideration in an assortment of application areas like business, money, space science, human services, telecommunication, and IoT. Among these regions, IoT is considered as a significant platform in bringing individuals, procedures, information, and things together so as to upgrade the nature of our daily living. However, the key difficulties are the manner by which to successfully extricate valuable highlights from the enormous measure of heterogeneous data produced by resource-constrained IoT devices and how to use this information insight in improving the presentation of remote IoT systems. In this chapter, we have tried to examine the issues of IoT big data analytics and have proposed the use of data fusion to deal with the problem of security in big data generated by the IoT devices deployed in the area of defense. However, for prospective research, there are certain significant obstacles that must be faced. All of which involves translating heterogeneous datasets into common storage formats.

References

1 Russom, P. (2011). Big data analytics. TDWI, 4th Quart. https://tdwi.org/research/2011/09/~/media/TDWI/TDWI/Research/BPR/2011/TDWI_BPReport_Q411_Big_Data_Analytics_Web/TDWI_BPReport_Q411_Big%20Data_ExecSummary.ashx (accessed 2 May 2020).

2 Golchha, N. (2015). Big data-the information revolution. *International Journal of Advanced Research* 1 (12): 791–794.

3 Tsai, C. (2015). Big data analytics: a survey. *Journal of Big Data* 2 (1): 1–32.

4 Zahra, S.R. and Chishti, M.A. (2019). Assessing the services, security threats, challenges and solutions in the internet of things. *Scalable Computing: Practice and Experience* 20 (3): 457–484.

5 Zahra, S.R. and Chishti, M.A. (2019). Ransomware and Internet of Things: a new security Nightmare. *9th International Conference on Cloud Computing, Data Science, and Engineering (Confluence 2019)*, Noida, India (10–11 January 2019). IEEE.

6 Marjani, M., Nasaruddin, F., and Gani, A. (2017). Big IoT data analytics: architecture, opportunities, and open research challenges. *IEEE Access* 5: 5247–5526.

7 Saleem, T.J. and Chishti, M.A. (2019). Deep learning for internet of things data analytics. *Procedia Computer Science* 163: 381–390.

8 Chen, C. and Zhang, C. (2014). Data-intensive applications, challenges, techniques and technologies: a survey on big data. *Information Sciences* 275: 314–347.

9 Saleem, T.J. and Chishti, M.A. (2019). Data analytics in the internet of things: a survey. *Scalable Computing: Practice and Experience* 20 (4): 607–629.

10 Zahra, S.R. and Chishti, M.A. (2020). A collaborative edge-cloud internet of things based framework for securing the Indian healthcare system. *International*

Journal of Sensors, Wireless Communications and Control 10 (4) https://doi.org/1 0.2174/2210327910666191218144157.

11 Zahra, S.R. (2018). MNP: malicious node prevention in vehicular ad hoc networks. *International Journal of Computer Networks And Applications* 5 (2): 9–21.

12 Ahmad, M.S. (2018). Alleviating malicious insider attacks in MANET using a multipath on-demand security mechanism. *International Journal of Computer Network and Information Security, MECS Press* 6: 40–51.

13 Dong, J. and Zhuang (2009). Advances in multi-sensor data fusion: Algorithms and applications. *Sensors* 9 (10): 7771–7784.

14 Hall, D. and Llinas, L. (2001). *Handbook of Multisensor Data Fusion. CRC Press.*

15 Alam, F. and Mehmood, R. (2017). Data fusion and IoT for smart Ubiquitous environments: a survey. *IEEE Access* 5: 9533–9554.

16 Zahra, S.R. and Chishti, M.A. (2020). Fuzzy logic and Fog based Secure Architecture for Internet of Things (FLFSIoT). *Journal of Ambient Intelligence and Humanized Computing.*

17 Showkat, S. and Qureshi, S. (2020). Securing the Internet of things using Blockchain. *2020 10th International Conference on Cloud Computing, Data Science & Engineering (Confluence)*, Noida, India (29–31 January 2020), 540–545. doi: https://doi.org/10.1007/s12652-020-02128-2.

18 Ding, W., Jing, X., Yan, Z., and Yang, L. (2019). A survey on data fusion in internet of things:Towards secure and privacy-preserving fusion. *Information Fusion* 51: 129–144.

19 Saleem, T.J. and Chishti, M.A. (2019) Data mining for the Internet of Things. *ICDM (Posters)*, New York;, USA (17–21 July 2019).

20 Bhatia, M. and Sood, S.K. (2018). Internet of Things based activity surveillance of defence personnel. *Journal of Ambient Intelligence and Humanized Computing* 9: 2061–2076.

15

New Age Journalism and Big Data (Understanding Big Data and Its Influence on Journalism)

Asif Khan[1] and Heeba Din[2]

[1]*School of Media Studies, Central University of Kashmir, Jammu and Kashmir, India*
[2]*Department of Mass Communication, IUST, Jammu and Kashmir, India*

15.1   Introduction

Data Journalism is a new exciting feature of information dissemination in the profession of Journalism. With all the latest technological revelations, the face of journalism is changing at a faster pace, and this is swaying more and more people. The amount of infographics, statistics, interactive online content via data visualization tools like maps, tables, graphs, microsites, and visual data is found to have more impact than the usual text-only method of storytelling. The analogy goes like if you can "correlate" you are indeed a genius. In today's journalism, data is a pressing and can enhance the impact of any report, broadcast, write-up, or analysis. It provides credibility to it, and the in-depth prefix can do wonders for any journalistic organization.

Although not new to the scene, the use of technology in journalism has a record as well. In the past, computers were used to draw correlations in elections or measure any social parameter. This has come to be known as computer-assisted reporting. But if we add the sense of comparative analysis and interactive data visualization to it the results can be helpful in not only deducing plausible references but engaging audiences with vast amounts of data without making it tedious and boring.

Big Data Analytics for Internet of Things, First Edition. Edited by Tausifa Jan Saleem and Mohammad Ahsan Chishti.

The ubiquity of big data and its subsequent methods is that it can make use of available data in any format and process it through the use of computer-based technology and thus serve as analysis which is more palatable in today's "no time" lifestyle of people. Comparing the data for solving any problem requires data or information first, and the fight for accessing factual or accurate information is always a never-ending process. There are myriad examples, which prove that facts are preserved through the epochs that have helped us in understanding the perceived situation in a much better way than ever. In short, the advent of data journalism has been an eye-opener as most often what meets our eyes is way beyond what is there actually. A mere comparative analysis sometimes or should we say most of the time opens up a new dimension of the facts different from what they might seem apparent.

Another aspect of data journalism is that in today's world, we have an abundance of information available. When we say accessible, it essentially means through every platform, which comes handy to users. Now if the same amount of data is customized for the use of consumers, the effect can be profound, which means filtering, processing, and then packaging the content for the end-users. And if the same is done using analytical tools or ways like use of infographics, comparative analysis, graphs maps, etc., the meaning of the content would amplify for the customized consumption. Journalism is the name of changes; the changes it adapts to and the changes technology brings in. The convergence of media is prompting these changes and relentlessly keeps on pushing the whole factory into a state of constant flux.

15.1.1 Big Data Journalism: The Next Big Thing

The advent of the digital revolution brought with itself alarms that journalism, as we knew, will never be the same and true to its warning it did change the face of journalism. The circulation and revenue generation of the newspapers have been on the decline for almost 20 years. In Canada, the situation has led to the Government authorizing a report, which hypothesizes as to what the country's democracy will look like in the world without newspapers. McLennan and Miles (2018) point out that, "the picture is grim in other major powers of world like UK & USA; where the former head of the state Theresa May has cautioned that the shutting of newspapers is a "danger to democracy;" The picture is similar in the USA where the weekday print circulation has decreased from a high of nearly 60 million in 1994 to 35 million for combined print and digital circulation today, amounting to a continuous 24 years of decline. On the other hand, advertising revenue has also taken a beating, falling from $65 billion in 2000 to less than $19 billion in 2016. Newsroom employment dropped nearly 40% between 1994 and 2014 [1].

However, while the digital turn announced the demise of print journalism, it also announced the onset of new-age digital journalism, which brought with itself the innovative set of practices that have changed the face of journalism. On the other hand, the ongoing trends further indicate that the picture of digital journalism is not all that rosy. To sustain the digital onslaught, newspapers launched their digital versions, but that has not curtailed the job cuts in the industry. It has been years since newspapers have launched their websites, but the transformation has not been all that profitable both in terms of revenue and readership [2].

The same thought is echoed by Iris Chyi in her book *Trial and Error: U.S. Newspapers' Digital Struggles Toward Inferiority*, where she believes that the digital change has not been all too successful as heralded. She further says, there is no indication that online news will ever be economically or culturally sustainable. They have killed print, their core product, with all of their focus online [2].

In addition to this the rampant spread of fake news, blurring lines between facts and post facts have further hit journalism in the twenty-first century [2].

At this critical juncture, for journalism to stay relevant, it has to engage its audiences with newer and exciting ways of storytelling that combine the richness and in-depth reportage like old school journalism and interactive, immersive nature of digital journalism. Data journalism does both perfectly; by encompassing an ever-increasing set of methods, approaches, and visualizations for storytelling and at the same time incorporating computational reporting along with increasing use of computer programs and software for data mining and analyzing data to come up with patterns. "The unifying goal is a journalistic one: providing information and analysis to help inform us all about the important issues of the day" [3].

The sheer ability of big data journalism or data journalism to tell a compelling story using vast amounts of digital information in multitudinous ways is what makes it stand out. It is imperative to mention here that digital turn has also changed our understanding of what we mean by data, hence the word, "Big Data" because it is available in such massive amounts. Data journalism uses computational software and tools to automate the process of gathering, sifting, and analyzing this vast amount of information in the digital spectrum and help find connections between them. A crucial thing to understand here is that having big data is not enough; it needs not only to be analyzed but also visualized in such a manner that audiences do not get bored or bogged down by all the numbers.

Data journalism helps a journalist to communicate a multifaceted and intricate story through engaging infographics that not only encapsulates the crux of the story visually but also reduces the reading time by increasing the time spent engaging with the story. Hans Rosling's story on envisaging world poverty [4] where he talks about the disparity in world income from over 200 countries, more than 200 years in just four minutes using visualizations, is a perfect example of how vast amounts of information can be presented in a visually engaging way and in a short time without compromising the data. Further, it can expand the process of news gathering itself, by open-sourcing and collaboration, which not only diversifies the story but also helps to mine through the massive amounts of data available. A perfect example for this is Panama papers, which consisted of 11.6 million documents amounting to 2.6 TB of data that was mined by 370 journalists from more than 100 media organizations across 70 countries [5]. "Data can thus act as a source of data journalism, or it can be the tool with which the story is told - or it can be both" [3].

Furthermore, the engagement with data has also meant that the job of journalists have transformed from mere reporting to uncovering, analyzing, and making the audience understand what a specific development means. The data journalist's job is not only to crunch numbers using big data analytics but instead connecting dots with the bigger picture as to how and what these numbers mean. This could interpret into identifying the next financial crisis or unearthing the misappropriation of funds, political scandals or scams; all told through immersive data visualizations that not only informs but at the same time explains [3].

In recent years, data journalism has garnered significant attention both in the academic literature and in the area of new developments in digital news production [6–8]. Similarly, the way social media journalism and citizen journalism were considered the future, data journalism is now being called the future of journalism [9], making it the next big thing. Data journalism has found journalists owing to the computational are still looking at growing acceptance in news organizations and newsrooms, but it with some amount of apprehension and the statistical methodologies involved. It does require the journalists to upgrade their skill sets, but due to the open-sourcing nature of the genre it further allows for collaborations, which can be fruitful for young journalists who are still wary of the style.

15.1.2 All About Data

The technological boom, along with other factors like a surge in the use of digital devices like smartphones, tablets, personal computers, etc., have been predominantly responsible for the growth of big data. It is currently doing wonders for the industry where advertising is slowly getting replaced by data analytics. This has proven to be more advantageous for business industries, companies, and news organizations with increased profitability and reach. Same is the case with the Journalism industry. The big data "story" has an unprecedented scope.

The amount of data that gets generated has reached humongous proportions. Humans have created more data in the last two years than in entire human history. We are constantly making data, even when on the go; all due to the digital turn. "On Google alone, we submit 40,000 search queries per second. That amounts to 1.2 trillion searches yearly. Each minute, 300 new hours of video show up on YouTube.

People share more than 100 terabytes of data on Facebook daily. Every minute, users send 31 million messages and view 2.7 million videos. Such unprecedented amounts of data have resulted in the creation of 8 million jobs in the USA alone since 2012, and 6 million worldwide. 79% of executives believe that failing to embrace big data will lead to bankruptcy. This explains why 83% of companies invest in big data projects. It is estimated, Fortune 1000 companies can gain more than $65 million additional net income, only by increasing their data accessibility with 10%"[10].

While enormous data is being generated, only 0.5% of the aggregate information that gets investigated. According to recent studies, big data not only means gathering large volumes of data but a careful and well thought of analysis of this data can lead to a focused assessment of future plans, etc. "In India, industry experts are of the opinion that the Big data analytics sector shall witness eight-fold growth to reach $16 billion by 2025 from the current $2 billion. The sector is expected to reach $16 billion by 2025 and register a CAGR of 26% over the next five years" [11].

According to Marco Lübbecke, refined data scrutiny can save lives and improve society. Citing a critical case study to showcase the importance of big data, Marco elaborates,

> In 2001, with massive amounts of data about polio cases around the world, the CDC (working with the Global Polio Eradication Initiative and consultants at Kid Risk, Inc.) faced an important choice: Direct $100 million in funding dollars and settle for controlling the outbreak of polio, or attempt to completely stop all new cases of polio. By developing sophisticated mathematical models that leveraged the best available scientific evidence and field knowledge, CDC officials became confident that it would be possible to prevent any further cases of wild polio from emerging. By coincidence, the CDC accepted its award in 2014 shortly after India celebrated three years in which its population of 1.27 billion people had not experienced a single instance of polio [12].

While no one can deny the impact of big data, the question lies in accessing, gathering, and analyzing the data, without which there is no value to the data. At the same time, Amazon web services, Spark SQL, Hive, and Hadoop Distributed File system are some of the biggest and most popular platforms for storing big data but also reading, managing, and analyzing large datasets. However, these are mostly being used by corporate Industries. When it comes to using data for journalism, a variety of different tools and methods come into play.

15.1.3 Accessing Data for Journalism

Accessing data is one of the most crucial aspects of data journalism and a stumbling block for many journalists. However, the age of big data has also opened the door for the era of open data, which gives free access to all the data in machine-readable format. Governments all over the world have taken the initiative of opening their databases to the public. An example of it can be seen when Barack Obama launched data. gov in 2009, where Government datasets were published. These open databases can be excellent sources for journalists giving them free access to vast amounts of data. Similarly, there are many different projects, which are working toward creating open data. Felle et al. [13] point "One such example of Opencorporates.com, which offers an open database of more than 110 million companies from all over the world with all the relative information about them that allows anyone to search and process it."

Data Hub – which allows access and reprocess of openly available sources of data, Scraper Wiki – an online tool which helps in extracting data so it can be reused, and Data Couch – a platform to upload, mine, share, and visualize data are some of the platforms providing easy access to open data.

According to Schellong and Stepanets, there are eight principles to categorize data. "First is the principle of completeness, which deals with the availability of data without any privacy restriction-second being the primacy, which means that the data is as extensive and restructured as possible. Then follows timeliness, which deals with sorting of data and keeping it available whenever it needs to be published in short time. It is followed by the next principle of accessibility, which revolves around the availability of data to a maximum number of people. While the fifth principle is of machine readability, that is related to the mining and reading of the data for further processing and analysis. The last three principles are the absence of discrimination, absence of compatibility and absence of license requirements; they determine that data is accessible to all, vis-à-vis open data movement" [14].

While dealing with open data, journalists also need to be aware of the type of license under which the data shared. The most common licenses encountered usually are creative-commons, open data commons, or any Government certificates. These licenses inform about what can or cannot be done with the data. While the entire dataset may be downloadable, it may not be shared. "In summary, there are three classes of open licenses: Public domain dedications, which also serve as maximally permissive licenses; there are no conditions put upon using the work; mPermissive or attribution-only licenses; giving credit is the only substantial *condition; mCopyleft*, reciprocal, or share-alike licenses; these also require that modified works, if published, be shared under the same license" [15].

15.1.4 Data Analytics: Tools for Journalists

Gavin Freeguard, a senior researcher at the Institute for the UK government, has given three ways to identify and use data. (i) Understanding data as pure raw data dealing with numbers. This allows reusing of data in various formats. (ii) Perceiving data as information and processing raw data into meaningful information via infographics and visualizations, which identify the key points in the dataset. (iii) Using data as evidence or proof. This is mostly used for investigative journalism for reporting scams or pressuring the state for reforms [13].

A major concern for data journalists is the requirement for or the proficiency in computer languages/coding to be able to sift through the data. This is labeled as a critical skill set for journalists and newsrooms all over the world that are equipping their systems to deal with big data. The International Consortium of Investigative Journalism (ICIJ) has given a list of tools that data journalists [16] can train themselves with to handle, analyze, and visualize the data (Figure 15.1). These include:

- *Spreadsheets + Pivot Tables (Google Sheets/Microsoft Excel):* It allows journalists to go through vast amounts of data within no time. By cumulating the spreadsheets and assigning a typical value to them, the tool enables the summarizing of large datasets. It is majorly used for finding inconsistencies or patterns.

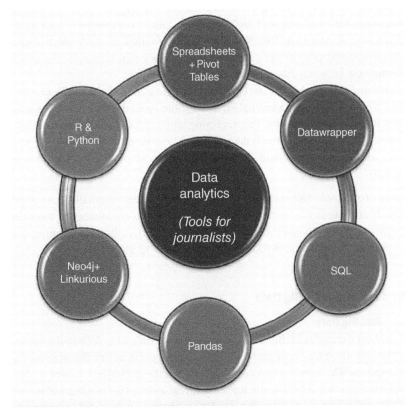

Figure 15.1 Various tools for journalists recommended by ICIJ.

- *Datawrapper:* An open-source tool which does not require any coding on the part of journalists and allows for importing data from a spreadsheet or an external link, and publishing it directly. It is an excellent tool for journalists to present their data using charts or maps visually.
- *SQL:* It fills in the limitations of the spreadsheets by allowing the journalist to accurately describe the type of data or identify the changes across the related datasets. With a simple command, it tailors the data mining according to the requirement of the journalist. Furthermore, the commands can be saved as a script, which can be used later. It can be used for filtering of data, finding patterns, or using it on relational datasets to look for similarities.
- *Pandas:* Using the language Python, this tool enables the data journalists to work like large datasets by indexing and categorizing the data, making it easier to investigate. It is useful in excluding duplicated data and describing data with a preferred label.
- *Neo4j + Linkurious:* ICIJ's investigations often expose previously unknown links between entities or characters in a dataset that seem unrelated at first glance. And had we been investigating a leak like the Panama Papers "by hand," many relationships between different individuals and shell companies would have gone unnoticed. Software like Neo4j allows journalists to contextualize the whole picture and present the data in such a manner that it exposes the interconnectedness between different sets of data.
- Programming Languages (R & Python): Learning to program will quickly expand your reach as a data journalist. Most of the time, the data is not given in the way the journalists require. Programming languages helps to scrape the data and manage it within no time. It does not matter so much which language you choose, although Python and R seem to be the current favorites among journalists.

Majority of the tools and software are open, which means journalists can copy the software and tweak it as per their requirements. Further, a number of them do not require journalists to be proficient in coding. Though, coding is becoming the need of the hour for data journalists (Figures 15.2 and 15.3).

15.1.5 Case Studies – Big Data

15.1.5.1 BBC Big Data

One of the best examples for big data comes from the BBC. BBC defines data journalism through the process of realizing three main points as per the manual on data journalism. The handbook reads, "BBC uses data to enable a reader to discover information that is personally relevant, reveal a story that is remarkable and previously unknown and help the reader to better understand a complex issue" [17].

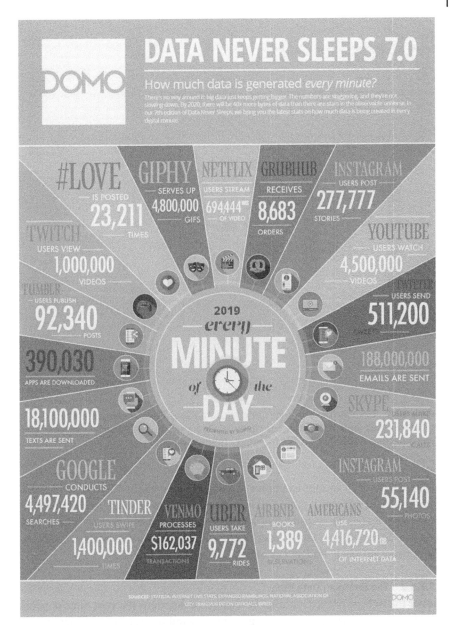

Figure 15.2 Illustration showing the amount of data generated per minute. (*Note:* From weforum.com, Image: Visual Capitalist.). *Source:* Data Never Sleeps 7.0, © 2020 Domo, Inc.

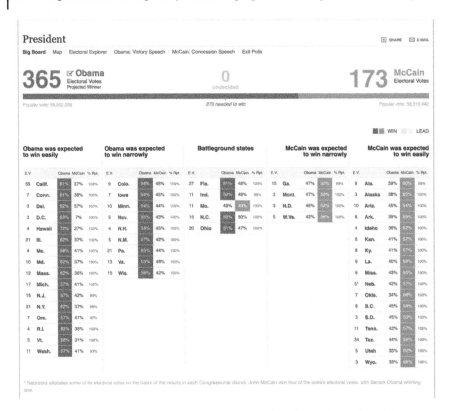

Figure 15.3 Example of case study using big data. (Note: From The Big Board for Election Results (New York; Times) as cited in datajournalism.com.). *Source:* The New York; Times, Election 2010. © 2010, The New York; Times Company.

There have been various examples of using big data by the BBC. Most of them were very noticeable with users and have helped the Government to take recourse in urgency. As per the website datajournalism.com, some of the cases which used data are "School league tables, every death on every road, The world at 7 billion: What's your number? Budget calculator" among many others.

15.1.5.2 The Guardian Data Blog

As per the website of the Guardian newspaper, it mentions the start of its popular data blog in 2009. The purpose is to present data and its usage in a format, which could be easily understood by its audiences. The main aim was to collect, publish, and analyze data. According to the statistics, the website has during the last decade published many stories and datasets on almost every topic which has immensely influenced the lives of people. Data blog (founded by Simon Rogers) is considered to be among the pioneers of data journalism with its operations starting as early as March 2009. Today, the newspaper has data editors positioned at its primary offices.

(Note: From The Big Board for Election Results (New York; Times) as cited in data-journalism.com*)*

The data blog provides statistics, maps, infographics, data and analysis, and has covered many important issues through this blog project. The significant stories, which have been quite influencing are Zero Tolerance project, Beyond the Blade, etc. They quote and feel inspired by the approach of Philipp Meyer who authored a book "Precision Journalism", way back in 1972, where he advocated the use of techniques from social sciences in researching stories in Journalism.

As real as it can get, the website reads "Data is not just about numbers, and behind every row, in a database, there is a human story. They're the stories we're striving to tell." Figure 15.4 shows an example of using data for mapping the wealth inequality estimates (Infographics from The Guardian).

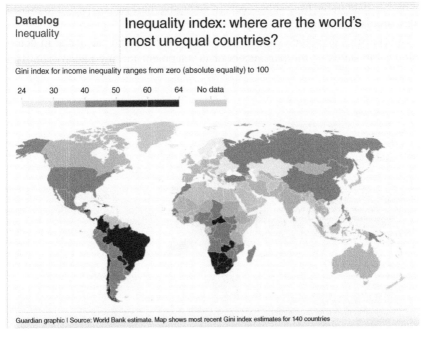

Figure 15.4 Map showing the financial inequality estimates for 140 countries. (*Note:* From World Bank estimate as cited in TheGuardian.com.). *Source:* Inequality index: where are the world's most unequal countries? © 2017, Guardian News & Media Limited.

15.1.5.3 Wikileaks

One of the most infamous document leaks in history is ascribed to Wikileaks, founded and managed by Julian Assange. Since 2006, this nonprofit organization has been working in the field of data journalism with some of the most impactful stories mainly based on document cache leaks/supply of files like the Afghanistan war, Guantanamo Bay detention camp, Iraq war logs, etc. It is a document archive and disclosure site and as per many studies has drawn criticism for the absence of whistleblowing and for criticizing the Panama Papers' and various other issues involving privacy, finances, etc. (Figures 15.5 and 15.6).

15.1.5.4 World Economic Forum

For Journalism, data can do wonders and in almost all the stories, data provides factuality, accuracy, and makes situations entirely predictable. In one of the crucial reports by the World Economic Forum, the data has been put as an important parameter to gauge the development status in any given region. The report published by WEF, compiled by Vital Wave Consulting says, "Sources such as online or mobile financial transactions, social media traffic, and GPS coordinates now generate over 2.5 quintillion bytes of so-called big data every day. And the growth of mobile data traffic from subscribers in emerging markets is expected to exceed 100% annually through 2015" [18].

The statistics show the essence of the data generated and calculated in determining the status or the parameters of development for any region. The data

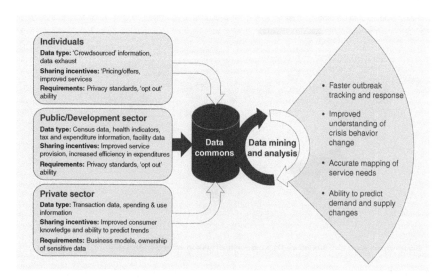

Figure 15.5 Illustration by World Economic Forum – diagram on Data Mining & Analysis and Data Commons.

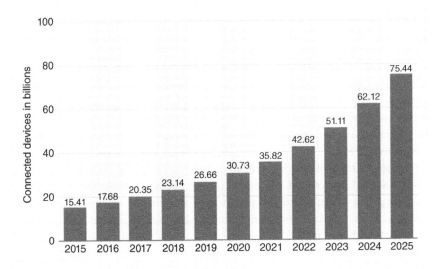

Figure 15.6 Internet of Things – number of connected devices worldwide 2015–2025. *Source:* From Statista Research Department, 17 November 2016, (statista.com), © 2016, Statista.

analytics can be used for the advantage of any administration, which can use the data for better positioning of its resources and related policies. The report talks about the importance of data, which if gleaned appropriately can be used in exploring the full potential in sectors, like financial services, education, agriculture, health, and general line administration.

15.1.6 Big Data – Indian Scenario

Just a decade before, the concept of big data, more specifically data journalism was not in the textbooks of journalists. There was no discussion of the topic in the newsrooms or in the editorial board meetings. But now since 2014, the concept is fast-changing newsrooms. Journalists are now able to ascertain the advantages of data, its analysis, correlations, analytics, and the same has helped in fact-checking and accurate predictions related to health, politics, elections, sports, development, economics, and many other important issues concerning our country. As the saying goes that brevity is the soul of wit, the big data has helped the journalists to tell stories in much fewer text and more illustrations, graphics, etc. A 1000-word article can now be made more specific and instead made more meaningful and understandable through infographics.

Sensing the importance of data and its usefulness in ascertaining the factual status of various initiatives, with increased transparency, the Government of India under the aegis of Digital India initiative started a portal data.gov.in. The portal (Open Government Data Platform India) is a joint venture of India and the US Government.

As per the website, "The portal is intended to be used by the Government of India Ministries/Departments and their organizations to publish datasets, documents, services, tools and applications collected by them for public use. It intends to increase transparency in the functioning of Government and also open avenues for many more innovative uses of Government Data to give different perspectives" [19].

Another example of data journalism picking up fast in India could be of "*Data Meet*", founded, "on 26 January 2011 by a handful of data enthusiasts who started a Google group where people can share tips for working with data. The intent was always to work on data in India and to find others doing the same." The group has an active presence (chapters) in five big cities in India, Delhi, Bangalore, Mumbai, Ahmedabad, and Pune. It is primarily involved in training, conducting workshops, and helping people, companies, and media practitioners in using data. There are also some small data-related initiatives where training and information are imparted to practitioners, students, and working journalists. One of such initiatives undertaken by DataLEADS which is a digital media and information venture aimed to build data-driven storytelling, media innovations through series of reporting, training, and media development initiatives, promote media research and creative educational interventions, and strengthen journalism in Asia.

The entire scene of data analysis and data usage for the advantage and progress of big small enterprises is flourishing at a faster speed in India. These data science helps companies, organizations in understanding data, analytics, digital campaign management, data management, social media strategies, and data-driven business decisions, which, in turn, supports their clientele in ensuring a better market presence and promotion. Likewise, reporters, media analysts, and media organizations have been using data for more in-depth insight into the day-to-day events which make a newsworthy content broadcast or published through multimedia channels, portals, etc.

15.1.7 Internet of Things and Journalism

Technological determinism or Media determinism is a theory, which describes technology as the leading force influencing our society. Technology is evolving every day, and so is its impact over the systems dependent on it. It has the power to permeate all the societal layers, provided resources are available. Among various technological advances, which have impacted the profession of journalism, is "Internet of Things."

From its applications for consumer use, healthcare, communications, remote monitoring, automation, manufacturing, agriculture, and many more to the ever-evolving media or journalism field, it certainly has future insinuations in store. The reason is simple: there is an unprecedented growth of devices, gadgets, and appliances controlled and connected through the Internet. As per some estimates, the number of devices, which will be connected to the Internet, shall be more than 50 billion by the year 2020. Although this comes with critical issues of security and privacy, which poses a formidable challenge, but the entire system is in constant flux. There is no end to it. According to statista.com, "The total installed base of Internet of Things (IoT) connected devices is projected to amount to 75.44 billion worldwide by 2025, a fivefold increase in ten years" [20].

15.1.8 Impact on Media/Journalism

Associated Press, an American news agency, on its website insights.ap.org reiterates what seems ubiquitous now,

> *According to Cisco, the number of connected objects is expected to reach 50 billion by 2020, equating to 6.58 connected devices per person. They are all controlled by tiny computers that communicate with each other, in an ecosystem commonly known as the Internet of things. It has implications for two distinct aspects of journalism – newsgathering and consumption. Smart devices connected can be used to provide better context to a story, such as data on traffic, weather, population density or power consumption* [21].

Already there is quite a lot of buzz about IoT in media and journalism circles. A lot of changes have been there at all possible levels, be it gathering, packaging, or dissemination of news. Even with such technological forces of media convergence and IoT, organizational, structural, and hierarchical changes are inevitable both inside and outside of newsrooms. The big data, IoT, newer technologies, social media, networking sites, media convergence, etc., all these are shaping up a new, competitive, and cutting edge journalism for people to consume. Reporters, correspondents, broadcast journalists, and stringers all have started using new technology-based devices and apps to tell their stories. With more analysis, decisive statistics, infographics, data correlation, etc., stories find more credibility and authenticity. This is the new face of purposeful storytelling now. With IoT set to be one of the predominant digital platforms, new paradigms and shifts will be evident in journalism. In today's networked society, the world is going to witness a new, innovative, collaborative as well as competitive, and different business models in this sphere leading to a smarter journalism activity.

It is undoubtedly the "Next Big Thing". The essence of "Big Data" and "Internet of Things" for media and journalism cannot be undermined and requires to be harnessed, promoted, and encouraged more so that the information we generate and circulate provides actionable insights. Both Big Data and IoT have tremendous potential to influence the economies of global markets, and at the same time change, the way content (information) is collected and produced for the audiences. As per statistics, all major companies are investing in this sector to secure their future. With more compatibility, interoperability, and power of integrating, the new systems and platforms have a lot to offer for users and businesses.

References

1 McLennan, D. and Miles, J. (2018). A once unimaginable scenario: no more newspapers. *Washington Post.* https://www.washingtonpost.com/news/theworldpost/wp/2018/03/21/newspapers/ (accessed 17 June 2020).

2 Rosenwald, Michael. (2016). Print is dead. Long live the print. *Columbia Journalism Review/Winter.* https://www.cjr.org/special_report/print_analog_comeback.php (accessed 19 June 2020).

3 Gray, J., Bounegru, L., and Chambers, L. (2012). *Data Journalism Handbook.* USA: O'Reilly Media.

4 BBC (2010). Han Rosling's 200 Countries, 200 Years-4 minutes. *The Joy of Stats* (26 November 2010).

5 McKeena, B. (2016). Panama papers revealed by graph database visualisation software. *Computerweekly.com.* https://www.computerweekly.com/news/450280758/Panama-Papers-revealed-by-graph-database-visualisation-software.

6 Appelgrena, E. and Nygren, G. (2014). Data journalism in Sweden introducing new methods and genres of journalism into "old" organizations. *Digital Journalism* 2 (3): 394–405.

7 Fink, K. and Anderson, C.W. (2014). Data journalism in the United States. *Journalism Studies* 16 (4): 467–481.

8 Mair, J. and Keeble, R.L. (2014). *Data Journalism: Mapping the future.* Abramis Academic Publishing.

9 Knight, M. (2015). Data journalism in the UK: a preliminary analysis of form and content. *Journal of Media Practice* 16 (1): 55–72.

10 Hostingtribunal.com (2020). 77+ big data stats for the big future ahead. https://hostingtribunal.com/blog/big-data-stats/#gref (accessed 1 July 2020).

11 NDTV Profit (2017). Big data analytics to become $16 billion industry by 2025: experts. *NDTV Profit.* https://www.ndtv.com/business/big-data-analytics-to-become-16-billion-industry-by-2025-experts-1719591 (accessed 4 July 2020)

12 Lübbecke, Mark. (2015). *Big Data Saves Lives*. CNN. https://edition.cnn. com/2015/02/12/opinion/lbbecke-big-data-saves-lives/index.html (accessed 22 June 2020).

13 Felle, T., Mair, J., and Radcliffe, D. (2015). *Data Journalism: Inside the Global Future*. Suffolk, England: Abramis Academic Publishing.

14 Aitamurto, T., Sirkkunen, E., and Lehtonen, P. (2011). *Trends in Data Journalism*. Next media.

15 Linksvayer, M. (2015). Using and sharing data: the black letter, the fine print, and reality. In: *Data Journalism Handbook* (eds. J. Gray, L. Bounegru and L. Chambers) (2012). USA: O'Reilly Media.

16 Kubzanksy, C. (2018). Nine essential tools from ICIJ's data journalism and programming experts. *International Consortium of Investigative Journalists* https://www.icij.org/blog/2018/08/nine-essential-tools-from-icijs-data-journalism-and-programming-experts/.

17 Leimdorfer, Andrew. (n.d). Data Journalism at the BBC. https://datajournalism. com/read/handbook/one/in-the-newsroom/data-journalism-at-the-bbc.

18 World Economic Forum (2012). Big data, big impact: new possibilities for international development. http://www3.weforum.org/docs/WEF_TC_MFS_BigDataBigImpact_Briefing_2012.pdf.

19 Government of India (2020). "Open government data (OGD) platform India. https://data.gov.in/about-us.

20 Internet of Things-number of connected devices worldwide 2015–2025. (2016). https://www.statista.com/statistics/471264/iot-number-of-connected-devices-worldwide/.

21 Marconi, F. (2016). *Making the Internet of Things Work for Journalism*. Associated Press https://insights.ap.org/industry-trends/making-the-internet-of-things-work-for-journalism.

16

Two Decades of Big Data in Finance
Systematic Literature Review and Future Research Agenda

Nufazil Altaf

School of Business Studies, Central University of Kashmir, Kashmir, India

16.1 Introduction

[1] defined "Big data analytics" as a combination of two words "big data" and "analytics." While big data refers to the voluminously large datasets, analytics refers to using these datasets to reveal patterns, trends, and associations that can be applied as a solution to various complex problems [2]. The past decade witnessed the growth in the application of big data analytics to the area of finance, particularly to the stock market finance. Finance practitioners viewed big data analytics as a powerful tool to improve the risk analysis process, enhance the fraud detection and prevention, and change the trade and investment paradigm from "of no moment" to "real-time" settlement [3]. Not surprising, many scholarly evidence also emerged that elaborated upon the application of big data analytics to the finance industry [3–8]. Specifically, these researchers viewed the application of big data to various finance businesses like crowd-funding, cryptocurrency, wealth management, SME finance, asset–liability management, trading platforms, payment and settlement system, and so on. These businesses generate an enormous amount of data and therefore the management of such data is an important factor in these businesses (Hasan et al. [3]). In line with this, big data has received overwhelming attention in the finance industry, where managers know that bits of information is an important part of the success of the enterprise.

Although the practitioners of finance believe that role of big data in the Internet of Things (IoT) is more about processing and gathering information about financial transactions that may be ideal for decision-making. Yet, many financial analysts believe that the application of the IoT in the finance industry as a technological

Big Data Analytics for Internet of Things, First Edition. Edited by Tausifa Jan Saleem and Mohammad Ahsan Chishti.

suite has a transformative effect on every dimension of financial and economic activity across the globe. The potential benefits of IoT in finance are seen in providing more comprehensive real-time data for decision-making. Additionally, innovative financial products like auto insurance telematics and customized banking services are examples of new products that became possible only because of IoT application to the finance industry. Therefore, the use of IoT in finance goes far beyond mere data communication and analysis and has been seen as a strategic gamification tool and incentivizes lower risk in the finance industry even if it is in the early stages of its development.

Beveled by the response of big data analytics as an information processing tool, the first illustration of the application of big data to finance was presented by [9]. In this study, the researcher investigated the textual content of a column in the Wall Street Journal "Abreast of the Market" as a prognosticator of market returns. Additionally, [9] also applied text analysis to predict companies' fundamentals. One more study on textual data was conducted by [10], who asserted that firms with a detailed description of products end up in a merger. Some studies, like [11], used Bayes algorithms for performing linguistic analysis. They suggested that disagreement in messages is associated with increased volatility. However, [12] criticized Bayes approaches and suggested that these approaches do not work in the finance industry. They proposed new positive and negative words that provided signals of volatility, trading volume, and so on, and were highly applicable to finance.

Some researchers like [13, 14] suggested that intraday data such as the flow of order types predict returns on a daily horizon. In a similar vein, [15] indicated that data from retail investor trades contain valuable information and they predict the returns positively with no reversals for a month. In follow-up research, [16] suggested that short sales of retail investors also predict stock returns but negatively. Using the ranking of stocks, [17] suggested that the higher the ranking of the stock, the lower will be the returns. Further, [18] assert that technical analysis tools are strong predictors of future stock returns, [19] suggest that interest in the company is conveyed through internet searches and search engines record these searches and sometimes such interest predicts returns.

Further, [20] suggested that product quality rating by customers of Amazon is one of the potential factors that affect returns, [10] assert that the proportion of negative words (opinions) posted by the investors on the investment blog acts as a significant determinant of lower returns on the stock. Additionally, [21] find that the volume of its posts predicts the company's abnormal returns.

The literature's critical review suggests that the studies on big data analytics in finance are mostly related to financial markets, internet finance, and financial services. Thereby keeping in view the increased acceptance of big data in the finance industry, this study attempts to contribute to the literature by conducting a systematic literature review (SLR) in an endeavor to collect the views of scholars, academicians, and industry practitioners related to big data and finance. This study will help gain deeper insights into the understanding of research conducted on big data in the area of finance and thereby discuss the findings of these studies as an implication to the finance industry. Additionally, the study attempts to provide directions for future research as a challenge to be faced by academicians and researchers.

The rest of the paper is organized as follows: Section 16.2 describes the methodology adopted for the study. Section 16.3 describes the article identification and selection process. Section 16.4 deliberates upon the description and classification of literature. Section 16.5 provides the content and citation analysis of articles and Section 16.6 reports the findings and research gaps.

16.2 Methodology

Following [22, 23], we conduct a SLR in this paper. SLR is recognized as a comprehensive methodology for conducting literature reviews. Additionally, SLR describes the procedures, processes by which all relevant material related to a particular phenomenon will be collected [24]. Accordingly, the adapted review process for identifying the issues and future avenues in big data for finance research is categorized into four phases, as shown in Figure 16.1.

16.3 Article Identification and Selection

The search of the published journal articles for big data in finance was made in the major academic databases like Scopus, Web of Science, and EBSCO. It must be noted that these three academic databases were chosen because they hold the

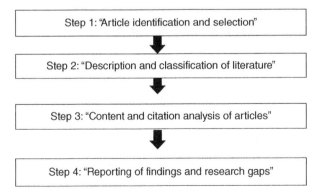

Figure 16.1 Procedure for systematic literature review. *Source:* From Singh and Kumar [22].

Table 16.1 Article identification and selection.

Database	Time-horizon	Total number of articles	Total number of selected articles
Scopus	2000–2019	85	46
Web of science	2000–2019	57	32
EBSCO	2000–2019	48	27
Total		190	105

academic might. The search began by a keyword (Big data) search that resulted in a huge amount of literature. The article that had the keyword "Big data" in the title, keywords, or abstract was initially selected. Following the keyword search, delimiting boundaries were setup for removing the unwanted articles. The delimiting boundaries' setup is as follows:

- Papers for which full-text was available.
- Papers were collected from 2000 to 2019.
- Papers disseminating links of big data for finance were only considered.

After the application of delimiting boundaries, the total number of articles available for reading was 190. However, after reading the articles, it was found that some articles were indexed in more than one database. Therefore, to remove duplicity, 85 articles were removed, leaving us with 105 articles as the final sample. The article availability and selection are given in Table 16.1. Perusing Table 16.1, the total number of articles found initially in the Scopus database was equal to 85, in Web of Science 57, and EBSCO 48. However, after removing duplicity or more than one indexing, the article that was solely indexed in Scopus was 46, in Web of Science 32, and EBSCO 27, making the total sample equal to 105 unique articles.

16.4 Description and Classification of Literature

The selected 105 published journal articles are further analyzed with regard to the research method employed, publications (year wise), and journal of publication. This exercise helps to identify the trends prevalent in big data for finance literature.

16.4.1 Research Method Employed

The published articles are broadly classified into three categories based on the research method employed; these include empirical, conceptual, and survey. The frequencies and percentage of articles within each group of research methods are given in Table 16.2. Perusing the table, it can be found that 75.23% (79) of articles employ empirical methodology and such methodology has remained most popular among the scholars working on big data in finance. The conceptual methodology is the second most popular among the researchers, while a survey methodology has not received much interest in the last two decades of research in big data for finance.

16.4.2 Articles Published Year Wise

Figure 16.2 presents the number of articles published from 2000 to 2019 across years. The visual presentation of the graph helps us to divide the time frame into three phases, the first phase (2000–2009), the second phase (2010–2013), and the third phase (2014–2019). The first phase can be regarded as a low growth phase because the number of articles published during these years amounts to be very less. It may be due to the low acceptance of big data in the area of finance. The second phase can be regarded as medium growth phase, during this phase the number of articles being published on big data for finance started growing, and

Table 16.2 Articles according to research method employed.

Research method	No of articles	Percentage of articles
Empirical	79	75.23
Conceptual	21	20
Survey	5	4.77

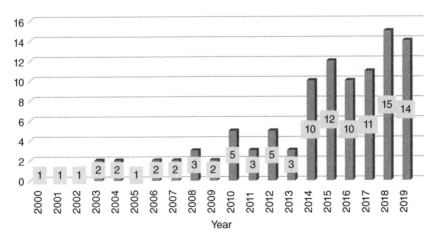

Figure 16.2 Articles published year wise.

lastly the third phase can be regarded as high growth phase, during this phase number of articles being published every year entered two digits, in fact, the maximum number of articles (15) were published in 2018, during this phase. The growth in the literature from the second phase may be due to the advent of the global financial crisis, where scholars and practitioners reacted by developing innovative methods to streamline crisis, big data is one such example.

16.4.3 Journal of Publication

The main aim of analysis by the journal of publication is to identify the journals that have taken the lead in disseminating information on big data in finance. Such information is presented in Table 16.3. It must be noted that from the long list of journals we have presented, only the journals that have published at least two papers on the topic during 2000–2019[1]. Among the long list of journals International Journal of Information Management has published a maximum of seven articles followed by the Journal of Econometrics and Journal of Business Research with six articles, respectively. The Journal of Big Data, Physica A: Statistical Mechanics and its Applications and Decision Support System have published five articles from 2000 to 2019. Further, a detailed list is presented in Table 16.3.

16.5 Content and Citation Analysis of Articles

In this section, we first take up the citation analysis followed by content analysis.

16.5.1 Citation Analysis

Citation is referred to as the reference of the work of one author made by another author. Citation analysis aims to identify the most influential and popular work among the researchers. The results of citation analysis are presented in Table 16.4. It must be noted that to save space, we have only presented the author(s) that have been cited at least 20 times. Additionally, it must be noted that we used citation information provided by Google Scholar for citation analysis. Further, it must be noted that among the 105 articles, 23 articles were not cited; however, these articles were published recently during 2018 and 2019. This may be the possible reason that these articles were not cited at all. Additionally, 105 articles have a total citation count of 2205, implying that the average citation per article was 21. The most

---

1 18 journals among the list of journals had published at least two articles during 2000–2019.

Table 16.3 Journal of publication.

S. No.	Journal name	Publisher	No. of articles published
1.	International Journal of Information Management	Elsevier	7
2.	Journal of Econometrics	Elsevier	6
3.	Journal of Business Research	Elsevier	6
4.	Journal of Big Data	Springer	5
5.	Physica A: Statistical Mechanics and its Applications	Elsevier	5
6.	Decision Support System	Elsevier	5
7.	Emerging Markets Finance and Trade	Taylor & Francis	4
8.	MIT Sloan Management Review	MIT	4
9.	Intelligent Systems in Accounting, Finance and Management	Wiley	4
10.	Journal of Behavioral and Experimental Finance	Elsevier	3
11.	Journal of Risk and Financial Management	MDPI	3
12.	Electronic Commerce Research	Springer	3
13.	The Journal of Corporate Accounting & Finance	Wiley	3
14.	Journal of Monetary Economics	Elsevier	2
15.	The Journal of Finance and Data science	Elsevier	2
16.	Australian Accounting Review	Wiley	2
17.	International Journal of Electronic Commerce	Taylor & Francis	2
18.	Wireless Personal Communications	Springer	2

influential and cited work is [25] with 395 citations, followed by [26] with 344 citations and [27] with 234 citations. These articles have received tremendous response and it may be because these articles are among the early works on big data in finance. The rest of the author(s) presented in Table 16.4 have citations in two digits.

16.5.2 Content Analysis

Through content analysis of the selected articles, we provide in-depth details where the research on big data in finance has been concentrated. It is worth noting that most of the research has been focused on big data for financial markets, internet finance, financial services, and other issues.

Table 16.4 Citation analysis.

S. No.	Author(s)	Citation
1.	Drake et al. [25]	395
2.	Einav and Levin [26]	344
3.	Dimpfl and Jank [27]	234
4.	Kshetri [28]	78
5.	Seddon and Currie [29]	59
6.	Begenau et al. [4]	49
7.	Chen et al. [30]	39
8.	Campbell et al. [31]	37
9.	Jin et al. [32]	36
10.	Choi and Lambert [33]	28
11.	Cerchiello and Giudici [34]	24
12.	Côrte-Real et al. [35]	22
13.	Fanning and Grant [36]	21
14.	Pejić Bach et al. [37]	20
15.	Pérez-Martín et al. [38]	20
16.	Blackburn et al. [39]	20
17.	Tian et al. [40]	20
18.	Xie et al. [41]	20

16.5.2.1 Big Data in Financial Markets

It is regarded that big data stimulate the financial markets by helping in returns prediction, valuations, forecasting volatility, algorithmic trading, and so on. In fact, it has been asserted that the efficiency of the financial markets is attributed to the amount of information it generates and its diffusion process. In fact, technological innovations that generate information are regarded to positively impact financial markets [3]. It is for this reason that the market overreacts to extremely negative news [42]. Additionally, opinions, ratings, and posts on social media also have an impact on the financial markets. For instance, [21] found that opinions and ratings predict company returns. Additionally, [17] show that stocks' aggregate rankings are negatively associated with future returns. Therefore, the amount of information generated acts as an influential force affecting the global financial markets.

More recently, [42] found that messages sent to the NASDAQ exchange for the S&P 500 were essential to establish a true market price. The increased electronic trading has generated interest among the fund managers for the use of big data [43], interpret the pricing screens [41], interpret the complex market data by using visualization software [15], and the use of technologies to facilitate electronic trading in the global coordination [44].

Further, given a large number of anomalies present in forecasting returns, big data presents promising upshots in forecasting stock returns even when the variables and stock returns change over time. Therefore, in forecasting future stock returns, big data can reduce uncertainty in investment outcomes. In fact, more data processing reduces the uncertainty that ultimately reduces the risk and the overall cost of capital [4].

16.5.2.2 Big Data in Internet Finance

Internet finance is regarded as an intersection of big data, cloud computing, social networking, and information technology over the Internet [45], making internet finance a new phenomenon in finance when compared to traditional finance. Internet finance includes electronic cash transfers, electronic payment and settlements, crowd-funding, peer-to-peer lending, and so on. However, these financial interactions take place over the Internet and thereby, internet finance is regarded as the integration of modern finance and technology. With the advent of internet finance, modern banking, online transactions, and banking applications produce a million pieces of data every day and the management of this data is important [3].

[41] recently explored the fundamentals of internet finance in abreast of explaining the relationship between information technology, e-commerce, and finance. They used factors like service variety, information protection, data volume, and predictive correctness as factors to explain this relationship. Further, it is contended that big data improve the risk management practices, alleviate information asymmetry problems, predict credit risk, and detect fraud [46, 47]. In fact, data mining technology is regarded as the chief factor in risk management and fraud detection and prevention [32, 37, 48].

Further, big data by way of information sharing has resulted in the formation of a transparent and competitive market where pricing processes are fair and smooth. This helps to further reduce the parties' financial disputes and improve dispute resolution [45]. Additionally, by way of data access, big data has impacted internet credit service companies. These companies are now able to access more borrowers, which was not possible with traditional models. Overall, big data has significantly uplifted financial institutions toward efficiency and approaching new consumers.

16.5.2.3 Big Data in Financial Services

The current landscape of the financial service industries' business model is changing rapidly with the advent of big data. In fact, many financial service firms are working toward developing novel business models that would consider the application of big data [3]. Further, it is asserted that such business models must incorporate big data application into risk control, creating finance sentiment indexes and financial market analysis for these institutions. As every financial services company receives and generates data in billions of pieces every day, their management is important for the overall risk management and profit maximization [49]. In line with this, [49] described four features of big data, volume (data of large scale), variety (data in different formats), velocity (data of different frequencies), and veracity (data is uncertain). These characteristics pose a challenge to the management of financial services firms for the application of big data and finding novel business models in handling big data.

Further, [33] assert that big data is increasingly important for risk analysis in financial service institutions. They suggest that big data enhances risk management by improving the quality of models and applying behavior scorecards. Big data also helps in interpreting the risk analysis information faster compared to traditional systems [50]. Additionally, [34] specified the risk modeling process that focuses on the interrelationships between financial institutions.

Further, [38] suggested that big data analytics measure credit risk in banking firms. They also suggest that the management of databases is important for automatic evaluation of credit status within a reasonable period. In fact, nowadays, big data techniques are applied in banking firms to segment risk groups and comply with legal and regulatory requirements. Overall, it is suggested that financial institutions must benefit from improved system bought up by big data.

16.5.2.4 Big Data and Other Financial Issues

Big data has also been applied to the management of personal finance [3]. Further, [4] asserts that big data in financial markets has enabled large firms to grow faster because big data helped large firms lower the cost of capital in a greater proportion compared to small ones. The significant relationship between information and the cost of capital is also supported by [51]. Further, [52] embeds big data into corporate finance and investment decision models. Additionally, [53] suggest a significant relationship exists between Internet message board activity and abnormal stock returns and trading volume. Moreover, [54] thrust upon the usefulness of big data analytics in financial statement audits.

16.6 Reporting of Findings and Research Gaps

This section is dedicated to the presentation of overall findings from the articles reviewed and identification of gaps thereof. It is worth mentioning that care has been taken to deliberate upon the major findings from the literature that are instrumental upon research gaps' identification.

16.6.1 Findings from the Literature Review

The critical review of the literature identifies the followings findings:

16.6.1.1 Lack of Symmetry

The studies conducted so far on big data in finance lack the symmetric theory development. In fact, the studies conducted so far deliberate upon different operational aspects of big data in finance, for instance, [4] considers big data in finance and growth of firms; [27] studies internet search queries as a predictor of stock market volatility; [8] intends to explain the value of big data for internet credit service companies; [26] deliberates upon the working of economics in the age of big data; [25] pondered upon google searches around earning announcements; [54] explained the usefulness of big data analytics in financial statement audits.

The lack of symmetry in the literature can be attributed firstly to the newness of big data in finance, secondly to the relatively smaller amount of big data in finance research. Given this finding in the literature, a future research opportunity emerges concerning the development of symmetric theory for big data in finance.

16.6.1.2 Dominance of Research on Financial Markets, Internet Finance, and Financial Services

Another key finding from the literature emerges with regard to the dominance of the research of big data in finance in the above-mentioned areas. Studies like [42, 43, 55] extensively researched on big data in financial markets, studies like [32, 37, 41] explored big data in internet finance, and some studies worked on big data in financial services, for instance [39, 42, 43].

The dominance of literature of big data in finance in these specific areas maybe because most of the journals publishing articles on big data in finance call papers in these specific areas. Additionally, these areas of finance have the direct application of the Internet or are internet-based industries, hence big data as an application in finance would naturally grow in such areas. Given this finding, future research can be conducted by extensively exploring other areas of finance like corporate finance, accounting, agriculture finance, and so on.

16.6.1.3 Dominance of Empirical Research

Based on the results mentioned in Table 16.2, the majority of the research articles were focused on conducting empirical research. In fact, 75% of the articles were such. These results should not be surprising as finance research is dominated by empirical setup. The other reason for the lack of qualitative and survey enquiry is that not many journals publish such research, at least not in the finance area. This finding puts thrust on conducting future research as a mixed enquiry or qualitative and survey-based research.

16.6.2 Directions for Future Research

Apart from the above-mentioned areas, future research can also be conducted by addressing technical problems with regard to data management. Additionally, there are a large number of datasets available, such as Eagle Alpha provides the dataset on customer receipts that can be used to forecast revenue, Twitter sentiment data on companies, and an attractive piece of the dataset is provided by iSentient. These datasets have not entered the academic space, thereby results gained after their analysis would be quite useful.

As mentioned earlier, big data in finance has been focused on a few specific areas; future research can be carried out by exploring its impact on a variety of organizational characteristics. Machine learning is also being considered as a promising area for academic work. Machine learning, coupled with big data, can do wonders; it can improve forecasts, risk management, prediction, and so on. Such an area can be a possible interest for academicians. In fact, software packages like R and Python have simplified these procedures and applications. Therefore, this area might find its space in the academic arena of finance.

References

1 Russom, P. (2011). Big data analytics. *TDWI Best Practices Report, Fourth Quarter* 19 (4): 1–34.

2 Suthaharan, S. (2016). Machine learning models and algorithms for big data classification. *Integrated Series in Information Systems* 36: 1–12.

3 Hasan, M.M., Popp, J., and Oláh, J. (2020). Current landscape and influence of big data on finance. *Journal of Big Data* 7 (1): 1–17.

4 Begenau, J., Farboodi, M., and Veldkamp, L. (2018). Big data in finance and the growth of large firms. *Journal of Monetary Economics* 97: 71–87.

5 Zetsche, D.A., Buckley, R.P., Arner, D.W., and Barberis, J.N. (2017). From fintech to techfin: the regulatory challenges of data-driven finance. *NYU Journal of Law and Business* 14: 393.

6 Fang, B. and Zhang, P. (2016). Big data in finance. In: *Big Data Concepts, Theories, and Applications*, 391–412. Cham: Springer.

7 Woodard, J. (2016). Big data and ag-analytics. *Agricultural Finance Review.* 76 (1): 15–26.

8 Zhang, S., Xiong, W., Ni, W., and Li, X. (2015). Value of big data to finance: observations on an internet credit Service Company in China. *Financial Innovation* 1 (1): 17.

9 Tetlock, P.C. (2007). Giving content to investor sentiment: the role of media in the stock market. *The Journal of Finance* 62 (3): 1139–1168.

10 Hoberg, G. and Phillips, G. (2010). Product market synergies and competition in mergers and acquisitions: a text-based analysis. *The Review of Financial Studies* 23 (10): 3773–3811.

11 Antweiler, W. and Frank, M.Z. (2004). Is all that talk just noise? The information content of internet stock message boards. *The Journal of Finance* 59 (3): 1259–1294.

12 Loughran, T. and McDonald, B. (2011). When is a liability not a liability? Textual analysis, dictionaries, and 10-Ks. *The Journal of Finance* 66 (1): 35–65.

13 Heston, S.L., Korajczyk, R.A., and Sadka, R. (2010). Intraday patterns in the cross-section of stock returns. *The Journal of Finance* 65 (4): 1369–1407.

14 Chordia, T. and Subrahmanyam, A. (2004). Order imbalance and individual stock returns: theory and evidence. *Journal of Financial Economics* 72 (3): 485–518.

15 Kelley, E.K. and Tetlock, P.C. (2013). How wise are crowds? Insights from retail orders and stock returns. *The Journal of Finance* 68 (3): 1229–1265.

16 Kelley, E.K. and Tetlock, P.C. (2016). Retail short selling and stock prices. *The Review of Financial Studies* 30 (3): 801–834.

17 Da, Z., Huang, X., and Jin, L.J. (2019). Extrapolative beliefs in the cross-section: What can we learn from the crowds? *Journal of Financial Economics* Available at SSRN 3144849.

18 Neely, C.J., Rapach, D.E., Tu, J., and Zhou, G. (2014). Forecasting the equity risk premium: the role of technical indicators. *Management Science* 60 (7): 1772–1791.

19 Da, Z., Engelberg, J., and Gao, P. (2011). In search of attention. *The Journal of Finance* 66 (5): 1461–1499.

20 Huang, J. (2018). The customer knows best: the investment value of consumer opinions. *Journal of Financial Economics* 128 (1): 164–182.

21 Tirunillai, S. and Tellis, G.J. (2012). Does chatter really matter? Dynamics of user-generated content and stock performance. *Marketing Science* 31 (2): 198–215.

22 Singh, H.P. and Kumar, S. (2014). Working capital management: a literature review and research agenda. *Qualitative Research in Financial Markets* 6 (2): 173–197.

23 Tranfield, D., Denyer, D., and Smart, P. (2003). Towards a methodology for developing evidence-informed management knowledge by means of systematic review. *British Journal of Management* 14 (3): 207–222.

24 Fink, A. (2019). *Conducting Research Literature Reviews: From the Internet to Paper*. Sage publications.

25 Drake, M.S., Roulstone, D.T., and Thornock, J.R. (2012). Investor information demand: evidence from Google searches around earnings announcements. *Journal of Accounting Research* 50 (4): 1001–1040.

26 Einav, L. and Levin, J. (2014). The data revolution and economic analysis. *Innovation Policy and the Economy* 14 (1): 1–24.

27 Dimpfl, T. and Jank, S. (2016). Can internet search queries help to predict stock market volatility? *European Financial Management* 22 (2): 171–192.

28 Kshetri, N. (2016). Big data's role in expanding access to financial services in China. *International journal of information management* 36 (3): 297–308.

29 Seddon, J.J. and Currie, W.L. (2017). A model for unpacking big data analytics in high-frequency trading. *Journal of Business Research* 70: 300–307.

30 Chen, Y., Chen, H., Gorkhali, A. et al. (2015). Big data analytics and big data science: a survey. *Journal of Management Analytics* 3 (1): 1–42.

31 Campbell-Verduyn, M., Goguen, M., and Porter, T. (2017). Big data and algorithmic governance: the case of financial practices. *New Political Economy* 22 (2): 219–236.

32 Jin, X., Shen, D., and Zhang, W. (2016). Has microblogging changed stock market behavior? Evidence from China. *Physica A: Statistical Mechanics and its Applications* 452: 151–156.

33 Choi, T.M. and Lambert, J.H. (2017). Advances in risk analysis with big data. *Risk Analysis* 37 (8): 1435–1442.

34 Cerchiello, P. and Giudici, P. (2016). Big data analysis for financial risk management. *Journal of Big Data* 3 (1): 18.

35 Côrte-Real, N., Ruivo, P., Oliveira, T., and Popovič, A. (2019). Unlocking the drivers of big data analytics value in firms. *Journal of Business Research* 97: 160–173.

36 Fanning, K. and Grant, R. (2013). Big data: implications for financial managers. *Journal of Corporate Accounting & Finance* 24 (5): 23–30.

37 Pejić Bach, M., Krstić, Ž., Seljan, S., and Turulja, L. (2019). Text mining for big data analysis in financial sector: a literature review. *Sustainability* 11 (5): 1277.

38 Pérez-Martín, A., Pérez-Torregrosa, A., and Vaca, M. (2018). Big Data techniques to measure credit banking risk in home equity loans. *Journal of Business Research* 89: 448–454.

39 Blackburn, M., Alexander, J., Legan, J.D., and Klabjan, D. (2017). Big data and the future of R&D management: the rise of big data and big data analytics will have significant implications for R&D and innovation management in the next decade. *Research-Technology Management* 60 (5): 43–51.

40 Tian, X., Han, R., Wang, L. et al. (2015). Latency critical big data computing in finance. *The Journal of Finance and Data Science* 1 (1): 33–41.

41 Xie, P., Zou, C., and Liu, H. (2016). The fundamentals of internet finance and its policy implications in China. *China Economic Journal* 9 (3): 240–252.

42 Blocher, J., Cooper, R., Seddon, J., & Van Vliet, B. (2018). Phantom liquidity and high-frequency quoting. *Journal of Trading*, Vol. 11, No. 3, 6–15.

43 Preda, A. (2007a). The sociological approach to financial markets. *Journal of Economic Surveys* 21 (3): 506–533.

44 Zaloom, C. (2003). Ambiguous numbers: trading technologies and interpretation in financial markets. *American Ethnologist* 30 (2): 258–272.

45 Yang, D., Chen, P., Shi, F., and Wen, C. (2018). Internet finance: its uncertain legal foundations and the role of big data in its development. *Emerging Markets Finance and Trade* 54 (4): 721–732.

46 Glancy, F.H. and Yadav, S.B. (2011). A computational model for financial reporting fraud detection. *Decision Support Systems* 50 (3): 595–601.

47 Ngai, E.W., Hu, Y., Wong, Y.H. et al. (2011). The application of data mining techniques in financial fraud detection: a classification framework and an academic review of literature. *Decision Support Systems* 50 (3): 559–569.

48 Hajizadeh, E., Ardakani, H.D., and Shahrabi, J. (2010). Application of data mining techniques in stock markets: a survey. *Journal of Economics and International Finance* 2 (7): 109.

49 Sun, Y., Shi, Y., and Zhang, Z. (2019). Finance Big Data: management, analysis, and applications. *International Journal of Electronic Commerce* 23: 1.

50 Hennessy, C.A. and Whited, T.M. (2007). How costly is external financing? Evidence from a structural estimation. *The Journal of Finance* 62 (4): 1705–1745.

51 Fang, L. and Peress, J. (2009). Media coverage and the cross-section of stock returns. *The Journal of Finance* 64 (5): 2023–2052.

52 Gomes, J.F. (2001). Financing investment. *American Economic Review* 91 (5): 1263–1285.

53 Tumarkin, R. and Whitelaw, R.F. (2001). News or noise? Internet postings and stock prices. *Financial Analysts Journal* 57 (3): 41–51.

54 Cao, M., Chychyla, R., and Stewart, T. (2015). Big Data analytics in financial statement audits. *Accounting Horizons* 29 (2): 423–429.

55 Subrahmanyam, A. (2019). Big data in finance: evidence and challenges. *Borsa Istanbul Review* 19: 283–287.

Index

Big Data Analytics for Internet of Things, First Edition. Edited by Tausifa Jan Saleem and Mohammad Ahsan Chishti.
© 2021 John Wiley & Sons, Inc. Published 2021 by John Wiley & Sons, Inc.

Printed and bound by CPI Group (UK) Ltd, Croydon, CR0 4YY

27/10/2024

14580268-0002